"Contrasting the functions of various actors and tools across a myriad of legal contexts, these authors present striking insights impacting legal cultures. The role of the citizen in law and the act of comparison in legal analysis are both key to untangling the threads of environmental responses. Offering an excellent range of environmental analyses, this book is a timely comparative tool to understand evolving environmental good governance. Practitioners, students, and a wide range of readers will find it an important contribution to gaining a deeper grasp of complex environmental legal themes."

Elizabeth Burleson, Co-Editor, Comparative Environmental Law and Regulation

"Whether a student, a lecturer, practicing attorney, an expert, or indeed just someone with a genuine interest you will find this text a profound study in comparative environmental law. This book goes way beyond comparing written norms and differences in practices, to include in its focus the fact that law in every societal circumstance is born of a concrete culture and its contents is therefore a consequent of the same. This text is at once historical and contemporary; covers from Canada to Chile and from India to Ethiopia and many others in between in its impressive geo-legal span. It makes the study of comparative environmental law very real through inter alia consideration of decided cases from different local, national and federal judiciaries in dissimilar legal systems in many parts of the world. From customary to statutory laws, and from constitutional principles to specific environmental laws of various countries are compared and contrasted not merely in search of differences and similarity; but to point out how justice is reasoned, defined and delivered based on history, culture, the requirements of modern life vis-à-vis the protection of the environment, and the meeting of human needs."

Dr. Richard A. Byron-Cox, International Law Specialist, Diplomat & Expert on Sustainable Development, United Nations

"This book is a refreshing and much-needed contribution to the area of comparative environmental law. It does not shy away from taking on the complex task of comparing the approach and practice of environmental law across contrasting legal cultures. The uniqueness of this book lies in its ability to effectively combine academic rigor and depth with learning from the practice of environmental law in different cultures as many authors are also practitioners from countries as diverse as India, China, and Germany to name a few. This is a must-have for anybody working or researching environmental law in diverse legal and political contexts."

Arpitha Kodiveri, Hans Kelsen Fellow at the European University Institute's Department of Law

ENVIRONMENTAL LAW ACROSS CULTURES

This book provides a practical, functional comparison among various institutions, tools, implementation practices, and norms in environmental law across legal cultures.

This is a new approach that focuses on the act of comparison, looking at legal practice from the ground up, including the perspective of citizens. Most literature on comparative environmental law either focuses on a two-way comparison of state jurisdictions or simply juxtaposes environmental features of two or more state jurisdictions without engaging in any analysis of the comparison. However, this book treats legal cultures as the objects of comparison as it provides practical comparisons among various institutions, tools, and norms in environmental law. The arrangement and organization of the material reverses the more traditional presentation of comparative environmental law as a series of countries within which separate descriptions are respectively presented. In this book the reader is presented with environmental legal themes, with examples and case studies drawn from various cultures that are compared in order to help understand the theme. Case studies draw on the authors' experiences in a range of legal cultures, including in Australia, Brazil, China, Chile, Ethiopia, Germany, India, Nigeria, Slovakia, and the USA. The comparative nature of the book allows domestic professionals to develop skills to enable them to understand and advocate broader contexts for clients, and helps students become more aware of specific legal systems while questioning why their own system functions (or does not function) as it does.

The book is aimed at advanced undergraduate and postgraduate students of environmental law as well as researchers and practitioners.

Kirk W. Junker is Professor of Law, Director of the Environmental Law Center, and Director of the International Master of Environmental Sciences Programme, Faculty of Law, University of Cologne, Germany.

ENVIRONMENTAL LAW ACROSS CULTURES

Comparisons for Legal Practice

Edited by Kirk W. Junker

Foreword by M.C. Mehta

Routledge
Taylor & Francis Group
LONDON AND NEW YORK

earthscan
from Routledge

First published 2020
by Routledge
2 Park Square, Milton Park, Abingdon, Oxon OX14 4RN

and by Routledge
52 Vanderbilt Avenue, New York, NY 10017

Routledge is an imprint of the Taylor & Francis Group, an informa business

British Library Cataloguing-in-Publication Data
A catalogue record for this book is available from the British Library

Library of Congress Cataloging-in-Publication Data
A catalog record has been requested for this book

ISBN: 978-0-367-02825-1 (hbk)
ISBN: 978-0-367-02826-8 (pbk)
ISBN: 978-0-429-39761-5 ebk)

Typeset in Bembo
by Deanta Global Publishing Services, Chennai, India

For all the persons who have insufficient air, water, or earth, but will never read this book. Some already know it; some do not. We hope the book will help someone to help you.

CONTENTS

ACKNOWLEDGMENTS

I would like to thank my colleague Mr. M.C. Mehta, whose public interest work is an inspiration for all lawyers working for improvement in the environment. My reading of his work, *In the Public Interest* (Prakriti Publications, Delhi, 2009), convinced me of the rewards of studying environmental law across cultures. I would also like to thank my former law practice colleague and engineer, Michael J. Heilman of the Pennsylvania Department of Environmental Protection in the U.S.A., for his constant willingness to keep me in contact with public practice concerns there. I am grateful to Robert D. Taylor, John Poulakos, Nicholas P. Cafardi, Frank Y. Liu, Klaus Klein, Erach Bharucha, Shamita Kumar, Sophie Robin-Olivier, Tsegaye Beru, and the late Edward Keynes for expanding my horizons and network. The production team who brought the book together began with Mrinalini Shinde, who organized all the authors, followed by Saskia Münster and Ryan Kraski, who painstakingly managed the work of organizing and sub-editing the texts. Emma Shensher, Lena Carduck, Laura Midey, Luka Vihuto, and Michelle Quindeau all deserve thanks for close proofreading. Marek Prityi, State Adviser at the Slovak Ministry of Environment, authored thoughts and ideas in nearly every chapter. Prityi and Münster both also provided valuable lectures to our students in order to give their written ideas life. The Environmental Law Center at the University of Cologne deserves credit for attracting and organizing environmental law authors from around the world. I would also like to thank Autumn N. Alko, the graphic artist who designed the cover image for the book. Thanks to Duquesne University Law Review for permission to include some ideas that I had previously published with them. Thanks also to the staff of Carnegie Library of Pittsburgh, South Side Branch, where the final manuscript was completed. And finally, I want to thank the students from the University of Cologne's Law Faculty, International Master of Environmental Sciences Program, and Culture and Environment in Africa

Master Degree Program, as well as the students of the Cologne University of Applied Sciences, all of whom have taken part in the more than fifteen years of comparative environmental law at the University of Cologne. The invaluable exchange of ideas among many of the authors and these students were vital for us to develop the concepts and text. We all learn from one another.

FOREWORD

This book, *Environmental Law Across Cultures: Comparisons for Practice*, reflects the principle of "*vasudhaiva kutumbkam*"—a Sansrkrit phrase that means "the world is one family." Air, water, and earth, necessary for all living beings, are not confined to artificial boundaries demarcated by different nations. Competition, greed, warfare, the race for arms, and erroneous development policies have blinded most of the affluent nations into believing that the Earth belongs to them and is merely an object and resource for exploitation. Despite all of our huge technological advancements and sophistications, we refuse to learn lessons from history or from other living species that have managed through the millennia to adapt to need-based survival, unlike us thinking, "superior" humans, who practice greed-based hoarding of natural resources. We have divided the Earth into continents, countries, and states and from that have shaped environmental laws and policies that are human-centric and exploitative, resulting in the merciless extinction of a large variety of both animal and plant species. As the climate crisis continues to expand, we become more heedless to the dangers looming before our very eyes. Pollution on land, in rivers, lakes, streams, and in the ground water; solid, nuclear, and hazardous waste of every kind; and even floating debris in space are all compounded by frequent warfare amongst one or more nations and are all contributing to a seemingly irretrievable situation for our beleaguered planet. We have reached a critical do-or-die juncture in our history. At any time, a catastrophe of major magnitude may erupt, and, when it does, it will have the potential to injure or destroy a large population.

Environmental Law Across Cultures is an excellent exercise in applied comparative law that gives the reader a bird's eye view of environmental law and policy, providing perspectives across a number of relevant topics, including green crimes and the use and abuse of enforcement mechanisms, while balancing perspectives from theory and practice in many parts of the planet. Kirk W. Junker and this network of authors from Australia, China, India, Ethiopia, Nigeria, Brazil, Chile, Germany, Slovakia, and the U.S.A. offer an opportunity for citizens, law-makers, lawyers, and judges, whether sitting in general jurisdiction, on environmental courts, or on environmental tribunals, to rise above parochial administrative or judicial boundaries or jurisdictional limits and provide environmental justice. Environmental justice is still possible and this book provides thoughtful provocation that incites and enables its readers to work toward providing it. Environmental justice recognizes that the natural environment does not practice discrimination, religion, or boundaries and speaks in a language that allows all beings to live and let live in their only home—the Earth.

M.C. Mehta, M. C. Mehta Environmental Foundation, Delhi, India

PREFACE

"Twelve lawyers, ten different legal systems, five continents, ten chapters; all comparing legal cultures in such ways as to include citizens as essential participants in environmental law." When a Greek mining compliance manager recently asked me what this book was about, that was my answer. Practice informs theory and theory informs practice. Having practiced environmental law, I now regularly teach environmental law to students of law or the natural sciences from Germany, France, Chile, Italy, Turkey, India, Ethiopia, Nigeria, China, Brazil, the U.K., the U.S.A., and other countries. In such a mixed group, the conversation often includes comparing legal cultures. But the practice of law before the courts of the world does not take place in such a mixed cultural environment. And for those practices one must address the question as to why one would want to compare legal cultures. I find the answer to the question in my same classroom. The more traditional approach to the category called "comparative law" would have the comparativists compare legal texts. As Bernhard Großfeld has warned, however; looking at legal texts alone is looking at something dead. One cannot know how law is practiced simply by reading legal texts. So we should shift our comparison to looking at law in practice. Therefore, one must also look at institutions of law-making, interpreting, and implementation. Even there, however, many comparativists look only at the lawyer, judge, police officer, or other professionally trained legal person. While it is true that these persons know what the texts say and how the institutions work for them, theirs is only half the story. It is the remainder of the people—the citizens—for whom the law is to be working. Those citizens might have expectations of, or practices in their legal system that are not exactly aligned with what a foreign lawyer would read in a text or expect from a court in that culture.

And so we are driven instead to the very difficult, but at the same time very telling, task of assessing the persons who live with the law in given legal cultures to provide a key insight into comparative law. That is why we call this book *Environmental Law Across Cultures* and not across jurisdictions. The enterprise of comparative law is most often conducted by lawyers with little or no experience in investigating culture, including legal culture. The enterprise of legal anthropology is most often conducted by anthropologists who are not lawyers and must treat the practice of law as something being witnessed from the outside; an observation of the other. Both perspectives have advantages and disadvantages. In this book we emphasize the advantages of comparing legal cultures as lawyers because the book is by lawyers who believe the practice of environmental law can benefit from comparing legal cultures. But at the end of the day, it is a book by lawyers for provocation of, and application by citizens and lawyers.

Texts and professionally trained law experts are of course included, but not to the exclusion of the culture in which they function, if they function. Historically, some would have given some attention to related ideas under the notion of the "legitimacy" of a legal system. So here, we can look to the example of the Organization for Economic Co-operation and Development's corruption index, and how a culture avoids its legal system in order to get the task of daily living accomplished, often through corrupt means because citizens disrespect, distrust, or cannot afford their professional law communities. So much for *what* one compares.

An equally important, and logically prior, question is "*why* compare?" In comparative law literature, the answer to this question is not often made explicit, and when it is made explicit, it frequently tends toward making a list like comparing football teams so as to enable the claim of who is best. As a result, we hear much too often the misleading label "best practices," which is just a thinly disguised version of "best management practices" or "best available technology"—terms that environmental lawyers know from legislative compromises. All these "bests" tell us something about why one is comparing—to determine that which is best. And then after we determine which is best, what do we expect? Are we going to transfer the best to a different legal community as if we were buying souvenirs to take home from the airport? "Best," ironically, is itself not an absolute term, although it might appear to be. Consider, for example, that best available technology will differ if different cultures have different economics, if they have different labor practices, different religions, and different expectations from the state. Best is relative to too many different cultural variables to be the absolute standard that it appears to be, and it is insufficient to justify neatly packaging legal texts or institutions for transfer. As soon as we properly interrogate "best" by asking "for whom?" and "to achieve what purposes?" we begin to see that one cannot be best for all, despite the superlative nature of the grammar. Worse still would be when "best" turns out to be the choice advocated by the powerful economy or political culture that can impose its way on other cultures under the brightly colored label of being "best." Speaking in superlatives slides too easily into speaking in absolutes.

In writing these chapters, our exigencies were felt both as pulls and pushes. We were pulled to explain and analyze some new things, and we were pushed away by more common approaches to comparative environmental law. The authors of this book are pulled to ask whether and how the law, as a human institution, might be used to help humans improve their relationship to the natural environment so that the relationship becomes at least sustainable, and, if possible, enjoyable. In that sense, when the authors succeed, they do more than provide detailed descriptions of legal regimes. We assume that cultures have different advantages and disadvantages in their legal regimes. A further assumption is that, if one were to compare those approaches, one might help to improve the human relationship to the environment. And because that relationship is to an interrelated worldwide system, improving the relationship through law in one culture may improve the relationship in other cultures as well.

We felt also pushed away from treating comparative law as a Eurocentric league table establishment exercise that would inevitably conclude with recommending practices that other cultures ought to adopt. First, we take the notion of comparativism seriously. Much of the impracticality of comparisons can be exposed by asking why the comparison is being carried out, why these select places, peoples, and practices are being compared, and through what method the author understands his or her comparisons.

Second, as mentioned above, we have added the word "culture" to our title because we include the perspectives of the citizens of the several states under comparison. To our way of thinking, law is not only of interest and concern to those who study or practice law, but to all citizens who wish to have access to the law in exercising their rights and acknowledging their liabilities both in their own society and in the context of other societies.

Third, we recognize that the study and practice of law contains a cultural element and cannot be known only from studying the rules of law enacted by a legislative body or the administrative rules thereunder. To understand how environmental law works in any particular culture requires that one see the law in practice, not just in theory. To that end, our authors include practitioners from all continents and at least ten different states, and have written not just from the perspective of the rational theorist, but from the experience of legal practice in each of their respective cultures.

The chapters of the book can be read in any order. Each chapter makes explicit references to other chapters that precede it and follow it, so some connections will be made for the selective reader who does not read from beginning to end. While at least ten states of the world are represented by authors in this book, our goal was not encyclopedism. Instead, we began by asking how comparative law might practically benefit the students and practitioners who read this book. Thus, **Part I** is designed to address basic questions of comparative law in general, not limited to environmental law. Then we turn to an arrangement that is partially traditional, and partially nontraditional. With our focus on culture, we might have begun with the persons who are subject to environmental law, given that they are in fact the theoretical source of power for having created environmental law. But in fact, we present these legal subjects in their position after the fact; that is, after they have transferred legal power to the state and now are subject to their own creation. That said, we do depart from treating the theoretical position of the text as being primary and instead begin with the institutions of the law. Thus, in **Part II**, "Institutions of Environmental Legislation, Implementation, and Dispute Resolution," we turn to first consider how environmental problems might be solved across a spectrum of cultures through various institutions, asking whether disputes are better solved in particular cultures through planning and administration, but in other cultures through litigation. One can read already at this early step that how one analyses a problem—in this case the environment—indicates a particular cultural approach to the law. We take the approach that there is no pan-legal approach to comparative law. Simply selecting the order of topics suggests that one topic is a base and the other are variations on it. As a simple technique to try to break students from getting stuck in the bias of their own culture, my colleague Prof. Heinz-Peter Mansel insists that students begin comparative law studies with law that is foreign to the author. This principle is simple and effective in illustrating that alone the order in which one compares things will guide one's thinking about solving the problem. Of course, beginning with persons and not with norms or texts may also expose my own common law heritage preferences.

Only in **Part III** do we arrive at the more common topics "Norms and Tools of Environmental Legislation, Implementation, and Dispute Resolution," the title of which needs little explanation to the reader. And then finally, in **Part IV**, we treat the area that makes most clear the emphasis on the concept of legal culture by featuring "Persons Subject to Environmental Law." Throughout all chapters, secondary literature is cited using Modern Language Association style, but primary law sources are cited according to the rules of the country whose law is being cited. One lesson is clear from all of comparative law—no type or degree of enforcement can surpass the quality of implementation that is possible when the persons subject to the law understand it and with understanding, also support it.

Kirk W. Junker
Cologne, Germany, December 2019

CONTRIBUTORS

Dennis Agelebe is a Partner at DAOS LP in Lagos, Nigeria. He holds an LLB from the University of Calabar, Nigeria and an LLM from Charles University, Prague in the Czech Republic in International Human Rights Law and Environmental Protection. He is currently a doctoral candidate at the University of Cologne in Germany. His research interests include sustainable energy, environmental friendly trade practices, and FDI with environmental ethics in developing countries.

Tsegai Berhane Ghebretekle is an Associate Professor of Law at Mekelle University College of Law & Governance, School of Law, in Mekelle, Ethiopia, where he teaches several courses including public international law, environmental law, land law, environmental theories and principles, environment and development, and climate justice, at the undergraduate and graduate levels. His research interests include sustainable development, corporate social responsibility, climate ethics and justice, and industrial pollution control management.

Vanessa Johnston is a Lecturer at Monash University in Melbourne, Australia. She is licensed to practice in Australia in both state and federal jurisdictions, and prior to lecturing at Monash University she practiced law for ten years in niche and mid-sized commercial firms. Her primary areas of expertise include property law, revenue, and taxation. She was awarded a Ph.D. from Monash University in 2016 for her thesis on carbon pricing for road transport activities. She teaches compulsory property law subjects to students at undergraduate and postgraduate levels, and climate change law electives for undergraduate and master's level students. Her research focuses on the impact that climate change has on land use and planning in Australia, as well as fundamental property law concepts such as ownership. In addition, she is involved in several interdisciplinary research projects on sustainable transport in Victoria.

Kirk W. Junker holds the Chair of U.S. American Law at the University of Cologne, where he is also Director of the Environmental Law Center, and is Director of the International Master of Environmental Science (I.M.E.S.) Program. He also serves as an expert witness before German courts. Before joining the Law Faculty at the University of Cologne, Prof. Junker was Associate Professor of Law and Director of International Programs at the Duquesne University School of Law in Pittsburgh, U.S.A. He held the first cross-border academic position after the Belfast Agreement in Ireland by being a full-time faculty member at both Queen's University of Belfast, Northern Ireland and Dublin City University, Republic of Ireland. Earlier, he was a member of the Faculty of Natural Sciences of the Open University in the United Kingdom. Prof. Junker is an Adjunct Professor of Comparative and International Environmental Law to the Law Faculty at Université Paris 1 Panthéon-Sorbonne, the Institute of Environment Education and Research, Bharati Vidyapeeth University, Pune, India, and Sts. Cyril and Methodius University, Skopje, Northern Macedonia. Prof. Junker is licensed to practice law in Pennsylvania and before the federal courts of the U.S.A. For nine years he practiced environmental law as litigation counsel to the Pennsylvania Department of Environmental Resources.

Sacha Kathuria is Counsel for Government Affairs at the American Coatings Association (A.C.A.) in Washington, D.C., where she focuses on sustainability and the paint industry's product stewardship program, PaintCare. Prior to joining A.C.A., she taught International Business and International Law at Saint Vincent College, worked in the oil and gas and trade industries, and on U.S. Department of Energy-funded electricity projects for the Midwestern Governors Association. She holds an LL.M in Energy and Environmental Law from The George Washington University Law School, a J.D. from Duquesne University, a Master in Public and International Affairs from the University of Pittsburgh, a Graduate Certificate from the European Union Center of Excellence at the University of Pittsburgh, and a B.A. in International Affairs and History from The George Washington University. She is licensed to practice law in Pennsylvania, the District of Columbia, and before the U.S. Court of Appeals for the Federal Circuit, and the U.S. Court of International Trade. The views expressed are those of the author and do not reflect any official position or policy of her current or past employers, clients, or associations.

Ryan Kraski is admitted to practice law in the Commonwealth of Pennsylvania and U.S. Federal Courts of the Third Circuit. He is currently a doctoral candidate at the University of Cologne, where he is a Research Fellow and a lecturer in tort law. He holds a B.A. in Business Administration and Political Science, *cum laude*, from the Washington and Jefferson College and a J.D. *cum laude* from the Duquesne University School of Law in the U.S.A.

Ana Miola is an environmental lawyer from Brazil and is a specialist in international and domestic Brazilian environmental law. She is a member of and former Secretary-General of the Environmental Law Commission at the Brazilian Bar Association Section, Rio Grande do Sul. She is currently completing master's degrees in environmental sciences and natural resource management and development, at the University of Cologne and the Technical University of Cologne, respectively, as scholar of the Heinrich Böll Foundation, Germany.

Saskia Münster is a researcher at the Chair of U.S. American Law, where she manages research projects for the Environmental Law Center at the University of Cologne. Her areas of interest include comparative environmental law and water law, especially focusing on comparisons between European and Latin American jurisdictions. She completed her law studies at the University of Cologne, where she is currently a doctoral candidate.

Jorge Ignacio García Nielsen is a partner at D.S.Avocats in Santiago, Chile, focusing on Environment, Urban Planning, Natural Resources, and Energy projects and a Professor in energy and pollution law at the Universidad del Desarrollo. He was previously a lawyer for the Department of Environment, Natural Resources, and Regulatory Law at the Moraga y Cía. law firm and a Partner at Green Chile, where his practice focused on environmental and city law, natural resources, and regulatory law. He has also served as the Head of the Department of Environmental and Criminal Law at the Araya y Cía law firm, and as Assistant Professor in administrative and criminal law at the Pontificia Universidad Católica de Chile. He holds a Master in Regulated Sectors Law from the University Carlos III of Madrid, a Diploma in Natural Resources and Energy Law, Natural Resources Law; and a degree with a Certificate of Specialty in Criminal Law, Minor in Ecology and Social Psychology and Law from the Pontificia Universidad Católica de Chile.

Marek Prityi is State Adviser at the Slovak Ministry of Environment. Prior to that, he completed his doctoral research in the field of environmental comparative law at the University of Cologne, Germany. He completed a traineeship at the German Parliament, and worked as a lawyer and contract analyst in Slovakia. He is a graduate of the Comenius University in Bratislava and Trnava University, Slovakia and collaborated as Research Associate with the organization gLAWcal-Global Law Initiatives for Sustainable Development. During his doctoral studies and in his capacity as Research Associate, he conducted research in the United States, India, and China. The views and interpretations are those of the author and do not necessarily represent the official position of the Ministry of Environment of the Slovak Republic.

Mrinalini Shinde is an environmental lawyer from India who has practiced law as an Advocate in India, notably in environmental legal disputes and public interest litigation, including before the National Green Tribunal. She is currently a Fellow under the Early Career Climate Fellowship Programme of the United Nations University and United Nations Framework Convention on Climate Change (U.N.F.C.C.C) secretariat, working in the Legal Affairs Programme of the U.N.F.C.C.C secretariat. Previously, she was a Research Assistant in comparative environmental law at the Chair of U.S. American Law at the University of Cologne. Her areas of interest include environmental crime and criminal enforcement, international climate law, biodiversity law, and the law of the sea. She holds a B.A. LL.B (Hons.) from National Law School of India University and an M.Sc. in Environmental Sciences from the University of Cologne, Germany. The ideas and opinions expressed by Mrinalini Shinde in this book are belong entirely her, and they do not, in any manner or form, represent the views and opinions of the UNUEHS, the UNFCCC Secretariat, or any of her affiliated bodies, organizations, or agents.

Yuan Ye is a Lecturer at Hohai University of Nanjing, China. Her areas of research include Chinese environmental law, including comparative law between the European Union and China on compensation for ecological damage, between the U.S.A. and China on effluent permit systems, and on China's administrative legal risks in environmental protection issues as well as natural protection areas of China. Dr.Ye has contributed to the German Ecological Relief Institution for the Chinese Ministry of Land and Resources, and to a project on the ecological legal system of China's Eco-Environmental Ministry. She is currently conducting research on German and E.U. water law as a potential model for development of a Chinese water law system.

PART I
Comparison

1

WHY COMPARE?

The biological, cognitive, and social functions of comparison for the human

Kirk W. Junker

Comparison is such a natural act for the human being that it would seem to require no investigation, no reflection, and no discussion. But that is precisely why it does need investigation, reflection, and discussion. Comparison is in part hard-wired in the human animal, as when for example it aids such things as visual depth perception, but for discussions of the law, comparison is largely as set of learned social practices. To practice useful and intelligent comparative environmental law across cultures, learned social practices must be examined and questioned. Comparisons may contain hidden ethical values, for example, and those should be discussed, chosen, and made explicit, and not remain implicit and assumed.[1] In conducting these examinations, theory informs practice and practice informs theory.

It is difficult to imagine human existence without its practices of comparison. We can begin the analysis with physical comparison, and then consider how much physical comparison determines or informs social and cultural comparison.

1.1 Physical comparison

The human experience of vision, for example, makes considerable use of comparison. Stereoscopic vision is possible because as the human animal evolved, our eyes moved further and further to the front of our heads, thus creating an overlapping field of vision.

1 See Brown, Donald A. *American Heat: Ethical Problems with the United States Response to Global Warming*. Rowman & Littlefield Publishers, 2002; Brown, Donald A. *Climate Change Ethics: Navigating the Perfect Moral Storm*. Routledge, 2013.

> As the primate nose progressively shrank in size over millions of years, there was a corresponding increase in visual capabilities. … All [primates] have binocular vision with fields of view that significantly overlap, resulting in true three dimensional (3-D) depth perception or stereoscopic vision. At the same time, the field of view for peripheral vision was reduced. Human eyes have an overlapping field of view of about 120°. It is only in this field that we have stereoscopic vision.[2]

The overlapping field of vision—that is, stereoscopic vision—enables us to perceive depth. And by perceiving differing depths, we can compare the closer to the further away and the larger to the smaller. The distance comparison enabled by stereoscopic vision was essential for the hunter,[3] and we should not forget that, despite the exponential increase in human population since industrializing, most humans who have ever lived have been hunters and gatherers.[4] "Depth perception is an invaluable tool for animals that need to move quickly. … [A]ccurate depth perception is a critical capability for hunters, such as cats and hawks, who need to accurately judge distances in the pursuit of their prey."[5] Using their stereoscopic vision, primates have been very successful animals due largely to the fact that they are intelligent and opportunistic in getting food. Other animals, such as koalas and giant pandas, can or will only eat a very limited variety of food, a practice that restricts where they can live.[6] Thus, one can see our ability to compare as an evolutionary advantage.

Not only has our vision changed over time, but so have other physical abilities. In the case of olfactory comparisons, most mammals have about the same number of genes that code for odor receptors in their nasal tracks. However, the majority of these genes in humans, and presumably other large primates, are no longer functional. Recent research has shown that of the approximately 1,000 human olfactory receptor genes, only 347 remain functional.[7] The rest have mutations that deactivate them. If we compare humans' ability to smell with other animals' abilities to smell, we find that we are poor at sensing short-chain chemical compounds but, by comparison, we are good at discriminating between complex long-chain ones such as those of many foods and flowers. By further comparison, dogs can detect the presence of something with fewer molecules

2 O'Neil, Dennis. "The Primates: Overview." *The Primates: Overview*, 1998, www2.palomar.edu/anthro/primate/prim_1.htm.

3 See Howard, Ian P., and Brian J. Rogers. *Perceiving in Depth.* Oxford University Press, 2012.

4 Hassan, Fekri A. *Demographic Archaeology.* Academic Press, 1981; Binford, Lewis Roberts. *Constructing Frames of Reference: an Analytical Method for Archaeological Theory Building Using Hunter-Gatherer and Environmental Data Sets.* University of California Press, 2001.

5 O'Neil, *supra* note 2.

6 *Ibid.*

7 Ibarra-Soria, Ximena, et al. "Variation in Olfactory Neuron Repertoires Is Genetically Controlled and Environmentally Modulated." *ELife*, vol. 6, 2017, doi:10.7554/elife.21476.

of the substance. For our work in comparative law, these facts serve to remind us that when the physical bases change, our physical comparisons change too. It should come as no surprise that since our cultural and social comparisons are constructed abilities,[8] our needs and attitudes toward social comparison change just as our physical abilities to compare change. These examples show that different species have different physical abilities and build comparisons differently therefor. When it comes to differing human cultural values, we must then consider that humans build comparisons differently, over time and across cultures.

1.2 Cultural and social comparisons

1.2.1 Generally

We have adapted our cognitive comparative practices from physical abilities, such as vision and smell, to our social world. And we have evidence in ancient and modern languages of the fact that our physical abilities to compare influence our abstract notions of comparison. Ancient Greek, modern English, and modern French all use a version of "I see" to mean "I understand."[9] Thus it is not difficult to understand that we bridge from our physical abilities and advantages of comparing attributes to our cognitive abilities to compare attributes. And we accomplish our abstract comparison through language. Language enables us to compare past and present, that which is not in our presence with that which is in our presence, and the imaginary with the material.[10] For comparative *environmental* law, however, we must take note of the fact that language may also be "mankind's first technology for letting go of the environment in order to grasp it in new ways."[11]

A master of language, the 1980 Nobel Prize winner in literature Octavio Paz, provocatively demonstrates that abstract concepts can be compared. His skill in comparison from outside of the law helps a reader to focus upon comparison, rather than law itself, and so I will use his observations to frame ideas throughout this chapter. As a result of his experiences and study in India, Paz offered that:

> The differences between Hindu and Christian asceticism is even more marked than those between their eroticism. The key word of Western eroticism—I am referring to the modern West, from the eighteenth

8 See Berger, Peter L., and Thomas Luckmann. *The Social Construction of Reality: a Treatise.* Penguin, 1979.

9 For a comparative study between English and German uses of "understand" and "verstehen," see Burke, Kenneth. *A Grammar of Motives.* Prentice-Hall, Inc., 1945.

10 Harari, Yuval Noah. *Sapiens.* Penguin, 2011, pp. 25–27.

11 McLuhan, Marshall. *The Classical Trivium: the Place of Thomas Nashe in the Learning of His Time.* Gingko Press, 2005, p. 14; Citing McLuhan, Marshall 1911–80. *Understanding Media.* Routledge, 2001.

> century to the present—is violation, which is an affirmation of the moral and psychological order. For Hindus, the key word is pleasure. Similarly, in Christian asceticism, the central concept is redemption: in India, it is *liberatio*. These two words encompass opposite ideas of this world and the next, of the body and the soul. Both point toward what has been called the "supreme good," but there the similarity ends: redemption and liberation are paths that lead from the same point—the wretched condition of man—but lead in opposite directions.[12]

Reading the work by Paz in its English translation, one is generally struck by how often he finds, or wants to find, ways of comparing India to Mexico. His reason is straightforward—he was the Mexican ambassador to India and he is interpreting India to fellow Mexicans. But aside from that coincidence of his own life (he tells the reader at the beginning that he was sent to India by his government) why does he go into such comparisons with Mexico in a book that is ostensibly to tell the reader about India? When one encounters "the other," is some comparison inevitable? If a human is transplanted to another culture, physically or mentally, can he or she do anything else but compare it to his or her home? And if so, once we begin comparison, where does it lead us? Literary critic Kenneth Burke insisted that all human comparison results in the creation of hierarchy.[13] Humans are "goaded by a sense of hierarchy"[14] says Burke. This question of human psychology will not be answered in this book, but we must remain aware of the potential dangers when conducting comparisons of any type, including in comparative environmental law.

While one can tell a convincing and simple story of the advantages of physical comparative abilities, we must examine whether there are advantages to cultural comparisons. Even the more *avant garde* theories of comparative law overlook (in a literal way, they seem to just see past the human practice they are employing) a scientific discussion of comparison itself. What can one hope to achieve? On the one hand, one wishes to say that the purpose of making a legal comparison is not to create some sort of list of football teams from which one can argue as to which one is better. On the other hand, if one stops at simply describing the basics, which is just step number one in the functional method of comparative law proposed by Zweigert and Kötz called "laying out the essentials," one has not gone far enough and has provided no comparison, but rather has only provided the basis for a juxtaposition. Here we see the core of a problem located in presumptions.

12 Paz, Octavio. *In Light of India*. Harcourt Brace, 1997, pp. 164–65.

13 Burke, Kenneth. *Language as Symbolic Action Essays on Life, Literature, and Method*. Univ. of California Press, 1966, pp. 15–16.

14 *Ibid.* at 16.

A lesson learned from the study of comparison is that although presumptions of similarity or difference are capable of identifying material to compare, they are just the beginning of the comparison process. Most comparison is difficult precisely because there are both differences and similarities between any phenomena under examination, and thus much more focus needs to be placed upon the degrees of difference and similarity,[15] rather than simply identifying that there is a similarity or difference. As Harvard Law Professor Mary Ann Glendon has noted, when it comes to comparison in law:

> most cases that reach the [U.S.] Supreme Court involve choices between positions that are supported by weighty moral and legal arguments, and the Court more often than not must make choices that, either way, will entail substantial individual or social cost.[16]

In other words, all the discussion about whether to proceed from the presumption of similarity *or* difference is misleading. Most, if not all, cases are both similar and different by matters of degree, and the decisions must be reasoned on grounds beyond simple categorical similarities or differences.

Evidence of the study of comparison is found at least twenty-four centuries ago in the Western tradition. Indeed Aristotle treated comparison (the greater and the lesser) as a *tópos koinós* that is, as a commonplace or common topic, to which orators gravitate when inventing arguments, like the common topics of relationship, circumstance, testimony and definition.[17] Aristotle develops his categories from observations of human behavior. From this practice of observation, we have extended the use of comparison to abstract social and cultural phenomena. We have successfully expanded our objects of study from social and cultural practices that enable choices among *physical* phenomena to social and cultural practices that enable choice among *social and cultural* phenomena themselves. One of these social and cultural phenomena is the law. One may well ask whether these conceptions of comparison have ever been directly applied to law. There are strong implications that they have been.

15 The principle of similarity and difference asserts that for any two things—material or abstract—one can always find a similarity or a difference and thus to simply say that things are the same or different is an empty assertion. This assertion is particularly pointed at academic conferences wherein papers presented are often limited to telling the audience that two things that appeared to be the same were in fact different, or two things that appeared to be different are in fact the same. See Stojkoski, Nikola, and Dragana Velkovska. *Essay on Human Reason: on the Principle of Identity and Difference*. Vernon Press, 2018; Heidegger, Martin. *Identity and Difference*. Harper & Row, 1969.

16 Glendon, Mary Ann. "Comparative Law in the Age of Globalization." *Duquesne Law Review*, vol. 52, no. 1, 2014, p. 1.

17 Early rhetoricians Protagoras, Gorgias and Isocrates made use of topics. Aristotle devotes an entire work to it, Τοπικά (*Topics*). and characterizes "topics" in his Ρητορική (*Rhetoric*) 1403a18–19. Cicero adapts much of Aristotle's *Rhetoric* in his own work, *Topica*, iii. 13–17; ix.41–xi.46.

1.2.2 Comparative practices made within the law

Before moving to comparisons *across* legal cultures, it should be noted how often comparison is used *within* a legal culture. Within the practices of law, humans apply their social skill of comparison. According to historian Harold Berman:

> The conceptualization of general legal terms, like the formulation of general principles underlying the legal rules, was closely related not only to the revived interest in Greek philosophy but also to developments in theology; and both the philosophical and the theological aspects were closely related to the great changes in political, economic and social life which constituted the Papal Revolution.[18]

Many of those ideas, including legal ideas that were received in Europe as Roman, could be traced to their Greek ancestors.

> "Comparison" arguments could be used by prosecutors (in the Roman empire of Cicero) showing that no one but the defendant could have committed the act, "or at least not so easily," while the defense would argue that the crime benefited others as well, or that others as well could have done the deed.[19]

Importantly, those Roman practices did not end with the fall of the Roman Empire.

> [T]he European jurists who revived the study of Roman law in the eleventh and twelfth centuries set out to systematize and harmonize the huge network of Roman legal rules in terms both of general principles and of general concepts, using methods similar to those which their colleagues in theology were employing to systematize and harmonize the Old and New Testaments, the writings of the church fathers, and other sacred texts. The jurists took as a starting point the concept of a legal concept and the principle that the law is principled.[20]

In constructing the general legal principles, they not only practiced comparativism, but made it the heart of their newly established legal science. In summary, history demonstrates that we have applied our physical abilities to compare social phenomena, including within legal practices. Furthermore, since the

18 Berman, Harold Joseph. *Law and Revolution*. Harvard University Press, 1983, p. 150.

19 Shapiro, Barbara. "Classical Rhetoric and the English Law of Evidence." *Rhetoric and Law in Early Modern Europe*, Yale University Press, 2001, p. 57; citing Cicero, M. Tullius, and Harry Caplan. *Ad C. Herennium De Ratione Dicendi: (Rhetorica Ad Herennium)*. 1964, pp. 2.3; 3.4–5; 4. 6–8.

20 Berman, *supra*, note 18, p. 150.

Enlightenment, we have tended to quantify degrees of difference and make our choices based upon numbers. That too might be a helpful comparative practice, but not all differences and similarities of degree are quantifiable. In the law, we are still often left with the task of choosing to say if and when qualities differ and defending our choices.

1.3 From comparative subjects to comparative legal cultures

This chapter focuses on addressing the question, "Why compare?" What normative force would suggest that if something worked well as a legal tool in Chile for Chileans or Australia for Australians, it should or could be adopted for example by Nigerians in Nigeria? There are many levels to explore in answering this question. On one level, one can be a mechanical practitioner and look for similar markers among the peoples and thus conclude the norm or practice is transferable. One can also be a colonizer and claim that the practice or norm to be borrowed is somehow in and of itself superior, and all humans ought to use it. This second approach is dangerous. It might be as innocent as a well-meaning philosophical abstraction or as predatory as a globalizing measure that has as its real goal to extend the legal influence of the globalizing state in other states for no other reason than to expand markets through norms and culture.

In an assessment of the style of writing court opinions, Sir Konrad Schiemann, former judge on the European Court of Justice (E.C.J.), and former judge of the High Court in England, made a subtle but scientifically important conclusion. Schiemann compared writing judgments in England to writing those of the E.C.J., using the criterion of style in his explorations and explanations. In the end, however, his comparison made a conclusion based not only upon style as an isolated marker, but upon social purpose *relative to the stated purpose of the court.* From that assessment, his refreshing conclusion was that:

> it seems to me that the E.C.J.'s practices have the advantage over English practices—at any event for the task that this court has to fulfill. I have the feeling that there is a genuine attempt to arrive at the best common solution that the brains of the court can reach.[21]

Where does the comparativist position himself or herself, such that the comparison can be conducted? Too often that which is labeled "comparative law" is in fact a discussion of foreign law, and one that lacks any self-reflective or critical element of comparison. Although the practice of comparison has been with us for some time, the practice is most often silent on the nature and workings of the very act of comparison itself. For example, my own survey of many well-known

21 Schiemann, Konrad. "From Common Law Judge to European Judge." *Zeitschrift des Europäisches Privatrecht*, vol. 4, p. 747.

scholars working today found that their institutions list their areas of research as including comparative law, but a review of the titles these scholars publish indicates little talk of comparison. Instead, the titles seem to indicate that the scholar is active in at least one foreign language and one or more foreign legal systems that practice law in that language, but nowhere do these titles talk about comparison *per se*. John Henry Merryman has noted that the study of *foreign law* is what "most comparatists do in fact most of the time."[22] Consequently the scholar explains legal system A which is practiced in language A^i to his or her native audience of culture B, working in language B^i without an explanation of the two necessary comparative interpretations,[23] the first of which is language and the second of which is cultural. While a comparison could begin without focus on its process, an uncritical juxtaposition will, in short order, employ a comparative theory, such as formalism or functionalism, consciously or unconsciously. In short, the study of any foreign legal system is *not* the same process as the study of one's own system and that study is vastly improved if one is conscious of the differences of cultural approach within the foreign legal system before one carries out the comparison.

So the scholar at home in Germany, for example, might know the Russian language, might have conducted research in Russia, or perhaps even studied in Russia, and thus can deliver a useful description of Russian law to a German audience. But if that same scholar has not explained and employed a theory or at least a model of comparativism, then he or she might well have delivered observations on Russian law, but will have left the work of comparativism for the German audience, with no guidance, instruction, or wisdom on the topic.

The result of overlooking the nature of the comparisons is to allow the assumption that there is some "natural" correct way to conduct comparisons. It would be more accurate and helpful to keep separate a category of scholarship called "foreign law," in which persons trained in one tradition and language serve the important function of reporting on other traditions and languages to persons for whom the others are foreign. Even then, the foreign law expert will need to translate culture *and* language to his or her native audience, but that will differ from comparative law because being a "foreign" law scholar means a one-way delivery of the other system to one's own audience. Comparative law should not be so parochial in its one-way interpretive construction. In their canonical treatment of comparative law, Konrad Zweigert and Hein Kötz have agreed that the "mere study of foreign law falls short of being comparative

22 Reimann, Mathias. "The Progress and Failure of Comparative Law in the Second Half of the Twentieth Century." *Am. J. Comp. L.*, vol. 50, 2002, p. 675; citing Merryman, John Henry. *The Loneliness of the Comparative Lawyer and Other Essays in Foreign and Comparative Law*. Kluwer Law International, 1999, p. 4. In that same footnote, Reimann notes that "[l]ooking through the volumes of the *American Journal of Comparative Law*, one quickly recognizes that almost invariably, the articles about foreign law outnumber (often by a huge margin) those explicitly comparing two or more systems."

23 Großfeld, Bernhard. *Core Questions of Comparative Law*. Carolina Academic Press, 2005, *passim*.

law."[24] Unfortunately, the majority of literature has historically failed to heed that admonition.

Perhaps the distinction between foreign and comparative law is blurred because anyone who is studying something that he or she would call "foreign" must be doing so while standing in his or her own legal system. While an author of such a study may not explicitly talk of comparison, he or she is of course comparing. When studying a foreign legal system, we cannot avoid comparing it to our own, at the very least. But it is important to become cognizant that we are comparing. Successful comparative law study must bring one's comparisons to consciousness and not act as though they are fixed and non-negotiable. We would improve our comparative law practice if these marginalia would be made explicit and brought to the core of the work by the author, rather than be ignored and left in need of searching by the reader or critic.

1.4 Comparing across space and time

We learn that at the great Paris Exposition in 1900, the International Congress for Comparative Law introduced comparative law in the form in which we know it today. The spirit of that age was "progress." The goal of the Congress, as recorded by its reporter, Edouard Lambert, was a *droit commun de l'humanité*. He went on to describe what appears to have been the noble goal that:

> comparative law must resolve the accidental and divisive differences in the laws of peoples at similar stages of cultural and economic development, and reduce the number of divergences in law, attributable not to the political, moral, or social qualities of the different nations, but to historical accident or to temporary or contingent circumstances.[25]

So at the height of the world's love affair with industrial progress, it was thought by Edouard Lambert and the others present in Paris at the International Congress for Comparative Law that it was not only possible to distill a universal private legal system from comparing all those already existing legal systems, but that it was desirable to do so. Two world wars, a cold war, and substantial decolonization followed by globalization have at each step changed these goals for comparative law.

24 Zweigert, Konrad, and Kötz, Hein. *Introduction to Comparative Law*, Translated by Tony Weir, 3rd ed., Clarendon Press, 1998. p. 6. Matthias Reimann notes that "[i]ndeed, foreign legal studies ('*Auslandsrechtskunde*') would be a more precise term." Reimann, *supra*, note 22 at 675.

25 Zweigert, *supra*, note 24, at 61. Note that the Congress sought a similar goal in the year 1900 as the author of the *Liber Augustalis* sought in the year 1231 "to resolve differences."

By the middle of the twentieth century, a catalogue of reasons *why* one might compare legal systems or legal families had developed and been mapped.[26] Summarizing the mid-century catalogue, Günter Frankenberg asserted that "[t]he ultimate aims of comparative law [are]—to reform and improve the laws, to further justice and to better the lot of mankind."[27] At the end of the century, Mathias Reimann renewed the catalogue of reasons why we compare from some of those same authors whom one might consider to be the canon of comparative U.S.–European comparative law—René David, Mary Ann Glendon, John Henry Merryman, Max Rheinstein, Rudolph B. Schlesinger, Konrad Zweigert, and Hein Kötz. Those reasons are:

- Foreign models may improve domestic law.
- Comparative law practice promotes international unification or at least harmonization.
- Comparative law study and practice reveals the common core of all law.
- Comparative law study teaches the basic skills of international legal practice.
- The study of comparative law provides an overview of law on a world-wide scale by introducing the student to the major legal families, or at least provides knowledge of foreign legal families.
- The study of foreign law familiarizes students with foreign rules, concepts, and approaches and thereby facilitates communication with foreign lawyers.
- By forcing students to compare foreign law with their own, it forces them to be critical of their own system.
- The study of comparative law helps students to understand law as a general phenomenon, in particular its contingency on history, society, politics, and economics. By providing critical perspectives and explaining alternatives, the study and practice of comparative law fosters tolerance toward other legal cultures and thus overcomes parochial attitudes.[28]

26 Comparativists René David and John E. C. Breierly summarized the reasons as follows: The present utility of comparative law can be analyzed under three complete heads: it is useful in historical and philosophical legal research; it is important in order to understand better, and to improve, one's national law; and it assists in the promotion of the understanding of foreign peoples, and thereby contributes to the creation of a context favorable to the development of international relations. David René, and John E. C. Brierly. *Major Legal Systems in the World Today*. 2nd ed., Legal Classics Library, 1978; quoted *in* Frankenberg, Günter. "Critical Comparisons: Re-Thinking Comparative Law." *Harv. Int'l L.J*, vol. 26, May 2017, pp. 411, 418.

27 Frankenberg, Günter. "Critical Comparisons: Re-Thinking Comparative Law." *Harv. Int'l L.J.*, vol. 26, May 2017, pp. 411–13; citing Zweigert & Kötz, *supra* note 22, at 12–14, 19–23; Tunc, André. "La Contribution Possible Des Études Juridiques Comparatives à Une Meilleure Compréhension Entre Nations." *Revue Internationale De Droit Comparé*, vol. 16, no. 1, 1964, pp. 47–67.; Stone, Ferdinand F. "The End to Be Served by Comparative Law." *Tulate Law Review*, 25, 1951, p. 325; David, René. "The Study of Foreign Law as a Contribution towards International Understanding." *Int'l Soc. Scie. Bull.*, vol. 2, 1950; Yntema, Hessel E. "Comparative Law and Humanism." *Am. J. Comp. L*, vol. 7, 1958, p. 493.

28 Reimann, Mathias. "The End of Comparative Law as an Autonomous Subject." *Tulane Eur. & Civ. L. F.*, vol. 11, 1996, p. 49.

One may note that the above list does not mention the daily practices of lawyers. Today, many lawyers in many countries find themselves often in need of serving their clients' interests abroad. Minimally, these practicing lawyers cannot proceed under the naive notion that another legal system is a projection of their own, but simply expressed in a different language. Thus, comparative law is not a luxury, but a necessity. For example, I was once asked to provide an expert opinion on U.S. law to a German court in a case concerning the tax ramifications of a corporate spin-off according to U.S. corporate law. I explained that the characterization would be a matter for interpreting the federal tax code in the U.S.A., and not the state law under which the parent company was incorporated. But in Germany, the same matter would be a matter for the law of corporations. One of the parties insisted to the court that I had to explain the characterization as a matter of corporate law, not as a matter of tax law. The problem was not one of linguistic translation, but one of cultural translation. The party just could not understand that the same problem would be resolved through the tax code in the U.S.A., not through corporate law.

On another occasion, I was asked by a German court to provide an expert opinion concerning child custody according to a U.S. state's law. I provided the likelihood, based upon the statute and prior court interpretations of the statute, of how the state court would resolve the matter. The German court said that was insufficient to constitute an expert opinion. As an expert, said the court, I was to provide the definitive answer, not a likelihood. I explained that the expert report in my legal culture's understanding is not the definitive answer—the judge's interpretation is—and that expert reports, in the words of Justice Oliver Wendell Holmes of the U.S. Supreme Court, *are* the law: "The prophecies of what the courts will do in fact, and nothing more pretentious, are what I mean by the law."[29] Again, there was no linguistic translation problem, but we struggled with the legal cultural translation.

1.5 Conclusions

In the century or so that comparative law has been researched and practiced as such, our goals and theories have changed. The optimism of that purpose eroded with the passage of time during which a world system was not forthcoming, soured by the discordance of two world wars, a lengthy cold war, colonization, and globalization.[30] Some have not been perturbed by these events of history,

29 Holmes, Oliver Wendell. "The Path of the Law." *Harvard Law Review*, vol. 10, 1897, p. 457.

30 Although it should be noted that, as recently as 1989, physicist Werner Heisenberg still wrote: If one looks around in history as to what great capacities human societies hold, next to the primitive same-race feelings that are prevalent already in the animal kingdom, is the shared language. But in addition to these strengths are two more still, which are stronger and can bring together even peoples of different races and languages: a common faith and, strongest of all, a common law. Heisenberg, Werner. *Ordnung Der Wirklichkeit*. Piper, 1990, p. 152.

however. Pierre Lepaulle, the French courtroom lawyer who became a Harvard lecturer, claimed in the *Harvard Law Review* in 1922 that the Great War was evidence of an even greater need for comparative law because "divergences in laws cause divergences that generate unconsciously, bit by bit these misunderstandings and conflicts among nations which end with blood and desolation."[31] After the Second World War, the European common market was established by the legal acts of treaties in order to prevent a third war. With each new "crisis" in the E.U., José Manuel Barroso, former President of the European Commission, was asked whether the E.U. was successful. He answered without hesitation that it was, and quickly continued to say that Europe had not started another world war since the E.U. began. The Organization for African Unity, followed by the African Union and now the revival of the East African Community, have not only substantially decolonized Africa, but, with explicit reference to the E.U., work toward legal harmonization to reduce conflict. Perhaps a similar, though lesser, claim could be made about the United Nations. Both examples echo Lepaulle's point that when cultures are in dialogue, through some institutionalized means, they are less prone to war.

Literature

Berger, Peter L., and Thomas Luckmann. *The Social Construction of Reality: A Treatise.* Penguin, 1979.

Brown, Donald A. *Climate Change Ethics: Navigating the Perfect Moral Storm.* Routledge, 2013.

Frankenberg, Günter. "Critical Comparisons: Re-Thinking Comparative Law." *Harv. Int'l L.J.* 26. (May 2017), 411–455.

Großfeld, Bernhard. *Core Questions of Comparative Law.* Carolina Academic Press, 2005.

Lepaulle, Pierre. "The Function of Comparative Law with a Critique of Sociological Jurisprudence." *Harvard Law Review* 35.7 (1922).

Reimann, Mathias. "The Progress and Failure of Comparative Law in the Second Half of the Twentieth Century." *Am. J. Comp. L.* 50 (2002), 671–700.

Zweigert, Konrad, and Kötz Hein. *Introduction to Comparative Law.* Translated by Tony Weir, 3rd ed., Clarendon Press, 1998.

31 Lepaulle, Pierre. "The Function of Comparative Law with a Critique of Sociological Jurisprudence." *Harvard Law Review*, vol. 35, no. 7, 1922, pp. 838, 857.

2

A TAXONOMY OF COMPARISON

The *Accesus ad auctores*

Kirk W. Junker

Social science disciplines use comparison and have formed comparative literature, comparative politics, comparative philosophy, comparative medicine, comparative media studies, and comparative architecture. In the natural science disciplines, we do not find comparative physics or comparative chemistry, for example, and that brings about an important point for comparative law. Physics and chemistry work from the foundational principle that human behavior is irrelevant to the truths known through these disciplines. At one point, any attempt at comparative law between any two cultures worked from a similar foundational principle—that some values or notions of justice were in some way "natural" to all societies and comparative law could proceed from that assumption. Legal positivists would disagree and insist that any legal system is made by the cultural values in which it is found. A third position could offer that there is in fact something common to various cultures, but only because we agree that it is. Rather than argue the truth value of one or the other assumption, we propose that one must simply make clear from which base assumption a comparison is proceeding.

Building upon this third position, two types of comparison are possible. In Type I, the lawyer is comparing environmental legal cultures that have no common legal bond, such as a convention or common legislation. Thus, while one has a great deal of flexibility in choosing Type I comparisons, one is challenged to establish the perspective from which one may approach either culture. Consequently, the question as to why one would compare these two environmental law cultures must be made explicit.

In Type II comparisons, there is in fact a common legal bond to the two or more legal cultures that are being compared such as the Convention on the International Trade in Endangered Species, or common legislation such as the ambient air quality Directive 2008/50/EC of the European Parliament and of the Council. Therefore, while one is constrained to considering cultures inside that bond, one is relieved of the task of justifying why one has chosen these cultures and can make a useful and well-founded comparison based upon the convention or legislation that binds both cultures. More will be said about these two types as they are applied in later chapters.

As we saw in Chapter 1, Aristotle offered a variety of ways in which comparison might be conducted, and Greek culture can be seen to have influenced European legal culture, but for legal practice in the present, we need something more concrete in order to apply comparison to legal practices. One may observe the role of abstract comparison for the human and then the role of comparison *within* specific legal practices. There are many ways in which comparison is used within the law, but not to compare one culture or state to another. Eventually one arrives at comparative law itself. If comparative law is practiced with purpose, organization, and method (one might say, "scientifically"), one is faced with some basic thought challenges that are also used in other disciplines. For example, when a physician is alleged to have committed medical malpractice, he or she is often judged by comparing what another physician would have done in the same community in the same situation.

And with law—what constitutes the work of comparativism? Mathias Reimann, after completing a survey of comparative law literature, concluded that:

> The problem is that these books, articles, ideas, and critiques do not add up to a sum that is larger than its parts. Instead, they constitute a potpourri of disparate elements that coexist side-by-side but rarely relate to any overarching themes.[1]

"But comparative studies are most valuable—and also most interesting—when they are used to answer questions. The nature of the questions, of course, depends on the interests of the student."[2] As the human experience of space compresses, the comparative law of the future may well be far more concerned with comparisons of time than of space. As an aid to focus on the concept of comparison, one can begin by looking at when and why comparative law came into being as a separate discipline of its own.

As a simple heuristic tool to organize our pursuit methodically, one may *return* to the highly developed practice of interpretation in Western antiquity (going back as far as exegeses of Homeric epics) that the Greeks used not only for oracles, dreams, myths, and philosophical and poetical works, but also laws and contracts. I use the word "return" because the *accessus* was also used by the ancients for analyzing many types of texts, including legal texts.[3]

1 Reimann, Mathias. "The Progress and Failure of Comparative Law in the Second Half of the Twentieth Century." *Am. J. Comp. L.*, vol. 50, 2002, p. 675, 686.

2 Bracey, Dorothy H. *Exploring Law and Culture*. Waveland, 2006, p. 5.

3 See Quain, Edwin A. "The Medieval Accessus Ad Auctores." *Traditio*, vol. 3, 1945, pp. 215–264; Copeland, Rita. "Rhetoric and Literary Criticism." *Oxford Handbooks*, 2014.

That practice is hermeneutics. According to Mantzavinos, this practice was carried forward to the Middle Ages, from which time we know it through its Latin terms, when "the most remarkable characteristic of the interpretative praxis was the so-called *accessus ad auctores*."[4] There were many versions of the *accessus*, but one of the more widely used was the following typology of seven questions. I have rearranged the order to suit the needs of the current text.

- Why (was the text written) (*cur/causa*)?
- Who (is the author) (*quis/persona*)?
- What (is the subject matter of the text) (*quid/materia*)?
- When (was the text written or published) (*quando/tempus*)?
- Where (was the text written or published) (*ubi/loco*)?
- By which means (was the text written or published) (*quibus faculatibus/facultas*)?
- How (was the text composed) (*quomodo/modus*)?

Like hermeneutics generally, the ***accessus*** provides us with a time-tested concrete method.

2.1 Why compare and who is to be compared?

The question of why one compares has already been addressed in Chapter 1. Here, we can turn directly to the second item in the *accessus*: who? When it comes to considering who one might compare in various legal families or systems, the obvious choices would be those trained in the law—lawyers, judges, prosecutors, and perhaps police officers. The assumption supporting such a list is that specialized knowledge defines the lawyer. But the difference between the citizen and lawyer is not limited to knowledge specialization. In insurance disputes, for example, the insurance representative that works daily with insurance law may know insurance law as well as or better than the lawyer, as may other non-lawyers in regular contact with other areas of the legal process. But unlike other disciplines in which one is defined as a member based solely on specialized knowledge, the law authorizes only one person—the lawyer—to represent and bind the rights and liabilities of another person. This "right of audience" that a lawyer has in the court of the state on behalf of another person is the defining feature of a lawyer—not his or her specialist knowledge. For context, one should also consider the related phenomenon of the legislator. In many of the world's systems, one is not required to be a lawyer in order to be a law-maker. Yet would one say a law-maker is not a legal actor?

4 Mantzavinos, C. "Hermeneutics." *Stanford Encyclopedia of Philosophy*, Stanford University, 22 June 2016, plato.stanford.edu/entries/hermeneutics/#Intr.

Much of today's discourse on comparative law is spent on the things we compare, and within that discourse, the list is too limited to sources of law (constitutions, statutes, regulations, and judicial decisions) and associated institutions. Lacking in the study are legal actors beyond the lawyer. Citizens are also "legal actors" and play the essential legal (not political) role of accepting and thereby legitimizing a legal system. Citizens must accept both their own law and foreign law in order for either to be practiced and legitimated. For example, in a recent discussion with a software developer from country X, the developer mentioned that he had a new idea that he wanted to bring to the market, but there were obstacles. I asked him what the obstacles were, and he said that, first, he did not know who to bribe. He was serious. While this story is only anecdotal, it does indicate an important, but too-little discussed part of comparative law practice. What is the reception of law in the country of interest? The Organization of Economic Co-operation and Development (O.E.C.D.) does make efforts to fight bribery and corruption and keep abreast of that which is measureable.[5] But for legal counselling, there is no substitute for inside information on the acceptance and reception of law.

In another example, a colleague reported that while working in country Y, she would tell co-workers there that her husband was a judge in the U.S., but the co-workers were not only unimpressed, they were demonstrably negative about the profession. After a while, she learned that in country Y, it is well-known that judges are all corrupt state officials and that they are disrespected and distrusted. The colleague did not realize that judges were disrespected in country Y, and her new colleagues in country Y did not know that judges in the U.S. *are* respected.

In yet another example, a constitutional law scholar in country Z, in considering the role of the state in village legal matters, opined that, for most rural citizens of country Z, the notion of state courts as the appropriate cultural institution to settle disputes is regarded as alien—one would far more often consult a local elder.

All of the preceding applies to legal practice. When it comes to legal scholarship, another person to be considered is the author. Because a comparison is not possible from some sort of person-less neutral observer position, it remains a task of comparative law scientifically to include one's subjectivity. To illustrate, one can consider a common example of comparative organization with an Indian menu. "Veg or non-veg?" asks the Indian waiter, in reference to the different menus available.[6] As comparativists we may already make several important observations, beginning with examining the base position of the subject—the menu's author. What does the author consider to be "normal"? The waiter's question began with "veg(etarian)," implying that it is the starting position. Then

5 "Bribery and Corruption." *OECD*, www.oecd.org/corruption/.

6 To complete my inquiry, I should also note that I am assuming the Indian menu to be "foreign" to the reader such that there will be an impact on the reader when considering vegetarian as the normal starting position.

one may observe that the printed menu itself begins not with a label "vegetarian," but instead just goes right into a list of the dishes, thus implying that is the normal starting point. Perhaps there is in fact no meat or fish at all on this menu or perhaps it comes second, under a category called "non-veg." We can again make two important observations—why is meat second and why is it described not as "meat" but as "non-veg"? From the first, we can infer that fewer people are interested in meat. More telling is the labeling of meat or fish as that which is "non-veg"! This is a striking statement about the primacy of the vegetable dishes—we understand meat as "the other" insofar as it is not vegetable!

Thereafter arises a subtler set of questions about what certain items on the menu mean. Does one begin to explain or understand meanings from usage, definition, or something else? So, for example, does the waiter explain to a customer whom he believes to be "foreign" when and how one would eat a particular dish and with what complements? Or does he "define" the dish by listing the ingredients and how they were cooked? Does he make an assumption about his foreign guest when he interprets for him whether the food is very "spicy" or not very "spicy"?

From these simple menu selection comparisons, we are prompted to consider similar questions in law by analogy. Does one begin a legal comparison by being positioned in one's home state or somehow from the perspective of a foreign state? What does one compare—sources of law, institutions of law, or something else, such as the persons subject to the law? If one begins from sources of law, which sources of law are paramount? If one begins from institutions of law, which institutions are paramount? Starting points make a difference.[7] Unconscious notions of home, family, normalcy, place, and familiarity will, without self-reflection, produce biases in the comparison. A tool to work against these tendencies would be to begin the comparison with "the other,"[8] and to build comparisons from that perspective. In the case of comparative law, that means beginning a comparison from the point of view of the "foreign" system, recognizing of course that one is doing so not as a cultural insider in that system, but as a comparative scientist.

2.2 What is to be compared?

Comparing India to his native Mexico, Octavio Paz wrote that:

> Indian sculpture is naturalistic, like that of Greece or Rome, and is the aesthetic opposite of the sculpture of ancient Mexico, that lover of terrible abstractions. But in their folk arts the Indian and Mexican sensibilities converge: fantasy, humor, bright colors, bizarre shapes.[9]

7 Großfeld, Bernhard. *Core Questions of Comparative Law.* Carolina Academic Press, 2005.
8 As is advocated by Prof. Dr. Heinz-Peter Mansel when teaching comparative private law at the University of Cologne.
9 Paz, Octavio, and Eliot Weinberger. *In Light of India: Essays.* Harcourt Brace, 1997, p. 138.

2.2.1 *Institutions of* techné *or* epistêmê?

To earn the qualification to practice law in most cultures, one must complete a long and difficult course of study, especially in comparison with some other disciplines. As a result, when a person completes legal studies, that person often feels as though he or she has accomplished an academic achievement. But what lawyers in fact have learned in the law is largely a set of practices, or what the ancient Greeks called *technê*, rather than abstract or absolute understanding, known to the Greeks as *epistêmê*.[10]

2.2.2 *Practice*

The most obvious way in which to implement that which legal education and training has to offer is through practice. From practice, not from education or training, one can then develop a more abstract wisdom. Theory informs practice and practice informs theory. One without the other is either an incomplete position from which to theorize law or an incomplete position from which to help clients. One might compare, for example, the German legal education and training system, which makes a clear split between the education a student receives in the classroom before the first state examination, and the various stations of practical training that the student receives between the first and second state examinations. In the U.S., during the same years that a student is learning in the classroom, he or she is required by the American Bar Association to have "experiential" training outside the classroom.

2.2.3 *Policy*

There are denotation problems with discussions of the word "policy" in the English language. Given the number of different ways in which "policy" is used, it seems to mean everything and therefore nothing. Some people, for example, interchange "policy" with "politics." But it is not the same word and does not mean the same thing. For others, "policy" means a wish list announced by politicians, perhaps even reduced to writing, such as a plan in a white paper. But such pronouncements are in fact just wishes, hopes, and theories. Why dress up a plan or a proposal with the word "policy"? Presumably, it is intentionally used to blur the distinction between that which has been done (and can be demonstrated to have been done) and that which is only a wish, a plan, or a promise. "It is our policy to be energy efficient" implies that we should believe it to be the case in the future because it has already been the case in the past. If it has not already

10 See Junker, Kirk "Rhetoric Demonstrates the Foundation of Law as *Technê*, Not *Empeiria*," *Philosophy, Art and Technology*, edited by K. Boudouris, Ionia Publications, 2011, pp. 99–114.

been the case in the past, why should we expect it to be the case in the future? Such promises do not merit being called "policy."

There is, however, the sense in which one looks backward, not forward, to say what a "policy" has been, not what one would wish it to be, and, in this sense, one can point to observable, recorded facts. In this sense, "policy" is that which has been done, not that which is promised or hoped to be done. An example was provided to me by a colleague in Ireland who pointed out that, for years, the Irish government said its policy was to keep all the brightest young workers and university graduates in Ireland. But if one observed the mass emigration of these young people, one could only conclude that the policy was not to keep them in Ireland.

Moreover, even when discussions of "policy" touch upon observable facts—that is, the *science* in "political science"—they are not just discussions of political theory. If one observes past practices and records them, one can claim to have observed a "policy." But we have no reason to believe that plans for the future will be actualized, and many reasons to believe they will not be actualized, so the planning sense of "policy" remains just that—plans. On another level of nuance, "policy" might not only mean a randomly observed set of past practices, but rather an observed set of acts of discretion by actors within the law, but not required to do so by the law. An example would be the permits chief of a state air quality program. The statutory law says what emissions limits may be. The chief must review permit applications and issue permits if they meet the criteria. But he must also be sure that there are not so many permits in his air corridor that overall exceedances reduce overall air quality, even though each facility would be within the law. His record of the quantity and quality of the limits that he imposes in the permit is his policy then. Not his plan, not his wish, but the summary of his recorded past actions of discretion within the law can then be described as his policy. The most important issue in such situations is whether the state may be held to maintain such a policy when that same chief all of a sudden reduces the emissions limitations in a new permit. Can that permittee appeal and say it may rely on the limits in past permits? *That* is a policy question—not newspaper talk about plans for 2050. It would seem that the only way in which one might proceed with some usefully scientific sense of public policy is to limit it to these observations of past acts of discretion by actors with legal authority of implementation, but no legal requirement to do so.

One may of course wish to speculate on future behavior as an extrapolation of past behavior, but given that the actors involved may use their discretion in the meaning that I offer, one has no guarantee that the discretion will be exercised in precisely the same way in the future—nor should it be. The persons charged with making decisions are given the right to exercise their discretion for a good reason, presumably, and not for the opportunity to hide unfairness, injustice, or corruption. Those who implement the law are not computer programs running on a mathematical algorithm.

2.2.4 Custom

The goal of the practitioner is of course to understand legal practice in a given matter. To do so, one needs to understand rules of law, perhaps expressed in texts, *plus* extra-textual practices. Seen anthropologically, all law begins in customary practices.[11] Some tend toward maintaining the role for custom, such as the common law "legal family" does, and others tend to construct systems outside of custom that vary in the degree to which they ignore explicit reference to their customary bases, such as the civil law family does. (I say "tend to" because neither family in the twenty-first century (if ever) is wholly and exclusively one or the other.) Arguments have been made that those that retain customary bases are therefore more supported by the culture in which they operate, rather than being seen as law for lawyers only. A basic function for any legal system is to resolve dispute among its members. Anthropology notes that until a group of humans is 150 or more, it does not really need a legal system.[12] Research shows that the common size of a chimpanzee group bonded by gossip is about 150 individuals. The threshold in human organization falls around the same number. Below the threshold, communities, businesses, social networks, and military units can maintain themselves based upon intimate acquaintances and gossip. Above that number, other means are necessary—formal ranks, titles, and law books.[13] It is only at the point of there being insufficient communication ability for a leader to make direct decisions among the group that other persons must make rules, interpret, and enforce them: in short, a legal system. Additionally, we expose a European bias when we become overly concerned with dividing the world neatly into common law and civil law jurisdictions, and even when we categorize other jurisdictions as a "mix," if they are constituted only as a mix of common law and civil law.

2.2.5 Texts

Bearing criticisms in mind, the fact remains that, if we want to compare legal cultures, there must be something we are comparing, and for the comparison to be useful and intelligent the things compared should in some way be justifiably comparable. Much of the work in comparative law compares civil law problems and solutions with common law problems and solutions. Here too, a type of legal ethnocentric bias is possible. Civilists, due to the structure of civil legal systems, might assume that all of law must be capable of a grand, unifying system not because comparative law must be able to make that possible, but because civil law behaves that way. As a result, civilists might conclude that:

11 Beru, Tsegaye, and Kirk Junker. "A Constitiutional Review and Customary Dispute Resoltion by the People in the Ethiopian Legal System." *NC.J. Int'L 1*, XLIII, 2018, p. 1.

12 Harari, Yuval Noah. *Sapiens*. Penguin, 2011.

13 *Ibid.* at 30.

"Only the Continental systems, with their tendency to abstraction and generalization, develop the grand comprehensive concepts, while the common law, with its inductive and case-by-case habits, produces low-level legal institutions especially adapted to solve isolated, concrete problems."[14] Alternatively, some authors would compare so-called "black letter law" (that is, texts) and conclude therefrom that civil and common law cultures are converging. The mixing of systems within the European Union contributes to this phenomenon. "It is worth noting, however, that English judgments nowadays, by reason of the incorporation of Community law and the European Convention of Human Rights, use such abstract concepts [as principles, freedoms, and rights] more than in the past."[15] But if instead of comparing the black letter law one compares the mentalities of the practicing lawyers, as does Pierre Legrand, one might well conclude that there is no convergence.[16]

Even at the height of critical legal studies, and after a long and convincing exposé of the weaknesses of comparative law in practice, Günter Frankenberg concluded that:

> To abandon comparative legal studies would be wrong-headed, I think, for it would freeze the tradition and current conditions into an eternal pattern. It would be equally wrong to go on with a comparative muddling-through. And from reading through the various approaches and from such highlights as Pound's *Comparative Law in Time and Space*, I infer that it is not just a more complex and longer process of comparison that is needed. Comparative law never had too little baggage in the overhead compartment. To this very day it is crammed with thoughts and thoughts, with aims and claims.[17]

A study of the history of comparative law yields some trends in method. And so the history of comparative law may be told as a history of the orthodoxies of method, whether they were consciously or unconsciously employed. Regardless of the reason why one compares parts or whole legal systems, one must proceed methodically in order to compare intelligently. As Paul Koschaker warned in 1935, "bad comparative law is worse than none."[18] For example, bad comparative law might simply look to compare texts (usually in some language of translation, perhaps for both countries).

14 Frankenberg, Günter. "Critical Comparisons: Re-Thinking Comparative Law." *Harv. Int'l L.J.*, vol. 26, 1985, pp. 441, 440.

15 Schiemann, Konrad. "From Common Law Judge to European Judge." *Europäisches Privatrecht (ZEuP) Heft*, vol. 4, 2005, p. 742.

16 Reimann, *supra* note 1, at 690.

17 Frankenberg, *supra* note 14 at 441.

18 Großfeld, *supra* note 7 at 14; quoting Koschaker, Paul. "Was Vermag Die Vergleichende Rechtswissenschaft Zur Indogermanenfrage Beizusteuern?" Vol. 1, 1936, pp. 145, 149, 150.

> Codes and treaties on the law of evidence are, at best, unreliable guides. Superficially, the rules of two systems may appear very similar. For example, some aspects of Dutch procedures bear a striking resemblance to recent proposals for new procedures in the Scottish courts. Yet we can be certain that if the Scottish proposals were to be implemented, the Scottish courts would still work very differently from the Dutch courts.[19]

To make useful and intelligent comparisons in the law, one must know more than the words of texts.

2.2.6 Reception of law—foreign and domestic

In considering the legal subject, one must include a consideration of the acceptance or reception of a legal system by the subjects as a necessary step toward legitimation of the system. In a review of three U.S. Supreme Court cases Professor Mary Ann Glendon of Harvard Law School lays out the disparate positions of U.S. Supreme Court Justices Stephen Breyer and Antonin Scalia on the reception of foreign law. She notes that the debate seems to be about the propriety of foreign norms in U.S. legal culture, but in practice might have much more to do with the balance of power between courts and legislative bodies. She points out that Justice Breyer ought to have noticed the fact that the foreign norms were legislated, not opined, by the courts, and that that difference mattered when it came to implementation or the "starting point," as Justice Breyer calls it, of foreign norms. She further cautions that foreign law should be used with attention to detail of how it functions in its own particulars, rather than transpose assumptions from one's own particular system. That is of course a solid critique. She could go further—the citizens of civil law jurisdictions do not expect authoritative interpretations of legislation from their judges. When they seek interpretation, they seek it from scholars in the field. So just as Professor Glendon points out in her study of U.S. Supreme Court reception of foreign law that Justice Breyer ought to have noticed that foreign countries' norms on juvenile death penalties or homosexual acts are legislatively determined, it should also be noted that in civil law countries one would not expect explication of norms by courts. This consideration of the reception of foreign law leads us also to consider what the attitude of a given society of citizens is to its own law, that is, the domestic reception of law itself.

19 Lever, Jeremy. "Why Procedure Is More Important than Substantive Law." *International and Comparative Law Quarterly*, vol. 48, no. 2, 1999, pp. 285–301; quoting Edwards, David "Fact-Finding: A British Perspective." in D.L. Carey-Miller and Paul R. Beaumont (eds.), The Option of Litigation in Europe (London: U.K.N.C.C.L., 1993), p. 44.

2.3 "When?"—comparative law has a history

Octavio Paz was comparing times, rather than places, when he wrote: "As one can see from these few examples, the poets of ancient India despite the centuries of differences in language, custom and belief are in many ways our contemporaries."[20] Legal comparison can be helpful over time in the same culture and over time in different cultures. Moreover, comparativist Paul Koschaker even went so far as to say that legal comparison is not possible, only comparative legal history is possible.[21] Where has the focus in comparative law been until now? Even though it was with the Paris Exposition of 1900 that modern comparative law became recognized in its present disciplinary form and name, practitioners of law looked outside their own systems long before. Already in the eleventh and twelfth centuries, the historical records indicate practices that we might today call "comparative." Historian Harold Berman tells us that, in the Kingdom of Sicily, "the jurists observed that *in all the various legal systems under examination* the question arose whether one who was forcibly dispossessed of his goods has the right to take them back by force."[22] The author of the 1231 *Liber Augustalis*, a codification of the law, was:

> moved by the spirit of scholasticism that informed the intellectual life of the age to resolve differences within the existing legal tradition of the *regno* and to distill his legal knowledge and that of his associates, probably practical men of the courts, into a unified body of law...[23]

Much later, we know that William Fulbecke wrote a *Parallele or Conference of the Civil Law, the Canon Law and the Common Law of this Realme of England* (1601–02), which like John Cowell's *Institutiones Juris Anglicani* (1605) and Sir Thomas Ridley's *View of the Civile and Ecclesiastical Law* (1607), compared and contrasted the rival codes.[24] In his 1614 *Eirenarcha*, William Lambarde recommended that a justice or judge was to "engage in the mental operation of 'comparison,' an act requiring analysis of whether 'none but he, or not so commodiously as he, could commit the fact.'"[25] After comparative law became established as a discipline, practicing lawyers and scholars could apply to individual areas of legal practice.

20 Paz, *supra* note 9 at 161.

21 Koschaker, *supra* note 18 at 150.

22 Berman, Harold Joseph. *Law and Revolution*. Harvard University Press, 1983, p. 153.

23 *Ibid.* at 427; quoting Frederick, and James M. Powell. *The Liber Augustalis: or, Constitutions of Melfi, Promulgated by the Emperor Frederick II for the Kingdom of Sicily in 1231*. Syracuse University Press, 1971, xxi. It is also worth noting here that, as with the common law, the knowledge of "the practical men of the courts" was part of the unified body of Sicilian law.

24 Prest, Wilfrid R. *The Inns of Court under Elizabeth I and the Early Stuarts, 1590–1640*. Rowman and Littlefield, 1972, pp.146–47.

25 Shapiro, Barbara. "Classical Rhetoric and the English Law of Evidence." *Rhetoric and Law in Early Modern Europe*, Yale University Press, 2001, p. 65.

Beginning with the onset of public "environmental law" in about 1970, the already seventy-year-old modern discipline of comparative law could be applied to environmental law.

2.4 Where?

Especially since the mantra of "think globally, act locally" has come on the scene in environmental thinking, one often speaks and reads of the role of local communities or cities,[26] and of tribal, native, or first peoples. Despite these discussions, when one enters the realm of comparative *law*, the discussion often reverts back to states. So on one's first pass, answering the question as to *what* one compares in environmental law, the answer is often "states." But even then, one must explain to the audience why these two (or more) states are being compared out of the nearly two hundred possibilities. Here we are again reminded of the comment by John Henry Merryman that the study of foreign law is in fact what most comparatists do most of the time. Matthias Reimann echoes this observation:

> outside of a small hard core, most of those engaged in comparative work of one sort or another do not even think of themselves (primarily) as comparative lawyers but mainly as Asia specialists, Russian law scholars, constitutional lawyers with comparative interests, etc.[27]

The motivation for the label in those cases might well be personal, such as when one is married to a person from a foreign legal culture, or one had the chance to study there years ago and holds that foreign culture dear to his or her heart. These are questions that only the author himself or herself can answer, but answer he or she must. And if it is personal, then that author must tell the reader that the states were, at least in part, chosen for personal reasons, and not exclusively due to scholarly or scientific reasons. Only then can a reader, particularly a beginner, not be misled into believing that a particular state has a particularly important role to play in comparative environmental law. If, on an enlightened day, the discussion can turn away from states and listen to the call—act locally—one might in fact compare provinces, regions, or cities[28] as other legal cultures.

2.5 By which means?

Journalists today would often be aware of, and perhaps even employ, the *accessus ad auctorem* as a quick checklist by which to investigate a news story. In the past, journalists' reflections on means might well have been a comparison of print or

26 See *ICLEI*, https://www.iclei.org/.
27 Reimann, *supra* note 1 at 687.
28 See, for example, the work of ICLEI. *ICLEI*, https://www.iclei.org/.

broadcast media. Those discussions treated the media as conduits for the substantive message. Then, in 1967, Canadian scholar and communication theorist Marshall McLuhan provocatively announced that the medium *is* the message. That observation was not only a mimetic label for the age, but became a prescient insight into the social media of today. The absence of trained, licensed journalists, the absence of state control, the absence of peer review, and the absence of editorial and publisher review, all joined by the ubiquitous ease through which anyone with internet access can publicize an opinion indistinguishable from evidence-based science, returns the production of knowledge—including environmental knowledge—to some of its most primitive nature.

2.6 How? Comparative law has methods

Like other lines of thought in this chapter, the notion of how one compares can be traced historically. What we might have once thought was the best or most useful method of comparison is no longer the case, and for those who are forward thinking and planning, we should realize of course that the present method(s) will not likely be the best or most useful in the future. German philosopher Martin Heidegger once voiced his concern that the more basic questions of "what," answered through ontological study, had been overtaken by simpler epistemological questions of "how?"[29] Heidegger's complaint finds resonance in U.S. law, where one speaks of "substantive due process of law" as a concept in which one is considered to have his or her substantive justice guaranteed by completion of a process.

The focus upon method consumes much of the comparativist's attention, even if in fact he or she is practicing comparison and not just describing a foreign legal system. In consideration of the fact that the human capacity and need to compare changes from time to time and place to place, we come to the question, "how is comparative law conducted currently?" A neatly-packaged history of the schools of thinking in comparative law cannot dispositively answer the question because there are not clean and complete breaks over time with any one "ism" (formalism or functionalism, for example). And besides that fact, few scholars, and even fewer practitioners, reside neatly in any one school.[30] Nevertheless, some approximations in both history and intellectual camps are helpful. To answer this question, it remains important to note that, as a discipline, comparative law cannot approach its topic from a neutral or omnipresent position. At a minimum, common law lawyers approach comparative law differently than civil lawyers do, because their point of departure differs.

29 Heidegger, Martin. *Being and Time.* Blackwell, 1997.

30 The full list, according to Frankenberg, would include formalism, encyclopedism, constructive historical reconstruction, juxtaposition-plus, functionalism, translational, and immersion.

Regardless of the means by which one arrives at a method of comparison, he or she must give serious consideration to the choice because a student or practitioner outside a particular legal society *cannot* study law in the same way that a person within that society would study its law. Learning language produces an apt analogy here. Even the person who does not study the law as a legal specialist has learned the values, norms, and procedures in that law, at some level, from everyday experience in his or her native culture. Some of the things a native of any culture learns about his or her legal system might even be regarded by legal professionals as being wrong, but if it is generally learned in that culture, it is a force that forms general cultural expectations and cannot simply be dismissed. By comparison, we learn our native language by usage, and learn that usage in the context of its native spoken culture. We learn additional languages through the methodical mechanics of grammar, syntax, and vocabulary, outside of the native cultural context. Therefore we should be wary of the mechanical artificialities, out of context, when learning additional legal cultures and making comparisons.

A form of weak comparison is that which is typical of colonization justifications, beginning with the assumption that what one already does is the norm by which all others are measured, and if another person or culture does things differently then that other person or culture is, quite literally, "abnormal." This seems to be more commonly accepted among non-comparativists and even among those who are neither comparativists nor lawyers, such as the German television journalist who telephoned me when Dominique Strauss-Kahn was arrested in New York and demanded to know why "American" law[31] allowed Mr. Strauss-Kahn to be shown in handcuffs and why "American" law allowed cameras in the courtroom. In the journalist's own culture both were prohibited, so to him the prohibition of either was "normal," he said.

> The similarities that surface in the course of such comparisons are mirror images of the categories of the conception of law in the comparativist's own culture. Ambiguities are defined away or adjusted to fit the model; thus the "home" law is positioned as natural, normal, standard...[32]

31 Substantive criminal law differs in the United States in each state; thus so long as it does not violate the US constitution, it remains a matter of state law, not federal, unlike German centralized federalism. In short, there is no one "American" criminal law.

32 Frankenberg, *supra* note 14 at 423. In the spirit of the critical legal studies relative, post-modernism, one can turn this critique back on Frankenberg and note that he too is too much wed to continental civil law when he says "In order to find, compare and evaluate these effects, the comparativist has to move back and forth between texts and their application." Such a characterization of law omits both customary law and the oral tradition, and assumes that at its foundation, law is a text.

Correcting for one's native bias is one of the thorniest problems in comparative law scholarship and practice. Can these biases be corrected simply by employing the correct method? Perhaps. In criticizing the status quo of comparative law, Günter Frankenberg wrote:

> The comparativist approaches her field of research purely as philosopher, historian, sociologist or legal scholar; her task is merely to collect interesting items, to systematize, to develop or unify the law and/or to bring about rational change ... They call for an objectifying methodological approach or trust that international collaborations will correct national biases. They modestly propose rules of comparative reason that, if every comparativist follows, will control subjectivity.[33]

The use of the *accessus ad auctores*, employed so far in this chapter, is an organizing principle that aligns with the functions of various aspects of comparison. There are of course alternatives, such as formalism, encyclopedism, translationalism, and immersion.

2.6.1 Formalism

Formalism in domestic law grows out of a sense that principles of law can be applied as (mathematical) forms to facts to determine just resolution of conflict or disputes. It therefore becomes a problem to extend formalism to comparative law because it involves using one's one domestic system as the form to which other systems must *con*form. An early attempt to correct for the bias of standardizing one's own norms was the practice of formalism. Given that formalism was a trend in other disciplines at the time, it should not be a surprise if one can trace a thread of formalist method in comparative law back to the beginning of the twentieth century.[34] "Formalism proposes a narrow conception of law that, in a comparative perspective, is informed by the domestic legal culture and then projected onto what in other historical or social contexts is, looks like or may be taken as law."[35] In comparative law, formalism theorizes law as a "set of rules and principles independent of other political and social institutions."[36] The limitations of formalism are demonstrated by considering whether one could compare Indian National Green Tribunal decisions with tribunal decisions rendered in another state by a court of a different form, such as general first or second

33 Frankenberg, *supra* note 14 at 424–25; Referring to Zweigert and Kötz as well as Ernst Rabel's statement, "If the picture presented by a scholar is colored by his background or education, international collaboration will correct it." Rabel, Ernst. "Deutsches Und Amerikanisches Recht." *Zeitschrift Für Ausländisches Und Internationales Privatrecht*, vol. 16, 1951, p. 359.

34 Schauer, Frederick. "Formalism." *Yale L. J.*, 97(4), 1988, pp. 19–58.

35 Frankenberg, *supra* note 14 at 422.

36 Garner, B. A. *Blacks Law Dictionary.* West Group, 1999, p. 913.

instance jurisdiction, or whether one could be restricted from comparing constitutional rights in the Republic of Ireland with those in the U.K., since the U.K. does not have a codified form to its constitution.[37]

2.6.2 Encyclopedism

As one would expect from the name, the legal encyclopedist strives to compile all legal jurisdictions, systems, and cultures in one descriptive work. However, in doing so, the encyclopedist must create categories that are common across cultures, jurisdictions, and systems. There has been at times great excitement around the method of encyclopedism.

> When the history of the *International Encyclopedia of Comparative Law* (I.E.C.L.) is written, it will be hard to avoid using superlatives. Monumental in scope, unique in conception, unprecedented in its worldwide cooperation of comparative law specialists, and based on tremendous amounts of research, expertise and technical effort, this unique scholarly enterprise, when completed, might well be known as the "work of the century" among comparative law scholars.[38]

Yet encyclopedism suffers from some of the same weaknesses as formalism. The legal encyclopedist, for example, needs to divide the objects of discussion in some consistent categories so as to organize the encyclopedia. The choice of divisions made by the author will not yield a neutral outline, but will follow more or less the outline of some form, often that of the author's home culture legal system. Here we return to the notion that if one is a legal naturalist, one might argue that there exists some pure legal forms and categories, natural to all human societies, to which the pieces of each legal system attach. The other extreme would be the legal positivist who would say there are no such categories natural to all systems. A third position would be to admit that, if any categories are common to all legal cultures, they are common by agreement, and not by some natural force. At that point, one can begin the useful and important work for determining when, where, and why these categories or norms were agreed upon.

How one categorizes and organizes the various elements of law are one of the first challenges of comparative law. René David proposed the classification of legal systems according to the different ideology inspiring each one, thus arriving at just five groups or families: Western Law, which was further subdivided

37 In addition to formalism, there are other schools of thought and theories of how comparativism ought to be done, including constructivism, historical reconstruction, and juxtaposition-plus. For a discussion of these, see Frankenberg, *supra* note 14 at 443.

38 Sprudzs, Adolf. "The International Encyclopedia of Comparative Law: A Bibliographical Status Report." *The American Journal of Comparative Law*, vol. 28, no. 1, 1980.

into the Romano–Germanic and the Anglo–Saxon subgroups; and, in addition, Soviet Law, Muslim Law, Hindu Law, Chinese Law, and Jewish Law.[39] David argued that the difference between the Anglo–Saxon Laws and Romano–German Laws is of a technical rather than of an ideological nature. David further noted that the characteristics that differentiate the Western legal family from the other four are its liberal democracy, its capitalist economies, and its Christian religion.

In the early 1950s, Pierre Arminjon, Baron Boris Nolde, and Martin Wolff organized seven law families or "groups," based upon common historical origins in sources of law and training of lawyers: the French group, the German group, the Scandinavian group, the English group, the Russian group, the Islamic group, and the Hindu group.[40] Like the system of David, this was an improvement over geographic boundaries or simple state identities.

Beginning from a basis in the Arminjon, Nolde, and Wolff system, Zweigert and Kötz arrive at a chapter on "Legal Families of the World" at the end of Part I of their work. Using their criteria, they categorized the legal systems of the world into seven families: the Romanistic legal family, the Germanic legal family, the Anglo-American legal family, the Nordic legal family, law in the Far East, Religious legal systems and the Socialist legal systems.[41] After the Soviet Union collapsed, the Socialist family was deleted from their list. Encyclopedism must not only make categories for the world's systems or families (depending on the encyclopedist), but also make categories within those families. So, for example, in *the International Encyclopedia of Comparative Law*, Volume I, State Reports, each state (not family) is divided into chapters on I. Constitutional System, II. Sources of Law, III. Historical Evolution, IV. Civil Law, V. Commercial Law, VI. State Direction of Trade, VII. Industrial Property Rights and Copyright, VIII. Principles of Judicial Procedure in Civil and Commercial Cases, and IX. Private International Law.[42] I had some personal experience contributing to the index for Volume I of the *International Encyclopedia of Comparative Law*. Beginning with Afghanistan, I worked in alphabetical order on state reports. Thus I came to the entry for the U.S.A. quite late in the exercise. The description looked foreign to me, a U.S. legal practitioner, because of the categories within the entry for U.S. law through the encyclopedist approach. That made me realize that although encyclopedism

39 David, René. *Traité Élémentaire de Droit Civil Comparé: Introduction à L'étude des Droits Étrangers et à la Méthode Comparative*. 1950.

40 Arminjon, Pierre, et al. "Traité De Droit Comparé Tome III." *Librairie Génerale De Droit Et De Jurisprudence*, 1951, p. 8.

41 Zweigert, Konrad, and Kötz Hein. *Introduction to Comparative Law*. Translated by Tony Weir, 3rd ed., Clarendon Press, 1998.

42 "International Encyclopedia of Comparative Law Online." *Brill*, referenceworks.brillonline.com/browse/international-encyclopedia-of-comparative-law-online.

does provide an organizational structure by using the first level, then selecting sub-categories, it has obvious and recognizable limitations, too.

2.6.3 Functionalism

From the formalist roots that often beset the beginning of attempts to systematize or scientize a discipline, through the organizational focus of encyclopedism, comparative law then arrived at functionalism. Not only does functionalism provide the theoretical advantages for legal science, but private practitioners who wish to represent the interests of their clients in negotiation or litigation are likely to use the functionalist approach. "Functionality becomes the pivotal methodological principle determining the choice of laws to compare, the scope of the undertaking, the creation of a system of comparative law, and the evaluation of findings."[43] Functionalism rejects formalism in part and incorporates formalism in part. The great proponents of functionalism, Zweigert and Kötz, reduce all of comparative law's method to functionalism: "The basic methodological principle of all comparative law is that of functionalism ... Incomparables cannot usefully be compared, and in law the only things which are comparable are those which fulfill the same function."[44]

At the center of functionalism is a presumption of similarity, which they acknowledge is a learned behavior: "The proposition rests on what every comparatist learns, namely that the legal system of every society faces essentially the same problems and solves these problems by quite different means though very often with similar results."[45] The *praesumptio similitudinis* is both the insight and the blind spot of the functionalist method. The weak point is that it begins from a presumption that legal systems of the world are sufficiently similar such that one can find similar functions among them. As a result of this presumption, Zweigert and Kötz go so far as to say that if at the end of research one:

> finds that there are great differences or indeed diametrically opposite results, he should be put on notice to go back to check again whether the terms in which he posed his original question were indeed purely functional, and whether he has spread the net of his researches quite wide enough.[46]

43 Frankenberg, *supra* note 14 at 436.

44 Zweigert, *supra* note 41, at 34.

45 *Ibid.* See Zweigert. "Die Praesumptio Similitudinis als Grundsatzvermutung rechtsvergleichender Methode," Inchiesete di Diritto Comparato—Scopi e Metodi di Diritto Comparato (M. Rotondi ed 1973), p. 735; Critiques of this presumption of similarity are found in Curran, Vivian Grosswald. *Comparative Law: an Introduction.* Carolina Academic Press, 2002.

46 Zweigert, *supra* note 41, at 31.

In other words, analyses can only be conducted if the questions used are capable of yielding similarities, regardless of how well those similarities represent the practices of lawyers in those systems.

It would be equally as logical to begin from the presumption that legal systems are dissimilar enough that no similarities can be found. Both assumptions put a rabbit in the old-fashioned magician's hat that pre-determines the outcome of the comparison. But as Zweigert and Kötz point out, assuming dissimilarity yields no results under the system. Perhaps that is where honest and open science should lead us. And here again we are reminded that, in order to adopt a suitable method, one must ask what the goal is to yielding such results. What claim can I make if I establish a similar function between two legal systems? Is it our goal to be able to say there are identifiable legal trains of thought common to all human societies? If one is attempting something like an objective or neutral science of comparison, it is indeed peculiar to advocate beginning from a perspective that is admittedly biased toward finding similarities, just as it would be to acknowledge and employ a bias in search of differences.

The Zweigert and Kötz *Introduction to Comparative Law* is designed to apply to private law, which, when the book was conceived, was a clear category within states of the civil law family. Indeed, functionalism seems especially to have an affinity for commercial law. In fact, "legal origins" scholarship, "produced primarily by economists, not legal scholars—has a close affinity with functionalist comparative law."[47] Commercial law and some other areas of the law are less culturally connected than other areas of the law.[48]

But other areas of the law are more intrinsically attached to the differences of a culture, such as constitutional law,[49] family law, criminal law, or, as we maintain, environmental law.[50] One of our themes in this book is that environmental law is not only about the material world, but is in fact culturally connected. These culturally-dependent areas cannot so easily be understood without a focus on the process of comparison. In the one hundred or so years that we have continuously recognized "comparative law" as an independent scholarly pursuit, many scholars have concluded that it has been unfortunately rather unproductive. A survey of comparative law literature, both among the authors who

47 Whytock, Christopher A. "Legal Origins, Functionalism, and the Future of Comparative Law." *BYU L. Rev.*, 2009, pp. 1879, 1880.

48 Gordley, James, and Von Mehren Arthur T. *An Introduction to the Comparative Study of Private Law: Readings, Cases, Materials.* Cambridge University Press, 2014.

49 See e.g. Dorsen, Norman, et al. *Comparative Constitutionalism: Cases and Materials.* West Academic Publishing, 2016; Tushnet, Mark. *Comparative Constitutional Law.* The Oxford Handbook of Comparative Law, 2006.

50 Discussions with practicing professional German lawyers who have obtained their master's degrees in law (usually the L.L.M. degree) have indicated that they chose their curricula based upon cultural interest—hence family law and criminal law—and not just practical, commercial interest. After all, is legal education not intended to be a complete education, not just formal training, *Bildung* and not just *Ausbildung,* in the words of Alexander von Humboldt?

are practice-oriented and those who are theory-oriented, makes two things clear: first, there is no agreed-upon set of practices or concepts by which one can denote "comparative law," and second, because of the failure to agree on the category, no method or theoretical framework seems to be happening that enables would-be researchers or practitioners to build a methodical, replicable discipline.[51] Mathias Reimann and others noted the unsurprisingly increased interest in comparative law in Europe as part of the Europeanization process of the Single European Act, the Maastricht Treaty, and the Lisbon Treaty, but they have also cautioned that the process is concerned largely only with black letter law, viewed positivistically, and advancing private interests. "As a result, there is no better theoretical framework in Europe than there is [for example] in the United States,"[52] where the bulk of publishing under the banner of "comparative law" has historically taken place. Perhaps it is therefore time to look outside this axis for a superior theoretical framework. Consequently, the successes of comparative law in the latter half of the twentieth century can largely be characterized as a tool for the "Europeanization of private law."[53] While new forays in comparative law may be happening recently in Asia, "Latin America continues to be understudied and Africa is almost ignored"[54] even by such flagship comparative law journals as the *American Journal of Comparative Law*.[55] And public law does not often get as much ink (even within Europe) as do Asia, Latin America, and Africa.

Functionalism has some advantages over other approaches, but is not without weaknesses. A first critique for consideration was launched already back in 1922, long before functionalism was named and introduced. Then, Pierre Lepaulle made a strong case that

> comparison restricted to *one* legal phenomenon [even if it does carry out the same function] in two countries is unscientific and misleading. A legal system is a unity, the whole of which expresses itself in each part; the same blood runs in the whole organism. Hence each part must necessarily be seen in its relation to the whole.[56]

51 Indeed, researchers in the field cannot even agree whether to call themselves "comparativists" or "comparatists."

52 Reimann, *supra* note 1, at 694.

53 *Ibid.*

54 *Ibid.*

55 One count of articles published from 1952–1997 of that journal contained 41.12 percent about Europe, 8.6 percent about Asia, and only 1.79 percent about Africa; See Upham, Frank. "The Place of Japanese Studies in American Comparative Law." *Utah L. R.*, vol. 1997, 1997, pp. 639, 641; Reimann, *supra* note 1, at 674.

56 Lepaulle, Pierre. "The Function of Comparative Law with a Critique of Sociological Jurisprudence." *Harvard Law Review*, vol. 35, no. 7, 1922, p. 853.

More recently, some of the criticism of functionalism has come from critical legal studies. Social unrest in much of the world in the 1960s eventually made its way to comparative law as well, and through the 1980s and 1990s comparativists in the critical legal studies movement in the U.S. were applying social thinking and post-modern theory to the law with the expressed aim of including marginalized and repressed persons' legal cultures among those to be compared, all the while questioning the orthodoxies that supported a notion of comparative law that one might easily describe as contributing to globalization.[57]

The problem of bias grew into allegations of ethnocentrism by proponents of critical legal studies. Günter Frankenberg offers as a corrective that:

> To cope with ethnocentrism we have to analyze and unravel the cultural ties that bind us to the domestic legal regime. A practical and rather fascinating beginning could be a deviant reading of comparative legal literature focusing on the marginal stuff that is normally skipped for lack of relevance. Forewords and prefaces have iterating stories to tell about how comparison, despite higher aims and claims, is inspired and organized, in part at least, by contingent factors that reveal perspective: the comparativist's legal education and exposure to specific legal cultures, honeymoons and travels, invitations to conferences, and so on. The marginal remarks indicate why and how the purportedly objective discovery and comparisons of the "compared" legal culture is undercut by the comparativist's assumptions.[58]

Where might one see evidence of ethnocentrism in functionality? If we apply an important lesson from comparative law itself to Zweigert and Kötz, we can legitimately examine the perspective from which they see the discipline of comparative law. Zweigert and Kötz divided their *Introduction to Comparative Law* book into public and private law parts, which is a fundamental civil law distinction, but not one so pronounced in the common law. They then organized Part II of the book, Private Law, according to the pattern of organization of the paragraphs of the *Bürgerliches Gesetzbuch* (German Civil Code): contracts, unjust enrichment, and torts. Most German readers will readily recognize these divisions as being borrowed directly from the organization of the German Civil Code. But rather than look for an introduction to comparative law written from one's own perspective, so as comfortably to ignore the fact that we are viewing the phenomenon from only one perspective, we can benefit from the perspectives of others who react to and analyze functionality. At the same time, one must, however, remain observant of the functionality principle, even while criticizing it.

57 See Mattei, Udo, and Mathias Reimann. "Comparative Law and Critical Legal Studies." *The Oxford Handbook of Comparative Law*, Oxford University Press, 2006, p. 815.

58 Frankenberg, *supra* note 14 at 443.

Thus in attempting to provide a science and a method for comparative law that is above or beyond any one system, even the powerful work of Zweigert and Kötz demonstrates that it is impossible to compare without having a point of perspective from which one compares. One cannot step out of the hermeneutic circle, but rather can at best become conscious that one is in the circle, that one sees the world from his or her place in the circle, and that one tries to make observations of other things in the circle knowing that his or her subjective perspective today is some other scholar's object of study tomorrow.[59]

2.6.4 Translationism

Bernhard Großfeld mobilizes translation as an alternative method for comparative law.[60] Two translations are always required: linguistic and cultural. Furthermore, Großfeld means more than simply to say that translation is *analogous* to comparative law. He notes that law

> arises much the way language does, as was so clearly pointed out by Friedrich Karl Savigny, who stated early in his classic work: In the earliest times to which authentic history extends, the law will be found to have already attained a fixed character, peculiar to the people, like their language, manners and constitution. Nay, these phenomena have no separate existence, they are but the particular faculties and tendencies of an individual people, inseparably united in nature, and only wearing the semblance of distinct attributes to our view. That which binds them into one whole is the common conviction of the people, the kindred consciousness of an inward necessity, excluding all notion of an accidental and arbitrary origin.[61]

Using an extreme example to make his point, Großfeld says that Eastern Law, like pictograms, is not comprised of alphabetized language, but is instead right brain thinking and therefore not comparable to other legal systems in which law is character language-based.[62] He also seems to allow the *praesumptio similitudinis*, but then offers both a critique of functionalism and an alternative that he builds through language study. Insofar as he allows for the *praesumptio similitudinis*, one

59 See Hegenbarth, Rainer. *Juristische Hermeneutik Und Linguistische Pragmatik Dargestellt Am Beispiel Der Lehre Vom Wortlaut Als Grenze Der Auslegung*. Athenaeum-Verl., 1982.

60 *Ibid.*

61 Hiller, Jack A., and Bernhard Großfeld. "Comparative Legal Semiotics and the Divided Brain: Are We Producing Half-Brained Lawyers?" *The American Journal of Comparative Law*, vol. 50, no. 1, 2002, pp. 175, 178–79; quoting Savigny, Friedrich Carl von. *Of the Vocation of Our Age for Legislation and Jurisprudence*. Littlewood, 1831.

62 *Idem*; See also Illich, Ivan, and Barry Sanders. *ABC: the Alphabetization of the Popular Mind*. Vintage Books, 1989.

could level a similar critique against him as that aimed at the functionalism of Zweigert and Kötz. Großfeld states that

> [W]e must not in comparative law first and foremost fix on "differences," and even exaggerate them as "exotic." It is more important to perceive commonalities. They form intercultural "bridges," enabling agreement across borders. If we total up the differences, we distort the picture. Comparative law first and foremost must be "bridge seeking" and "bridge building."[63]

The "must" in that statement is of course a matter of choice.

He further reminds us of the process of learning any second language. It is mechanical—one learns the structure of grammar and sentence syntax, and then inserts as much vocabulary as one can memorize. That is not how we learn our native languages. Like natural languages, we all begin our lives doing everyday things within our own legal system. We typically spend years in that system, accumulating knowledge of it in the same way as other laypersons before we ever begin to study it scientifically or professionally. Thus the categories in our mind called "language" or "legal system" are in fact biased by the fact that we have learned *our* language and *our* legal system, not some neutral category called language or legal system, and we thus learn any second language or legal system by translating it into our own. This is not a disastrous state of affairs, so long as we remain conscious of it. But if we assume that our legal system is in some way *the* legal system, with some sort of natural or objective connection to the world, then we have made the same error as if we had assumed the same about our spoken language.

We learn our native languages by usage before formal language education can even begin. In addition, when learning languages after the mother language is learned, we do so by comparison to our own first language. But beyond my analogy in which learning another legal system is *like* learning another language, Großfeld maintains that learning another legal system *is* learning another language, for we accomplish this practice of comparison by *creating*—not finding—relationships that we say are equal. In the same way that we cannot learn a second language as we learned our mother tongue,[64] we must at least become conscious of these realities of cultural translation, and not fool ourselves into believing that we can learn another culture, even a legal culture, simply by reading the same texts that a native reads. By seeing the act of comparison as an act of translation, comparative law goes beyond textual comparison into order and relationship comparison.[65]

63 Großfeld, *supra* note 7 at 12.

64 With the notable exception of course of those who have learned a second language at an early age and are truly bilingual.

65 Großfeld, *supra* note 7 at 125.

> [T]he reading of a foreign legal text often gives us a false picture. Misunderstanding is programmed. In order to avoid this, we need to recognize the context of the text. Yet we are not able to observe foreign contexts in an unbiased way, for each system responds to observation, observes itself. Just as in the natural sciences, social reality is different if we observe it from when we do not.

Großfeld borrows a concept of translation from the natural sciences, citing physicist Werner Heisenberg: "the connection between two different closed conceptual systems always must be a very painstaking analysis."[66] He observes that comparative law "is not for 'clever boys' or 'clever girls.' It requires patience and the capacity to empathize. As in the natural sciences."[67] He then adds: "Only rarely does one read anything about the execution [of law] in practice. We remain thus always at risk of taking the representation for life, and dead law for living law."[68] We take this critique seriously and look at environmental law practice in the remaining chapters.

Großfeld not only offers his own method, but presents a critique of the orthodox comparison method, while addressing what he calls the "core questions."[69] In his critique, he unequivocally states that, "To give order to connections and to render them comprehensible also is the scholarly duty of comparative law."[70] He says that "we should not take the functional method too narrowly."[71] Großfeld begins the final chapter of his work with a critique of the focus on method:

> We have dealt with some basic problems of comparative law; we found ways to approach them, but we did not find sure answers: Great tasks lie ahead for comparative law; it must let itself be challenged if it wants to win the future in a world that is not European in imprint. A technical–functional comparison without a study of national customs, without cultural-research and without loving empathy (Bernhard von Clairvaux … *Res tantum intelligetur quantum amatur.* A matter is understood to the extent that it is loved.) remains a study of words, letters and numbers, touches only the superficial level and leads into error.[72]

Among the many texts on the topic of comparative law, Großfeld's disquisition on the core questions of comparative law are the closest that one comes to a focus, or re-focus, on issues of the act of comparison, rather than a focus on the laws themselves or ideological problems.

66 Heisenberg, Werner. *Physik Und Philosophie: Mit e. Beitr. Von Gu'nther Rasche u. Bartel L.van Der Waerden*. Hirzel, 1984, p. 77.
67 Großfeld, *supra* note 7 at 110.
68 *Idem*. at 95.
69 *Idem*.
70 *Idem*. at 93.
71 *Idem*. at 7.
72 *Idem*. at 245.

2.6.5 Immersion

Among comparativists, Vivian Curran also expresses discontent with functionalism. Like Großfeld's translationalism, Curran's notion of immersion offers an alternative to functionalism. She points out that in relying upon functionality, Zweigert and Kötz assume away the object of inquiry. The question "*Is* there a similar function?" is not asked, but rather it is assumed there is and the method launches itself from this assumption into determining *what* function is in operation. More importantly, she says Zweigert and Kötz are assuming that it is possible to carry out the old goal of comparative law—to distill a global, common, and unified system of private law—and to do so, they begin by assuming they will find similarities. Like the old-time magician's rabbit, the functionalist's similarities are unsurprisingly always there.

In addition to an approach across time, the historical record demonstrates that there are lessons to be learned from immersing oneself in comparison across space. French court lawyer who became a Harvard lecturer Pierre Lepaulle noted:

> If I may state a personal experience, I never completely understood French law before coming to the United States and studying another system. History of law seems inadequate to give the student this sense of relativity, because in history we often deal with forces which are not yet dead, which still unconsciously bend the mind of the student in a certain direction. To see things in a true light, we must see them from a certain distance, as strangers, which is impossible when we study any phenomena of our own country. This is why comparative law should be one of the necessary elements in the training of all those who are to shape the law for societies...[73]

The work of Curran guides things in the direction of addressing the language questions of comparative law, and away from pure discussions of method or ideology. Curran's critique is related to a more general critique of the identity–difference principle. In the identity–difference principle, it is said that for any two phenomena one can of course find one basic similarity for example, simply the fact that they exist, they are both made of molecules, and so on, and also at least one difference (even two pencils, mass-produced and side-by-side, do not exist in the same physical space, for example).[74] Curran's preferred alternative mode of comparison is through immersion in the new legal culture. If one looks to the academic

73 *Idem.* at 858.

74 Here one must be careful not to engage in the sophomoric practice of "compare and contrast," as a substitute for comparative method. Simple juxtaposition is not the comparative method. I am reminded of an observation made by Dr. Deborah Dysart-Gale, Chair of the Centre for Engineering in Society at Concordia University, Montreal, who noticed that the "analysis" offered by most academic conference papers could be boiled down to one of two themes: "X and X1 may appear to be similar, but I will show you how they are really different," or "X and X1 may appear to be different, but I will show you how they are really the same."

requirements for students willing to become professional ethnologists or comparative anthropologists, one finds that knowing a culture, including a legal culture, requires that one be immersed in that culture. A great advantage of immersion is that the scientist can observe directly how legal systems operate for persons in practice in the culture that is being studied. The disadvantages are obvious—the money and time cost of travelling to that culture of study, not to mention the sad irony of air travel's carbon-loading to facilitate the study of environmental behavior.

2.7 Conclusions

By the new century, comparative law might well have reached a new era. Mathias Reimann avers that "comparative law has moved way beyond … relatively rudimentary models in at least three regards."[75] First, according to Reimann:

> it is understood today that all classifications are mere approximations to, not accurate reflections of, reality. … Second we have learned to look beyond legal systems and families as static and isolated entities … and have come to think, in a more dynamic fashion, primarily of legal traditions. [And third,] we have developed at least some understanding of the *interactions* between these legal families, traditions, and cultures.[76]

While functionalism has come under criticism—much of it justified—a single, new ship sailing forward upon which to fly the flag of comparativism has yet to emerge.

Literature

Arminjon, Pierre, et al. "Traité De Droit Comparé Tome III." Librairie Génerale de Droit et de Jurisprudence, 1951.

David, René. *Traité Élémentaire de Droit Civil Comparé: Introduction à L'étude des Droits Étrangers et à la Méthode Comparative*, Librairie Générale de Droit et de Jurisprudence, 1950.

Großfeld, Bernhard. *Core Questions of Comparative Law*, Trans. Vivian Curran Carolina Academic Press, 2005.

Quain, Edwin A. "The Medieval *Accessus ad Auctores*." *Traditio* 3 (1945), 215–264.

Schauer, Frederick. "Formalism." *Yale L. J.* 97.4 (1988), 509–548.

Whytock, Christopher A. "Legal Origins, Functionalism, and the Future of Comparative Law." *BYU L. Rev.* 6 (2009), 1879–1906.

Zweigert, Konrad, and Kötz Hein. *Introduction to Comparative Law.* Trans. Tony Weir, 3rd ed., Clarendon Press, 1998.

75 Reimann, *supra* note 1, at 676.

76 *Idem.* at 677–78. (emphasis added).

PART II

Institutions and bodies of environmental legislation, implementation, and dispute resolution

3

LOCATING ENVIRONMENTAL LAW FUNCTIONS AMONG LEGISLATIVE, JUDICIAL, AND IMPLEMENTATION BODIES

Marek Prityi, Ana Miola, Yuan Ye, Ryan Kraski, and Saskia Münster

3.1 Preliminary issues: experience or rationality?

We might well begin Part II of the book asking what the function of the state is in environmental problems. To answer this question, there are several different conceptual dichotomies that will run throughout this chapter, each of which is different from the other but is often confused with the other. The first dichotomy distinguishes public law from private law. The second dichotomy distinguishes common law from civil law. A third dichotomy, discussed in section 3.2.1, is of less categorical impact than the first two. It distinguishes the role of administration in a presidential system, with a full third branch of the state dedicated to execution of the law, from the role of administration in a parliamentary state that incorporates execution of the law into the administration of legislation.

Many of the harms that we would today call "environmental" were historically only actionable by one individual against another individual in private law. In those instances, the state's role was only to establish legally recognized norms of behavior either through legislation or, in common law jurisdictions, perhaps additionally through case precedent. In addition, the state would interpret those norms as applied to particular disputes through the process of private litigation and subsequent enforcement of court judgments. The determining factor for whether a constitutional right, statute, or regulation is "environmental" is whether its implementation benefits the environment. To make that determination, one must look to the remedies and damages available. If the remedy benefits the public and the damages, criminal penalty, or civil penalty are paid to the state, then one might characterize the law as "environmental." If it benefits individuals and the exercise of the remedy concerns only private parties, and payment of the damages is to a private party, then the matter is simply an exercise in the private law of obligations, not environmental law.

Eventually, however, the imbalance of power in favor of industry, which was in turn responsible for many of the harms that one could call "environmental," meant that the state had to step in on behalf of the environmentally injured individual in order to balance the excesses of economic, social, and political powers in the legal system. This re-balancing of imbalanced power did function to stop environmental injuries to some individuals, but the state's presence also came at a price. In many situations, the state's new environmental authority, derived from constitutions, statutes, or regulations, meant that the individual was no longer a party to the action, but rather a witness giving evidence to the state. As with criminal law, the state was expected to act to benefit all citizens, but the damages paid by the wrongdoer went to the state, not to the harmed individual. An additional change that made at least as much impact was the fact that once the state was empowered to enforce environmental law, individuals often felt that environmental problems were no longer their concern, but rather only a concern for the state.

With these comparisons over time in mind, we turn to current comparisons. Much of comparative law begins by distinguishing what René David called the two big "legal families" of the world, thus introducing the second dichotomy for this chapter. When it comes to comparative environmental law, however, this distinction may not be very significant.[1] We say "may," because the significance depends upon the distinctions featured. An important but often overlooked distinction that runs throughout all thinking that one might say distinguishes the two cultures is the distinction between featuring experience or rationality, or what philosophers would call the distinction between empiricism and rationalism. Seen constructively, common law thinking built itself upon the experience of past adjudications of real disputes with actual facts. Seen constructively, the civil law legislature has used the collective rational thinking of its legal minds to anticipate harms and address them through rational rules arrived at with contribution from the citizen-elected legislature. Even the judges themselves, when empowered, exhibit the features of rationalism or empiricism. The German judge, for example, is invited to the bench after he or she has demonstrated exceptionally high rational problem-solving ability on the state examinations to become a lawyer. But the U.S. judge, for example and by comparison, is either appointed by the president (if a federal judge) or governor (if a state judge) or voted upon (some states elect, rather than appoint, judges). In all these U.S. scenarios, a common appeal is the judicial candidate's experience as a trial lawyer.

If one looks only to sources of law, for example, one would find that common law jurisdictions, like civil law jurisdictions, address environmental matters through constitutions, statutes, and regulations. The much-talked-about role of

1 From the perspective of comparative law in general, Matthias Siems emphasizes the need to concentrate on characteristics typical for "Western legal tradition" rather than accentuating the "common/civil law divide." Siems, Matthias. *Comparative Law.* Cambridge University Press, 2014, p. 68.

case precedent in common law jurisdictions, when it comes to environmental matters today, serves largely only as binding interpretations of environmental rights already expressed in constitutions, environmental statutes, or environmental regulations, not as exclusive, new norm developments by the courts. There are of course matters of private obligation in the common law, such as torts, in which a private individual may take an action against another private individual for harms that arise through what one might call an "environmental" medium, such as destroying one's water supply, but as such they remain just private law torts and do not typically fall under the disciplinary category of environmental law. Thus whereas the starting point for comparative law is often to distinguish common law from civil law, the greater difference for practice in either family is in fact between private law and public law.

Matters that were historically remedied, if at all, by private legal actions were turned into a topic of direct state control through public law. One must consider not only the formal differences between public and private law remedies, but the function of those remedies and how one arrives at them. For that consideration, traditional comparative law often begins by distinguishing civil law from common law. That distinction is not a difference limited to the legal text, statute, or judicial opinion. It is a difference that involves the entire *accessus ad auctores*. Beginning with who the citizens are, one finds cultures with different expectations of their legal systems, beginning with a *laissez faire* or planning mode of legal control. This fact leads to the next question—how were these expectations created?

The critique of legislation from a common law perspective can also address the issue that the legislative process, as such, is not wholly capable of establishing balance between the interests of competing groups, as it promotes "the interests of certain groups either too strongly or too weakly."[2] By contrast, some scholars argue that common law, coupled with equity, is in a better position to establish balance between competing interests.[3]

In many states, as the transition from private law remedies to public law administration took over, the theory of the state's interest in environmental matters became couched in terms of the "public trust." The influential public trust doctrine can trace its roots back to ancient Rome.[4] The public trust doctrine is built upon the premise that natural resources, including water resources, are in "government ownership and must be held in trust for the use and benefit of the public as well as

2 Green, Marlow. "Common Law, Property Rights and the Environment: A Comparative Analysis of Historical Developments in the U.S.A. and England and a Model for the Future." *Cornell Int'l L. J.*, 30 (2), 1997, pp. 541, 543.

3 *Ibid.*; see also Smith, Lionel. "Common Law and Equity in R3RUE." *Wash. & Lee L. Rev*, 68, 2011, pp. 1185, 1193, 1195.

4 One of the Roman law maxims can be interpreted as stipulating that "the conservation of resources is intrinsically good and necessary for the continuation of society." See Klass, Alexandra. "Modern Public Trust Principles: Recognizing Rights and Integrating Standards." *Notre Dame L. Rev.*, 82 (2), 2006, pp. 699, 711.

future generations."[5] The public trust doctrine represents in this regard essentially the "judicial recognition of a trustee–beneficiary relationship between the government and public."[6] Its importance in the common law world and its role in environmental matters, mainly in the context of the U.S.A., was reiterated in the 1970s by one of the first persons to call himself an "environmental law professor," Joseph Sax. He argued for the extended application of the public trust doctrine in environmental matters and its ability to compel governments to protect natural resources from development and related threats.[7] Gradually, the public trust doctrine in the U.S.A. has been incorporated and reiterated in common law, constitutional law,[8] and statutory law.[9] The line of jurisprudence[10] that has emerged since the 1970s has clearly interpreted the public trust doctrine in the light of its environmental and aesthetic dimensions, reflecting the thoughts of Professor Sax and the raising of environmental awareness across the world.[11] The public trust doctrine should not be perceived as an isolated matter, but also in the context of an intricate interrelationship among common law, constitutional law, and statutory law. It can serve as a technique for courts to remedy the shortcomings in legislative and administrative process.[12] The doctrine reiterates the need for enhanced accountability in decision-making concerning natural resources and it can serve as a mechanism that can be used by the public to remedy the deficiencies associated with the application of planning law or environmental impact assessments.[13] Moreover, the echo of the public trust doctrine might be heard when discussing the need to address

5 Alexandra Klass, *Fracking and the Public Trust Coctrine: A Response to Spence*, 93 Tex. L. Rev., 47, 49 (2015). In fact, the principle of intergenerational equity also represents one of the constituent and fundamental parts of the Public trust doctrine. See Treves, Adrian. "Predators and the Public Trust." *Biol. Rev.*, vol. 92, 2017, pp. 248, 251; see also Klass "Modern Public," *supra* note 4 at 699–754.

6 Ripley, Wayne E., et al. "Environmental Land-Use Control: Common Law and Statutory Approaches." *U. Miami L. Rev.*, 28, 1973, pp. 135, 151.

7 *Ibid.* at 152 (1973); see also Razzaque, Jona. "Application of Public trust doctrine in Indian Environmental Cases." *JEL*, 13 (2), 2001, pp. 221, 228.

8 Article 1, para. 27 of the Constitution of the Commonwealth of Pennsylvania stipulates that the state (Commonwealth) of Pennsylvania shall act as a trustee of natural resources to "conserve and maintain them for the benefit of all the people." The state's obligations in this regard are reinforced through provisions enabling mobilization of citizens and the state to enforce their rights associated with environmental protection. See Constitution of the Commonwealth of Pennsylvania; see also Klass "Modern Public," *supra* note 4 at 751.

9 Klass "Fracking," *supra* note 5 at 50.

10 See for instance the Wisconsin Supreme Court's decision in the case *Just v. Marinette County* (201 N.W.2d 761), the California Supreme Court's decision in the case *Marks v. Whitney* (491 P.2d 374 Cal. 1971) or the Illinois Supreme Court's decision in the case *Scott v. Chicago Park District* (360 N.E.2d 773 111. 1977). See Klass "Modern Public," *supra* note 4 at 708–10.

11 Klass "Modern Public," *supra* note 4 at 708–10; see also Razzaque, *supra* note 7.

12 The Public trust doctrine can act as a "check upon administrative action by providing a mechanism for judicial review or resource allocation decisions [as well as] an additional tool for environmental protection particularly where administrative discretion has been abused." See Redgwell, Catherine. *Intergenerational Trusts and Environmental Protection.* Juris, 1999, pp. 178–79.

13 Razzaque, *supra* note 7 at 229.

climate change. Some states claim to own the natural resources themselves. But if governments hold natural resources in trust for their citizens and have obligations to protect them for future generations, it is worth considering "atmospheric trust litigation" with the objective to compel the governments to comply with carbon emissions reduction targets.[14]

Apart from the U.S.A., the public trust doctrine has played an important role for instance in India, where the courts treated it as a concept embedded in its national common law system.[15] Indian jurisprudence also shows the tendency to apply the public trust doctrine when there is a lack of legislation to protect natural resources and in cases of a breach of municipal law, while linking it to the right to life under Article 21 of the Indian Constitution.[16] The application of the public trust doctrine in the Indian context can be also useful in triggering the "atmospheric trust responsibility."[17]

The development of comprehensive environmental statutes as well as the establishment of administrative agencies in the U.S.A. took place along with the rediscovery of the public trust doctrine in the environmental field. The Clean Air Act[18] or the Endangered Species Act[19] might be listed as examples.[20] Gradually, the developments started in the 1970s evolved, and the modern public trust doctrine was accompanied by the evolution of both constitutional and statutory law.[21] The enactment of statutes coupled with the development of a corresponding institutional framework were significant steps forward in the field of environmental protection, as they introduced substantive air and water quality standards and equipped agencies with the basis for an interdisciplinary approach to the solution of environmental problems and to the evaluation of environmental impacts of planned projects.[22]

14 Wood, Mary, and William Burns. "Atmospheric Trust Litigation." *Adjudicating Climate Change State, National, and International Approaches*, 2009, p. 99.
15 Razzaque, *supra* note 7 at 229.
16 *Ibid*. at 233 (2001); see also Supreme Court of India, M.I. Builders Pvt Ltd v. Radhey Shyam Sahu (AIR 1999 SC 2468); Supreme Court of India, M.C. Mehta v. Kamal Nath and Ors. (2000) INSC 334.
17 Blomquist, Robert F. "Comparative Climate Change Torts." *Val. U. L. Rev.*, 46 (4), 2012, pp. 1053, 1062–1063.
18 The adoption of the Clean Air Act in 1970 and related laws need to be seen in the light of institutional developments as well. Congress passed the act shortly after President Nixon established the Environmental Protection Agency by executive order. See Melnick, R. S. *Regulation and the Courts: The Case of the Clean Air Act: the Case of the Clean Air Act*. Brookings Inst., 1983, p. 5; see also Ripley, *supra* note 6 at 136.
19 As reiterated by Mathews, the Endangered Species Act, adopted in 1973, has been significant mainly for two provisions: Sec. 7, which prohibits federal agencies to harm enlisted species, and Sec. 9, which prevents the taking of endangered species. See Mathews, Jud. "Turning the Endangered Species Act inside out?" *The Yale Law Journal*, vol. 113, no. 4, 2004, pp. 947, 948–49.
20 Klass "Modern Public," *supra* note 4 at 720.
21 *Ibid*.
22 Ripley, *supra* note 6 at 164–65.

The relevant case law serves to explain the interrelationship and interdependence among constitutional law, statutory law, and common law and their interplay in the U.S.A. For instance, in 2004 the Hawaii Supreme Court decided a case concerning the application of the public trust doctrine when granting a permit to reserve water for instream use.[23] When making the decision, the court underscored the public trust character of the case,[24] building its decision not only upon the public trust doctrine, but also on the relevant constitutional and statutory provisions.[25] The interplay between "constitutional, statutory and regulatory requirements" was also stressed in the case *Save Ourselves, Inc. v. Louisiana Environmental Control Commission*,[26] in which the Louisiana Supreme Court concluded that the decision to construct and operate a hazardous waste disposal plant violated constitutional and statutory law as well as the public trust doctrine.[27] In that case, the public trust doctrine was interpreted as a constitutional–statutory concept with the state acting in its capacity as a trustee for all the people.[28] The Louisiana Supreme Court reiterated the need for all state agencies with environmental protection responsibilities to integrate the public trust doctrine, emphasizing the need for a balanced approach in decision-making that weighs economic and social factors against environmental costs, thereby also echoing the essence of sustainable development.[29] The cases demonstrate that to ensure the implementation of the public trust doctrine in environmental matters (and matters relevant from the perspective of environmental protection in general), the courts in common law jurisdictions such as the U.S.A. make use not only of common law, but also of constitutional and statutory laws.[30] These respective mechanisms need to be seen as mutually supportive.

3.1.1 Private environmental obligations

The migration of legal actions and remedies for environmental harms from the private sphere to the public sphere is not a one-way, irreversible path.[31] With

23 *In Re* Wai'ola O Moloka'i, Inc., 83 P.3d 664 (Haw. 2004).

24 In this regard, it is also interesting to note that the public trust duty extended also to protect traditional and customary gathering rights of native indigenous people of Hawaii. See *In Re* Wai'ola O Moloka'i, Inc., 83 P.3d 664, 671–72, 682–83 (Haw. 2004).

25 Klass "Modern Public," *supra* note 4 at 730–31.

26 452 So. 2d 1152 (La. 1984).

27 Klass "Modern Public," *supra* note 4 at 731.

28 Houck, Oliver. "Save Ourselves: The Environmental Case That Changed Louisiana." *La. L. Rev.*, 72 (2), 2012, pp. 410, 434–35.

29 As noted by McCowan, the court in the case called for a balanced approach without specifying any fixed criteria or introducing any fixed specifics of a balanced process to all aspects of regulatory mechanism in the field of environmental protection. See McCowan, Charles S. "The Evolution of Environmental Law in Louisiana." *La. L. Rev.*, 52 (4), 1992, pp. 918–20.

30 Klass "Modern Public," *supra* note 4 at.745.

31 The different approaches to rights and remedies can serve as examples. While the common law prefers the approach "where there is a remedy there is a right" (*ubi remedium ibi ius*), continental

public attention to climate change litigation leading the way against industry-bound governments, citizens suffering environmental harm can and do return to private redress. This raises an important question for research in which one can make a comparison over time or space: from the perspective of citizens, do citizens of a particular legal culture enjoy a cleaner environment with only private law of obligations enforced by them against individuals and corporations that harm them, or do they enjoy a cleaner environment with the state taking public actions on their behalf?

From the perspective of legal science, the common law fosters case resolution and development of law by the means of adjudication of both public and private interests. From a historical perspective, environmental problems in the common law system of the U.S.A. were addressed mainly by the means of tort law before the regulatory framework of statutory environmental law expanded in the 1970s.[32] In the common law, private obligations are understood to be two distinct bodies of law—contracts and torts. This larger, clearer separation than that found in the law of obligations in civil law cultures is due to the historical fact that both common law torts and crimes originated in the Germanic system of compensatory fines for wrongs, where no clear distinction was made between crimes and other wrongs.[33] It was not until the Norman invasion that the influence of European continental civil law thinking began to change this.

In the U.S.A. today, solving environmental problems by means of tort law might be seen as a form of regulation "through deterrence by imposing liability for prohibited actions."[34] However, it is an age-old problem to determine exactly which "actions" are "prohibited;" common law gas developed an extensive body of jurisprudence to determine the duties and obligations that are owed. The main purpose of tort law rests in the realization of corrective justice, providing peaceful means for solving a dispute and restoration of the original conditions for harms caused by wrongful conduct, i.e. it is a fault-based compensation system.[35] Corrective justice, in terms of legal remedies, means that injured plaintiffs may recover compensatory damages for past injuries. Furthermore, equitable remedies provide relief from ongoing and potential future harms. The view of tort law emphasizing deterrence as well as corrective or distributive justice sees

---

approach prefers the sequence "from rights to remedies." See Nowag, Julian. "Article 11 TFEU Environmental Rights." *Environmental Rights in Europe and Beyond*, Hart Publishing, Bloomsbury Publishing Plc, 2018, p. 156.

32 Latham, Mark. "The Intersection of Tort and Environmental Law: Where the Twains Should Meet and DepArticle" *Fordham L. Rev.*, 80 (2), 2011, pp. 737, 739.

33 Malone, Wex S. "Ruminations on the Role of Fault in the History of the Common Law of Torts." *La. L. Rev:* vol. 31, 1970, p. 1.

34 Yang, Tseming. "Environmental Regulation, Tort Law and Environmental Justice: What Could Have Been" *Washburn L. J.*, vol. 41, 2002, pp. 607, 617.

35 Latham, *supra* note 32 at 746; see also Lin, Albert C. "Beyond Tort: Compensating Victims of Environmental Toxic Injury." *S. Cal. L. Rev.*, vol. 78, 2005, pp. 1439, 1452–53; Abelkop, Adam. "Tort Law as an Environmental Policy Instrument." *Or. L. Rev.*, vol. 92, 2013, pp. 381, 428.

tort law also as a mechanism to foster certain social objectives.[36] This might prove to be particularly useful when addressing the cases concerning members of marginalized groups[37] whose interests and concerns can appear to be too insignificant to be subsumed under general environmental legislation that focuses on the interests of the public at large.[38] Considering this, the oversight of regulatory authorities in such situations can often be unsatisfactory and less effective than the law of private obligations which in contrast to administrative law and criminal law provides people in environmental cases with both standing to sue and private remedies.[39] Thus, if it is true that marginalized groups' interests are not represented in environmental legislation and regulatory oversight, private causes of action are invaluable. The law of private obligations alone would be insufficient and inadequate as a sole regulatory system, particularly due to individual and *ex-post* nature, but it does serve as a good complement to other statutory and administrative approaches.

3.1.2 Public environmental obligations

Environmental legislation introduces environmental norms and standards to address environmental harms before they occur.[40] Private law remedies only provide relief when private parties bring a private action. The private duties and obligations may inhibit further harms through economic disincentives. These economic disincentives may dissuade anti-social behavior, including harmful intentional acts and negligence. Within liberal economic systems, this can be thought of as being as much of a deterrent toward anti-social behavior as is environmental legislation, but the key differences remain the available remedies and the source of the obligations. Furthermore, duties and obligations under tort, such as the "reasonable prudent person standard," allow for flexibility in remedying actions that were, so to say, unreasonable by social standards. This flexibility frees judges (and juries in common law systems) from pre-formulated statutory standards that may not anticipate particular harmful acts.[41]

36 As noted by Shmueli, "corrective justice begins where the rationale of deterrence ends." See Shmueli, Benjamin. "Legal Pluralism in Tort Law Theory: Balancing Instrumental Theories and Corrective Justice." *U. Mich. J. L. Reform*, vol. 48, 2015, pp. 745, 750, 761.

37 The role of judges and courts as unbiased actors and institutions deciding tort law cases involving different parties has been also seen as a relevant and important factor in comparison to alternative compensation mechanisms for victims of pollution cases, such as private insurance schemes. See Abelkop, *supra* note 35 at 429.

38 Yang, *supra* note 34 at 616–17.

39 *Ibid.* at 623.

40 Yang, *supra* note 34 at 623-624.

41 For a general discussion on tort law, reasonable person standard, and remedial action in common law, see Hogg, Martin, "Liability for Unknown Risks: A Common Law Perspective." *J. E. T. L.*, 7 (2), 2016, pp. 113, 142.

The difference between addressing environmental problems by the means of environmental legislation and private law rests in the differing theories underpinning each. While the purpose of environmental legislation with respect to environmental protection is rather straightforward, the law of private obligations may serve to deter environmental damage through "the threat of financial liability."[42] These aspects should be considered when encouraging private remedies as a public regulatory tool.[43] The inherent shortcoming of the law of private obligations lies in its inability to provide a remedy when causation cannot be proven for a direct injury.[44] In such cases, environmental legislation can prove to be more useful.[45] Public environmental law and the law of private obligations are mutually supportive when their objectives overlap,[46] such as when environmental law fosters corrective justice or deterrence. In theory, the probability of clean-up and disposal of hazardous substances at a particular site could be higher if it is addressed by private and public law after the fact, as compared to being generally addressed by public environmental law's objectives of prevention or conservation before the fact.[47] Scholars generally give more credit to activism and regulation through public law than to the decentralized private tort law system in addressing the problem of pollution in the second half of the 20th century.[48] Also, the number of environmental cases based on environmental statutes and regulations has been gradually increasing in comparison to private law cases.[49]

A private action may make use of a scientific expert, but, to a much larger degree, administrative regulatory systems employ specialists and experts who are in a position to assess large amounts of information and conduct studies.[50] Still, the mutually supportive relationship between an administrative regulatory system of public law and private litigation remedies can be highlighted by the ability of the private law system to compensate for eventual gaps in environmental information as well as for insufficient enforcement of environmental laws on an *ad hoc* basis. In common law jurisdictions especially, tort remedies may act

42 Lin, *supra* note 35 at 1453. As also noted by Abelkop in this regard, tort law as an economic tool is useful only when the threat of litigation serves as an incentive for "risk-takers to adopt cost-internalizing precautionary measures." See Abelkop, *supra* note 35 at 427.

43 *Ibid.* at 382 (2013).

44 Latham, *supra* note 32 at 750.

45 *Ibid.* at 754.

46 *Ibid.* at 749.

47 *Ibid.*

48 This conclusion can be, to a certain extent, attributed also to the fact that for instance nuisance law fosters economic development in the early stages, while it addresses environmental problems only later, when they occur, and the demands for a better environmental quality are articulated. On the other hand, there are also voices arguing that tort law in the U.S.A. has also undergone certain development and the factors limiting its effectiveness at the times when these observations were made are no longer present. See Abelkop, *supra* note 35 at 456–57.

49 "Note: Causation in Environmental Law: Lessons from Toxic Torts." *Harvard Law Review*, vol. 128, 2015, pp. 2256, 2263.

50 Lin, *supra* note 35 at 1465.

as a safety net behind legislation and regulation, making up for the lack of state enforcement.[51] Moreover, the flexibility of tort law can prove to be relevant and useful even in cases when the economic or scientific foundations of environmental laws and regulations become outdated or discredited.[52]

3.2 Public law: can sufficient planning replace dispute resolution?

As we construct legal systems, there arrives a point at which customary cultural values are concretized by an institution that has sufficient power to make rules for the entire society. The implication in such a social environment is that this institution has the ability to anticipate any disputes that might arise and legislate in advance how such conflicting interests should be resolved. One would hope, thinking ideally, that the legislative body has anticipated every harm. Experience shows that this is not the case, however. Especially in areas of environmental harm, where the harm and the evidence of harm may require complicated scientific explanations, it is even more difficult for law-makers to anticipate all harms and address them in general legislation.

Statutory law is gaining more prominence in common law jurisdictions in most fields of law, with environmental law being a prominent example.[53] The approach of both the common law judge and the civil law judge rests in building their reasoning upon the analysis of legislation, while the extent and depth of their analysis dedicated to case precedent differs. Considering that judicial decisions represent a binding source of law in common law jurisdictions, common law judges can be expected to pay more attention to the developments in the field of case law than their counterparts in civil law jurisdictions, including developments through private law. Research indicates that civil law judges are increasingly paying more attention to developments (and interpretation of the legislation) in case law as well.[54] Even the lawyers in civil law jurisdictions are known to make reference to prior case law.[55] Aside from the mechanics of the

51 *Ibid.* at 1469.

52 In this regard it is also relevant to mention that the fact that scientific matters are (or are not) correct has to be proven also by the means of expert testimonies. However, even expert testimonies can be sometimes biased and not correct. As noted by Perez, tort law provides only limited guarantees and chance for correction in this regard. Eventual bias of expert testimonies would probably still have to be reviewed by courts, which are not the most adequately equipped bodies to conduct such a task, considering the often very technical and specific nature of expert testimonies. See Perez, Oren, "Can Experts be Trusted and What Can Be Done About It? Insights from the Biases and Heuristics Literature", in *Nudging and the Law: A European Perspective*, Alberto Alemanno and Anne-Lise Sibony eds., Hart, 2015; and Abelkop, *supra* note 35 at 443.

53 Tort law represents an exception in this regard. See Junker, Kirk W. *Legal Culture in the U.S.A.: an Introduction*. Routledge, 2016, p. 61.

54 Dainow, Joseph. "The Civil Law and the Common Law: Some Points of Comparison." *Am. J. Comp. L*, 15 (3), 1967, pp. 419, 427.

55 Junker, *supra* note 53.

sources consulted, the approach to reasoning also differs to a certain extent. While deductive reasoning is more common for civil law systems, the common law builds its norms inductively from the resolution of prior disputes.[56]

3.2.1 The implementation of environmental law through administrative law

The notion of legal administration as a separate legal function, complete with its own institutions, is a relatively new notion. In civil law culture, with the agent of law enforcement integrated into the legislative body as a minister, execution and enforcement of the law are integrated into the legislative branch of the state. Administration complements enforcement to accomplish execution and implementation of the law. By comparison, in a presidential system such as that of the U.S.A., three independent and separate branches of the state carry out Montesquieu's notion of a balanced division of state power, with execution accomplished through a fully separate branch of the state. Thus, if a separate executive branch already exists, why would it need separate institutions of administration?[57] The invention of additional institutions to *administer* statutory law, beyond the fully independent branch designed to *execute* statutory law, was regarded as necessary in the twentieth century in part due to the specialist knowledge needed for execution.[58] Many of these agencies were thus seen as specialist groups within the executive branch and both their substantive regulations and institutions were only needed at the pleasure of the executive bodies and answered to those executive bodies, not to the legislatures.

While the separation of institutions has occurred in various systems—executive agencies in the federation and several states in the U.S., for example, as well as environmental protection offices (*Bundesumweltamt* in Germany, for example)—in parliamentary systems, the separation is not as clean and clear as separating the judiciary from the legislative branch, for example. In the U.S., several attempts have been made to elevate the federal U.S. Environmental Protection Agency to the level of a Presidential Cabinet position, but each time the effort has failed. In Germany, one may compare the several attempts that have been made to unify the various statutes that concern air, water, and other environmental matters into one comprehensive *Umweltgesetzbuch* (environmental code),

56 Observable differences in the creation of meaning exist as well. Civil law lawyers have a tendency to look for a meaning in definition, whereas common law lawyers are open to deriving meaning from usage. See Junker, *supra* note 53 at 143.

57 Sources of administrative law in the U.S.A. may be described as rule-making and adjudicatory powers upon organizations in government outside of the judicial branch. The body of administrative law encompasses statutory law, quasi-regulatory provisions, quasi-legislative rules, and quasi-judicial orders. See Prettyman, Barret. "The Nature of Administrative Law." *Va. L. Rev.*, 44 (5), 1958, pp. 685, 687–88.

58 Lin, *supra* note 35 at 1467.

efforts which have also failed. (Here one might note evidence of thematic differences between common law and civil law systems—the common law made its efforts through establishing an institution, and the civil law made its effort through establishing a comprehensive source of law.)

Administration within civil law culture is regarded as integrated into the legislative system, and, therefore, anywhere one finds statutes being made, administration of these statutes is inherently present.[59] By comparison, and rather ironically, in the U.S.A., where the separation of the executive branch is complete, there are additional, administrative agencies that are independent of the executive. Nonetheless, the federal and state environmental agencies *are* executive, answerable to the president and the governors, respectively. The U.S. E.P.A., the U.S. Forest Service, and the U.S. Fish and Wildlife Service are all under the executive branch of the state.

Administrative law has evolved differently in Germany, Brazil, the U.S.A., and China. By comparison, in a parliamentary system, in which executive functions are part of the legislative branch, there are additional reasons beyond specialization which seem to make it clearer why separate administrative institutions would need to have been constructed.[60] By implementing the Zweigert and Kötz *praesumptio similitudinis*, one may adjust the level of generalization in order to identify certain underlying differences having their roots in the different historical evolution in these regions. In the context of countries such as Germany or Austria, whose legal culture influenced legal systems in the wider Central Europe to a large extent,[61] the state historically took over the role of a sovereign in determining what constitutes public interest and making clear the distinction between interests of the state on the one hand and private interests on the other.[62] A historical frame of reference sheds light on these differences. When we take the example of Germany, administrative law was associated with the traditionally strong bureaucracy, which served the interests of the state.[63]

59 In Germany, the hierarchy of norms lists for instance *Verordnungen* (regulations), *Satzungen* (bylaws), and *Verwaltungsvorschriften* (administrative rules). See Peine, Franz-Joseph, and Thorsten Siegel. *Allgemeines Verwaltungsrecht: Mit Ebook: Lehrbuch, Entscheidungen, Gesetzestexte.* Müller, 2018, p 29.Similar hierarchies of norms are characteristic also for other countries, for instance Slovakia. The non-statutory, derived legislation in Slovakia consists *inter alia* of government regulations, generally binding regulations of ministries and bodies of state administration.

60 Saunders, Cheryl. "Apples, Oranges and Comparative Administrative Law." *Acta Juridica*, 2006, pp. 427–28.

61 Szegedi, Laszlo. "The Eastern Way of Europeanisation in the Light of Environmental Policymaking: Implementation Concerns of the Aarhus Convention-Related EU Law in Central and Eastern Europe." *Elte L. J.*, 2014, pp. 117, 121.

62 Greve, Michael S. "The Non-Reformation of Administrative Law: Standing to Sue and Public Interest Litigation in West German Environmental Law." *Cornell Int'l L. J.*, vol. 22, 1989, pp. 197, 232.

63 Bignami, Francesca. "From Expert Administration to Accountability Network: A New Paradigm for Comparative Administrative Law." *Am. J. Comp. L.*, vol. 59, 2011, pp. 859, 864.

On the other hand, the roots of administrative law in the U.S.A. suggest that it evolved at the beginning of the twentieth century in a manner consistent with common law thinking, as "the law of independent agencies, whose shape and structure was more a matter of legislative improvisation and experimentation than of systematic thought or action."[64] Essentially, the purpose of administrative law was to allow public administration to pursue the common good.[65] Even though it was later systematized also in the U.S.A., there were still differences reported in the terms of resilience to political and public pressure, for example, in comparison to German administrative law.[66]

Similarly, the legal system of Brazil presumes that the state is responsible for overlooking public welfare, health, education, and the environment, thus bringing these topics under public interest. In this system, administrative law is used to regulate the behavior of private individuals when dealing with public affairs. The agencies and institutions act as law enforcement bodies and are the main recipient of private complaints. As is the case with several other Latin American countries that were colonized by Iberic governments, Brazil has an independent body that is responsible for protecting the public interests before any harmful actions, the Public Prosecutor's Office. This institution is part of the country's legal system and is normally the author of court actions against public and private offenders. At the same time, the legislation also allows private persons to pursue the application of law in courts without going through the institutions—the instrument used is called "popular action."[67]

The functions of agencies and bodies of state administration in this regard might differ. In countries such as Germany where public administration falls largely under the executive branch, only few independent agencies exist[68] and they do not possess rulemaking powers comparable to the ones of the government and ministries. The status of independent agencies in the U.S.A. has been historically more independent, as they possess rule-making powers and enforcement competences.[69] In contrast to the tendency toward robust bureaucracy in continental Europe, the creation of independent agencies in the U.S.A. can be seen in the light of the historically often prevalent small-government ideology in this part of the world.[70] Brazil presents an (almost) hybrid system in which the executive is comprised of agencies and institutions that are capable of participating in decision-making, independent of the executive power.[71]

64 Greve, *supra* note 62 at 234.

65 Bignami, *supra* note 63 at 866.

66 Greve, *supra* note 62 at 235.

67 Pietro, Maria, Direito Administrativo, 17th ed., (São Paulo: Forense 2014).

68 For instance, the Federal Environment Agency (*Umweltbundesamt*) in Germany.

69 Bignami, *supra* note 63 at 882.

70 Boughey, Janina. "Administrative Law: The Next Frontier for Comparative Law." *Int'l & Comp. L, Q.*, vol. 62, 2013, pp. 55, 83.

71 Gomes, Jose, Direito Constitutional Ambiental Brasiliero (Constitutional Environmental Law) (São Paulo: Saraiva 2015).

In China, the central government and local governments have authority to make regulations and ordinances to concretize the statutes for their implementation.[72] The adoption of the first Environmental Protection Law in 1979 was followed and accompanied by the establishment of specialized agencies in central and local governments and even corresponding agencies at lower levels.[73] However, the functioning of administration in the field of environmental protection has not been without problems. State implementation in general is criticized as being incoherent. Specific issues include insufficient law enforcement, unclear division of administration and supervision competences, compliance with administrative regulations and laws, and the authority to deliver "original interpretation on the process of law administration and supervision by courts and the Public Prosecutor."[74] Environmental protection competes with other interests and priorities.[75] Administrative officials in Chinese environmental protection have expressed the need to have a vertical system of supervision of control functioning under the central government to resist the pressures from local governments.[76]

In China, environmental law has experienced a transition from a looser system to a more rigorous one. This transition can be traced back to the 1970s when the rapid industrial development started to take a visible toll on the state of the environment.[77] It has been explicitly stated that developments in the U.S.A. and Japan influenced the development of environmental law in China, including the Chinese Environmental Protection Law of 1979.[78] The problems with the enforcement of environmental legislation by the bodies of state administration in China have been to a certain extent compensated by promising developments in the fields of litigation and judiciary.[79] These developments are to a great extent associated with the revised Environmental Protection Law that came into force in 2015.[80] The inclusion of registered social organizations among subjects with the right to initiate environmental public interest litigation is among the main points addressed by the revised law.[81] This development can be seen in the light of the amendment of the Civil Procedural Law from 2012, which granted the

72 Article 65 and Article 73 of Legislation Law of the People's Republic of China.

73 Wang, Jin. "Evolution of Environmental Thought and Enforcement of Environmental Protection Legislation in China: The Status Quo." *Environmental Policy and Governance in China*, Springer, 2017, p. 21.

74 Kitagawa, Hidetiki. "Environmental Policy Under President Xi Jinping Leadership: The Changing Environmental Norms." *Environmental Policy and Governance in China*, Springer, 2017, p. 11.

75 Wang, *supra* note 73 at 23–24.

76 Kitagawa, *supra* note 74 at 13. Kitagawa interprets Jin Wang's 2009 research.

77 Wang, *supra* note 73 at 18–20.

78 *Ibid.* at 20.

79 *Ibid.* at 42.

80 *Ibid.* at 18.

81 *Ibid.* at 36.

possibility to initiate environmental public interest litigation to "the authorities and related organizations prescribed by the law."[82]

From the perspective of judicial enforcement of environmental laws in China, developments in the field of public interest litigation supplement the enforcement of environmental laws through administrative litigation or class actions.[83] Citizens have the possibility to initiate lawsuits against the administrative actors if their rights have been violated by administrative acts.[84] Hence, concrete administrative acts in China are subject to judicial review for various reasons. The latest reform of China's judicial system has even subjected abstract administrative acts to judicial review. Although China has a special situation with respect to the relationship between courts and administrative organs due to the role of the Communist Party, the new round of reform strives to balance the judicial authority and administrative power, as illustrated in Article 52 of the Administrative Law revised in 2015.[85] Moreover, China started the "Plan for the Pilot Reform of the Ecological Environment Damage Compensation System" by the State Council in 2015. With this plan, provincial-level governments have the authority to raise lawsuits against those who harm the environment including enterprises and private citizens.[86] Moreover, the concept of "ecological civilization" has the potential to have a positive impact on the country's economic structure change and environmental protection development, and even on the improvement of certain human rights.[87]

Environmental law in different corners of the world has been, to a certain extent, permeated with various forms of transparency requirements and participatory decision-making, which will be discussed at length in Chapters 7 and 9. These and similar developments create to a certain extent a bridge between various countries and systems, minimizing the differences between them.[88] The role of information disclosure and public participation in the field of environmental governance has demonstrably increased in China.[89] The evolution of the Chinese legal system reflects changes in the society. Moral rules and customs reflected in traditional beliefs, such as Daoism or Confucianism, as well as the socialist ideology of the Communist Party, are no longer enough to maintain social stability in rapidly evolving China. Hence, legal rules have taken over the role of one of the

82 *Ibid.* at 31.

83 McElwee, Charles R. *Environmental Law in China: Managing Risk and Ensuring Compliance.* Oxford University Press, 2011, p.257.

84 Article 2 of Administrative Law of People's Republic of China.

85 Changshan, Ma. "Possibility and Limit of New Round Judicial Reform." *Tribune of Political Science and Law*, 2015.

86 Haisong, Chen. "The Formation and Normalization of Rule of Political Parties in Eco-Environmental Law." *Law Science*, 2019.

87 Guohua, Jiang. "The Entering of 'Ecological Civilization' into the Constitution and New Development of Environmental Law." *Journal of Nanjing U. of Tech*, 2019.

88 Bignami, *supra* note 63 at 884–85.

89 Kitagawa, *supra* note 74 at 12.

main guarantors of stability.[90] Legal mechanisms providing public participation and access to justice possibilities for citizens in China need to be seen in cultural and political contexts. For example, the notion of social stability has led to a tendency to rely on mediation as a preferred alternative to litigation.[91]

Some concepts might at first glance remind one of Western-style rules, but the underlying cultural concepts are adjusted to the Chinese realities. For example, the Western understanding of the rule of law, with its emphasis on matters such as separation of powers or independence of judiciary, essentially contradicts the functioning of a regime ruled by the Communist Party. Hence, to underscore the need to maintain social stability, in the context of China it is more appropriate to refer mainly to the rule *by* law rather than rule *of* law.[92]

Addressing environmental issues through constitutions, statutes, or judicial opinions separately provides only partial responses and solutions to environmental problems. All of them have certain limitations, and for this reason it is beneficial to perceive them as interrelated and interdependent, to an extent defined by their respective geographical and legal contexts.[93] When making a comparison, one needs to be aware of the place of environmental law in civil law and common law countries. While environmental law in civil law countries falls to a large extent in the category of administrative law, and in a broader sense in the category of public law, environmental law in common law countries typically involves both public and private interests,[94] with many cases brought by administrative agencies against private citizens.[95] Also, when thinking about resolving environmental disputes in a common law system, such as that of the U.S.A., one needs to be aware that civil litigation is often seen as a universal tool to address social problems, i.e. the first instinct when confronted with a problem is to try to address it by means of private action. Only if this approach is not enough should public law and state enforcement assume the leading role in resolving the problem.[96] For countries like Germany it is more characteristic that environmental disputes are prevented and addressed in advance, usually at the planning stage. Hence, there is a desire to prevent litigation, if possible.[97] This is again one element indicating the existence of cultural differences when compared to the

90 Siems, *supra* note 1 at 81.

91 See Minzer, Carl. "China's Turn Against Law." *American Journal of Comparative Law*, vol. 59, 2011, pp. 935–984.

92 Ruskola, Teemu. "Legal Orientalism." *Michigan L. Rev.*, vol. 101, 2002, pp. 230–31.

93 Klass "Modern Public," *supra* note 4 at 726.

94 As noted by Houck, the process of solving environmental problems can stretch "from court-like trials with formal rules and contested testimony to the show-up-at-the-microphone affairs of a city planning board that basically allow people to blow off steam." See Houck, *supra* note 28 at 426.

95 Junker, *supra* note 53 at 58.

96 *Ibid.* at 59.

97 Beru, Tsegaye, and Kirk W. Junker. "A Constitiutional Review and Customary Dispute Resolution by the People in the Ethiopian Legal System." *NC.J. Int'L* 1, XLIII, 2018, pp. 62–63.

countries of continental Europe or Latin America with strong bureaucracy and the historically omnipresent role of a state as the representative of public interest.

As another example of private litigation supplementing public action, in 2015 a large dam collapsed in the Brazilian city of Mariana, in which the deposited material ran off and contaminated more than 680 km of riverbed and allowed 62 million m^3 of toxic mud into the River Doce. The company responsible for the dam had an environmental license/permit emitted by the state's environmental body and was simultaneously subject to law enforcement agents from the local, state, and national agencies. Shortly after the collapse, the state's Environmental Prosecutor's Office filed a lawsuit against the company in order to revoke all the environmental licenses of the company in the surrounding area, since there were indications of fraud and negligence during the licensing process of similar enterprises. Parallel to this, all three levels of environmental executive bodies (local, state, and national) imposed several monetary administrative penalties, because the toxic mud crossed the administrative boundaries of several districts, municipalities, and federal property. After the losses and damages were assessed, the Environmental Prosecutor's Office filed a new lawsuit against the company, in which it required compensation for the environmental damages caused in the River Doce, and requested immediate environmental recovery measures and collective moral damages in the name of the affected communities. At the same time, several families affected by the accident (e.g. who lost houses, lost income source, lost family members, etc.) also filed private lawsuits against the company to request a compensation for their private losses.[98] In summary, what we can derive from this example is that environmental damages are treated in Brazil as a public interest, and therefore will be protected by the Prosecutors' Office and law enforcement agencies—a private person is not allowed to file lawsuits in the name of this institution to protect public interests, unless in the exclusive case of "Popular Actions." The private individual may, however, file a lawsuit to pursue the private losses and damages derived from an environmental harm.[99]

3.2.2 Permits

Permits are of course not a source of law per se, but they form the basis for so much of the work of environmental lawyers and engineers that they must be included in this chapter's reflection on the allocation of environmental governance functions. Although the theoretical science of environmental law might begin with constitutional authorization, from which legislatures make laws and delegate powers to agencies, which in turn make regulations and issue permits,

98 Meneghetti, Diego. "Quais as Consequências Do Desastre De Mariana (MG)?" *Superinteressante*, Superinteressante, 4 July 2018, https://super.abril.com.br/mundo-estranho/quais-as-consequencias-do-desastre-de-mariana-mg/.

99 See generally Machado, Paulo Affonso Leme. *Direito Ambiental Brasileiro*. Malheiros Editores, 2016.

a citizen and his or her practicing lawyer are likely to encounter environmental law in reverse. That is to say, a citizen is most often faced with needing to know first whether a statute in an area of concern, like wastewater discharge, requires a permit. If it requires a permit, the citizen will then be required to follow the permit's requirements, once a permit application is approved by the state. And even if a permit is not required, the technical detail needed to know how to comply with the law will be found in the regulations, not the statute. Briefly stated, an academic researcher might discuss sources of law from constitution to statute to regulation to permit, but a practitioner will first encounter the permit then the regulations then the statute and only on rare occasions a constitution. The permit is a bridge between regulations and compliance. In many cultures, when environmental law was introduced that imposed emission and effluent standards on industries, the industries responded by saying that they could not continue to produce under such limitations. If the industry was producing something considered to be essential to the culture, such as electricity or refined oil, the state would allow the industry to continue under the promise that it would redirect research to reduce pollution. With this promise made, the state permitted limited amounts of pollution for limited periods of time. The ability of the state to issue a permit was typically created in legislation, the details of which were then constructed by state permit engineers in environmental regulations. Operators were allowed to apply for a permit, as long as their application complied with the requirements in the regulations. The state's engineers reviewed these applications and either denied or granted them, or required different or new materials, designs, or processes, in order to work on schedule to meet pollution reduction goals of the state on behalf of the citizens who drink the water and breathe the air. Historically, no state created a right to a state permit, nor was the concept of the permit created as an indefinitely repeatable exercise in allowing pollution to harm citizens, the environment, or both.

A permit may be a general environmental impact permit or a permit specific to a pollutional medium such as an air quality permit or a water quality permit. There are many features on which that which is called a "permit" can differ. For example, in China, an air pollution permit is transferable from one operator to another,[100] but in the U.S.A. a permit is not transferable, just as a license is not transferable. And in Germany, a building permit[101] is often the tool used to address environmental issues, which is why jurists in Germany often think of environmental law as a topic proper to the area known as *Bau- und Planungsrecht* (building and planning law). The German building permit is issued if a proposed

100 Yuan, Ye. A Comparative Study: China's Water Pollutants Discharge Permit System with Reference to the National Pollutants Discharge Elimination System of the U.S.A. Dissertation, Universität zu Köln. 2016.

101 As the legislative power for building permits is with the states, whenever talking about such a permit in this text the reference is made to § 74 *Bauordnung NRW* (Building Regulations in the State of North Rhine Westphalia).

project, if completed, would not conflict with public law. As well as other norms of public law, this concerns environmental law norms as they are mainly part of public administrative law. In Germany, an exceptionally large variety of projects may be realized under the concept of "building," and all are therefore regulated by the building permit. An example is a building permit in the context of a development plan.

There are several planning stages for the development plan applicant during which the environment must be addressed. In the first stage, before a building permit can be issued, land-use planning must be considered.[102] The function of this is to prepare and control the use of land within a municipality, for buildings or other purposes, in accordance with the General Urban Planning Legislation. This is the first stage where environmental matters must be considered. Section 1 VI of the *Baugesetzbuch* (Building Code) names different requirements for the preparation and the control of the use of land within a municipality such as the social and cultural needs of the population (§ 1 VI Nr. 3 of the Building Code). Furthermore, it puts a special focus on the interests of environmental protection, including conservation and landscape management (§ 1 VI Nr. 7 of the Building Code). Section 1 VI Nr. 7 of the Building Code then mentions different environmental concerns such as the avoidance of emissions and proper handling of waste and wastewater. This is where all of the different environmental laws must be taken into account, such as the *Bundesimmissionsschutzgesetz* (Federal Pollution Control Act) or the *Bundesnaturschutzgesetz* (Federal Nature Protection Law). For the sake of environmental protection, the municipality must issue an environmental audit, and an estimate for the significant environmental impacts can also be made. A second stage, the Land Use Plan, is the basis for the development plan, which contains the legally binding regulations for the town planning regulation. Ultimately, whether a building permit is issued will depend on the requirements of the development plan. If the building permit is in accordance with public law, it will be issued.

By turning one's focus to India, one notices that the different legal culture results in changes in the terminology and procedural pathways used for the permitting process. For instance, in India, the permitting process involves the applying for and granting of a "license." Statutes contain "schedules" with several different kinds of forms which proprietors of industrial activities must duly complete with details such as water use, effluent outputs, and presence of treatment facilities in order to successfully obtain licenses from the appropriate pollution control regulatory body or municipal authorities. Therefore, at the stage where pollution can occur or can be prevented, often it is the interaction between industry and authorities based on regulations and forms that plays a stronger role in the march of environmental law, as opposed to the broader remit of constitutional law or statutory law.

102 See § 1 of the federal *Baugesetzbuch* (Building Code).

3.3 Common law obligations

We return now to private remedies and common law obligations to see how those might supplement public environmental law, rather than considering private law as a historical predecessor to public environmental law. In common law, actions for nuisance could be taken because an individual has interfered with quiet use and enjoyment of one person's property (private nuisance), or an entire population or sub-set of the population's quiet use and enjoyment of property (public nuisances). As we shall discuss in greater detail in Chapter 10, public nuisances are also a crime. And it is through this connection of private obligations and crime that we see why common law separates tortious obligations from contractual obligations so fundamentally.

3.3.1 Public nuisance

The concept of public nuisance has been utilized in air pollution cases. Thus, it is a natural evolution to apply public nuisance to climate change issues.[103] Private obligation lawsuits have been filed with the intention to lower the level of greenhouse gas emissions.[104] The incentive for steps in this direction can be seen in the need to enforce compliance with existing statutes and regulations as well as in the frustration and dissatisfaction with existing statutes and regulations, or a lack thereof,[105] or a government's intentional disregard for climate so as to make expedient political gain. However, the utilization of tort law to address climate change requires the overcoming of certain limitations inherent in the nature of tort law. Yet because a tort claim requires evidence of causation, geographic jurisdiction over the subject matter, and the possibility to calculate damages, its application to a global problem such as climate change is problematic.[106] Greenhouse gas emissions "diffuse and disparate in origin, lagged and latticed in effect … represent the paradigmatic anti-tort."[107] One of the main problems is the need to demonstrate that a defendant's emissions directly contributed to climate change, which is caused by multiple persons and atmospheric reactions.[108] The issue with multiple entities in climate change-related cases cannot be seen analogically to the cases in an asbestos context, which the courts in the U.S.A. tend to see as indivisible, triggering joint-and-several liability for the harm caused by defendants to plaintiff.[109]

103 Blomquist, *supra* note 17 at 1056.
104 Latham, *supra* note 32 at 760.
105 Kysar, Douglas. "What Climate Change Can Do about Tort Law" *Envtil. L.*, 41, 2011, pp. 1, 9.
106 Latham, *supra* note 32 at 761.
107 Kysar, *supra* note 105 at 4.
108 Grossman, David, and William Burns. "Tort-Based Climate Litigation." *Adjudicating Climate Change: State, National, and International Approaches*, 2009, p. 217.
109 See Kysar, *supra* note 105 at 51-2.

Climate change issues have already been litigated before national courts in common law and civil law jurisdictions, including the U.S.A., Netherlands, and Germany. One of the main underlying obstacles to all the litigation is acceptable evidence of causation—or a lack thereof—between climate change and injury.

The plaintiffs in *Native Village of Kivalina v. ExxonMobil Corp*[110] argued before a U.S. federal court that diminishing sea ice that protects the Kivalina coast from winter storms would lead to the costly relocation of Kivalina's residents.[111] The plaintiffs sought damages from twenty-four companies. Their claim was based on a common law claim of nuisance from the defendants' carbon dioxide and greenhouse gas emissions, which contributed to global warming in general and, more specifically, to the deterioration of living conditions of the people on the Kivalina shore. Certain parts of the plaintiffs' claims, especially the possible infrastructure damage owed to storm exposure associated with the loss of sea ice, formed the base of the causation in the claim.[112] One of the core issues in the case was whether such a nuisance claim would fall under the court's subject matter jurisdiction or should be denied because it is merely a political question.[113] In the U.S.A., the political question doctrine interprets the separation of powers principle to say that if issues brought before a court are political in nature, then they should not be adjudicated but rather addressed by state branches other than judiciary.[114] The court determined that a decision attributing the cost for global warming just to the defendants named by plaintiffs would be of a political nature. The court further stated that if the defendants'

110 Native Village of Kivalina v. ExxonMobil Corp., 663 F. Supp. 2d 863 (N.D. Cal. 2009).

111 The US Army Corps of Engineers estimated the cost of relocation to be between US $95 and 400 million. See Native Village of Kivalina v. ExxonMobil Corp., 663 F. Supp. 2d 869 (N.D. Cal. 2009).

112 Kysar, *supra* note 105 at 28.

113 Native Village of Kivalina v. ExxonMobil Corp., 663 F. Supp. 2d 869 (N.D. Cal. 2009), at 868, 869.

114 The determination on whether the claim is non-justiciable and political in nature is based on the fact whether the court needs to make a policy decision of legislative nature in contrast to legal and factual analysis, which is normally required and expected from the court to resolve a dispute. Native Village of Kivalina v. ExxonMobil Corp., 663 F. Supp. 2d 871 (N.D. Cal. 2009); see also *E.E.O.C. v. Peabody Western Coal Co.,* 400 F.3d 774, 785 (9th Cir.2005). More specifically, the so-called Baker doctrine reiterates the above-mentioned by emphasizing the need to respect the constitutional provisions on separation of powers in this regard. To detect whether the claim involves mainly a question of political nature, the court needs to assess whether:a textually demonstrable constitutional commitment of the issue to a coordinate political department; or a lack of judicially discoverable and manageable standards for resolving it; or the impossibility of deciding without an initial policy determination of a kind clearly for non-judicial discretion; or the impossibility of a court's undertaking independent resolution without expressing lack of the respect due coordinate branches of government; or an unusual need for unquestioning adherence to a political decision already made; or the potentiality of embarrassment from multifarious pronouncements by various departments on one question. See Native Village of Kivalina v. ExxonMobil Corp., 663 F. Supp. 2d 871 (N.D. Cal. 2009), at 871, 872; see also *Baker,* 369 U.S. at 217, 82 S. Ct. 691.

contribution to climate change was more significant than the contribution of others, it remains for the executive and legislative branches, not for the court, to determine this.[115]

In determining whether the nature of the claim sufficed to categorize it as a public nuisance, the court reiterated the need to consider whether the defendants' action represented "an unreasonable interference with a right common to the general public."[116] The court reasoned that in applying the political question doctrine, the finding of a public nuisance in the case would have required a political decision by the court to set limits on greenhouse gas emissions.[117] With respect to causation, the court distinguished other cases of water or air pollution from the claim made by the plaintiffs in the case. While in other cases it was possible to satisfy the requirement of causation and trace the specific sources of pollution to specific areas, the claims involving climate change and greenhouse gas emissions were sufficiently different insofar as the whole planet is affected by pollution sources that cannot be isolated as easily as in other cases of water and air pollution.[118]

In the end, the trial court dismissed the *Kivalina* case, thus thwarting the attempt to use tort law to address climate change. On appeal, the plaintiffs again lost and the court emphasized that plaintiffs need to address climate change through the legislative or executive branch, not the judiciary.[119] Moreover, the Court of Appeals concluded that such claims should rely upon the public source of law in the Clean Air Act for redress and not upon a common law claim of public nuisance.[120] But the Clean Air Act does not address climate change nor offer remedies for individual private citizens.[121] Of course the court was also aware that for political reasons the U.S. Congress had failed to enact climate change legislation, so it left the plaintiffs without a remedy, an act one might also characterize as "political."

Another example from the U.S.A. is the climate change litigation *Washington Environmental Council v. Bellon*.[122] In that case, environmental groups filed a claim to compel the state authorities to regulate greenhouse gas emissions stemming

115 Native Village of Kivalina v. ExxonMobil Corp., 663 F. Supp. 2d 871 (N.D. Cal. 2009), at 876, 877.

116 To conclude whether an interference is unreasonable, it is necessary to assess "the gravity of the harm against the utility of the conduct." See Native Village of Kivalina v. ExxonMobil Corp., 663 F. Supp. 2d 871 (N.D. Cal. 2009), at 874.

117 Native Village of Kivalina v. ExxonMobil Corp., 663 F. Supp. 2d 871 (N.D. Cal. 2009), at 876, 877.

118 *Ibid.* at 876.

119 U.S. Court of Appeals for the Ninth Circuit, Native Village of Kivalina v. ExxonMobil Corp., No.09–17490 D.C. No. 4:08-cv-01138-SBA Opinion by Judge Thomas.

120 Davies, Haydn. "Native Villagers' Demand for Climate Justice Exposes the Limitations of Federal Common Law." *Env. L. Rev.*, 15 (1), 2013, p. 1.

121 *Ibid.* at 8.

122 Washington Environmental Council v. Bellon, 732 F.3d 1131 (9th Cir. 2013).

from oil refineries.[123] The plaintiffs claimed that the injury included the risk of future harm, making their lives less enjoyable and satisfactory due to environmental degradation.[124] The court found that the plaintiffs' arguments failed to identify any meaningful contribution of oil refineries' emissions and failed to demonstrate causation between the lack of state controls and the injuries claimed by plaintiffs.[125] The analysis of causation in the case revolved mainly around the issue of whether there was sufficient evidence to prove causation.[126] The court said that the problem associated with general and specific causation in climate change cases rests in the fact that, even in a situation when a plaintiff succeeds in demonstrating a general causation between injury and greenhouse gas emissions, it would be hardly possible to demonstrate a specific causation linking the injury to particular defendants.[127] The court in the case did not dispute the reality of climate change and its basic premise was that greenhouse gas emissions are contributing to climate change.[128] However, one of the requirements establishing the applicants' standing in the case rested also in their ability to prove that the alleged injury was causally linked or, eventually, "fairly traceable" to the alleged misconduct on the part of state authorities to regulate emissions.[129] To have had success, the plaintiffs would also have needed to prove that the remedy would redress the harm, and the court found it would not.[130]

Court rulings in *Urgenda Foundation v. the State of Netherlands* are a breakthrough in climate change litigation.[131] The case was not brought against an industry or corporation, but against the Dutch government for its failure to mitigate climate change. Had Urgenda and the 900 Dutch citizens instead taken their action through the administrative process, the only remedy available would have been to evaluate the government's action for its compliance with existing environmental statutes and regulations.[132] The claim in this case was twofold. On the one hand lies an international dimension based on human rights claims under the European Convention on Human Rights, namely its Articles 2 and 8.[133] The court considered also the obligations emanating from European Union

123 Washington Environmental Council v. Bellon, 732 F.3d 1131, 1142 (9th Cir. 2013); see also "Note: Causation," *supra* note 49 at 2256, 2267.

124 Washington Environmental Council v. Bellon, 732 F.3d 1131, 1140 (9th Cir. 2013).

125 *Ibid.* at 1142; see also"Note: Causation," *supra* note 49 at 2256, 2267.

126 "Note: Causation," *supra* note 49 2256, 2267.

127 See Abelkop, *supra* note 35 at 47; and "Note: Causation in Environmental Law: Lessons from Toxic Torts." *Harvard Law Review*, 128, 2015, pp. 2256, 2267.

128 Washington Environmental Council v. Bellon, 732 F.3d 1131, 1135-36 (9th Cir. 2013).

129 *Ibid.* at 1141.

130 *Ibid.* at 1146.

131 Urgenda Foundation v. The State of the Netherlands, C/09/456689/HA ZA 13–1396, judgment from 24 June 2015.

132 Cox, Roger. "The Decision of the Hague District Court in the Climate Case Urgenda Foundation v. The State of the Netherlands." *J.P.L.*, vol. 4, 2016, pp. 323, 326.

133 Urgenda Foundation v. The State of the Netherlands, C/09/456689/HA ZA 13–1396, judgment from 24 June 2015, para. 3.2, 4.3.5.

legislation and international law obligations.[134] With respect to the human rights claims, the court rejected any direct claims based on this ground.[135] The court underscored the developments in the jurisprudence of the European Court of Human Rights demonstrating the trend of human rights law and environmental law as "mutually reinforcing."[136] On the other hand, it targeted the government's failure to adopt proper mitigation measures by the means of national tort law, arguing that there was a breach of duty of care.[137]

When discussing the claim concerning the duty of care, the court's supportive approach toward the tort law claim could be detected already when it concluded by relating the doctrines of hazardous negligence to hazardous climate change, articulating the need to expand the understanding of what it means to act negligently toward society.[138] While recognizing the power of the state to determine national climate policy,[139] the court stated that the objectives and principles embedded in the documents and agreements regulating the topic of climate change, as well as those in the European Union law, should be considered and adhered to when implementing the scope of policy-making and duty of care in the context of climate change.[140] Considering that the Netherlands participated in climate change agreements and policy actions undertaken under the auspices of the United Nations and the European Union, the state had to have been aware of the risks associated with climate change.[141]

According to the court's reasoning, to comply with the duty of care in the context of the case, the state must undertake mitigation measures (as a priority before adaptation measures) in the greatest possible extent and as soon as possible.[142] The mitigation reduction target for 2020 of less than 25 percent compared

134 Urgenda Foundation v. The State of the Netherlands, C/09/456689/HA ZA 13–1396, judgment from 24 June 2015, para. 4.38–4.44.

135 Lambrecht, Jesse, and Claudia Iuarte-Lima. "Legal Innovation in National Courts for Planetary Challenge: Urgenda v State of Netherlands." *Environmental Law Review*, 18 (1), 2016, pp. 57, 61.

136 Urgenda Foundation v. The State of the Netherlands, C/09/456689/HA ZA 13–1396, judgment from 24 June 2015, para. 4.48.

137 *Ibid*; see also Jodoin, Sébastien, and Sébastien Duyck. "Look Before You Jump: Assessing the Potential Influence of the Human Rights Bandwagon on Domestic Climate Policy." *The Routledge Handbook of Human Rights and Climate Governance*, Routledge, 2018.

138 *Ibid.* at para. 4.54.

139 It is interesting to see this approach of the Dutch court in comparison to the *Kivalina* case and the political question doctrine mentioned above. The Dutch court did not see its judgment as infringing the principle of separation of powers by entering the area of policy-making, but as a step for which "it had a democratic legitimacy to provide legal protection, also against government authorities." See Lambrecht, *supra* note 135 at 61.

140 Urgenda Foundation v. The State of the Netherlands, C/09/456689/HA ZA 13–1396, judgment from 24th of June 2015, para. 4.55.

141 Heiskanen, Heta, and Sébastien Duyck. "Climate Change and the European Court of Human Rights: Future Potentials." *The Routledge Handbook of Human Rights and Climate Governance*, Routledge, 2018.

142 Urgenda Foundation v. The State of the Netherlands, C/09/456689/HA ZA 13–1396, judgment from 24 June 2015, para. 4.73, 4.75.

to the year 1990 was considered as negligent.[143] In this regard, a sufficient causal link was established between global climate change and its impact on the living conditions in the Netherlands and Dutch greenhouse gas emissions. The court further noted that even a limited amount of emissions produced in the Netherlands contributes to global climate change and needs to be adequately addressed.[144]

The original judgment of the District Court in the Hague in the case was confirmed by the Court of Appeals in October 2018.[145] Even though the *Urgenda* case represents progress and a step forward when evaluating the approach and the role of courts at a national level in the context of global efforts to combat climate change, it needs to be considered that the court considered only the lower threshold of efforts required from states by international treaties. Such a minimalistic approach would be insufficient to achieve the goals of the Paris Agreement.[146]

3.3.2 Private nuisance

Historically, nuisance claims based upon environmental degradation did not represent a serious obstacle for industrialization.[147] However, there are some instances when the law of nuisance was able to offer certain environmental protections of air quality or water quality, or against excessive noise.[148] A public nuisance can be characterized as "the unreasonable injury to a public right," and an action to address it can be initiated by both public authorities as well as private citizens who suffered an injury different from the one suffered by the general public.[149] Some have disputed the potential of public nuisance to address issues such as climate change on the ground that it is not a tort, but a form of public action which requires the authority and guidance from legislature to determine its limitations and boundaries.[150] Private nuisance represents "an unreasonable interference with another's right to the private use and enjoyment of land," which concerns also environment-related matters, such as air-polluting

143 *Ibid.* at 4.93.
144 *Ibid.* at 4.90.
145 Urgenda Foundation v. The State of the Netherlands, 200.178.245/01, C/09/456689/HA ZA 13–1396, judgment from 9 October 2018.
146 Robiou du Point, Yann. "Warming Assessment of the Bottom-Up Paris Agreement Emissions Pledges." *Nature Communications*, 2018, p. 2.
147 Dewees, Donald. "The Role of Tort Law in Controlling Environmental Pollution." *Canadian Public Policy*, 1992, pp. 425, 428.
148 Chalifour, Nathalie, and Sébastien Duyck. "Climate Change. Human Rights and Private Remedies." *The Routledge Handbook of Human Rights and Climate Governance*, Routledge, 2018.
149 Latham, *supra* note 32 at 751.
150 Davies, Haydn. "Native Villagers' Demand for Climate Justice Exposes the Limitations of Federal Common Law." *Env. L. Rev.*, 15 (1), 2013, pp. 1, 7.

activities concerning one's land.[151] The operative word in a private nuisance case is "land," meaning that the cause of action is dependent on a landowner's enjoyment thereof being affected. Clearly being an anthropocentric approach, private nuisance claims have a very limited reach, not to mention the difficulty in proving factual and legal causation.[152] Despite its limitations, public nuisance appears to represent a more feasible option to address also more remote and complex harms and dangers of public nature.[153] However, today most common law causes of action for public nuisance are codified, mainly by criminal statutes.[154] From an ecological perspective, one of the limitations of the private nuisance rests in its restrictive view of "neighbor."

A cross-boundary matter involving nuisance was brought before German courts in the Huaraz case. The plaintiff in the case was the owner of a property in the Peruvian city of Huaraz, which is located in proximity to the glacial lake Palcacocha. The risks associated with progressing climate change include the possibility of rising sea levels and avalanches that could damage the plaintiff's house.[155] The plaintiff filed a claim against the company R.W.E., which together with its affiliated companies scattered across the world was alleged to have contributed to the global anthropogenic greenhouse gas emissions caused by electricity produced from burning fossil fuels.[156] The claim was based on § 1004 of the *Bürgerliches Gesetzbuch* (German Civil Code) which enables one to demand the removal of an interference with use of his or her property and to seek an injunction to prevent further interferences.[157] The plaintiff argued

151 Latham, *supra* note 32 at 752; see also Blomquist, *supra* note 17 at 1056.

152 Factual causation requires that a plaintiff prove that something has a causal link with an injury. Here, the "but for" test is applied, asking, for example: but for the defendant's CO2 emissions in Slovakia, would the forest fire have damaged my house in California? Assuming that a causal chain can be linked back to the defendant, the plaintiff still has to show legal causation. Legal causation requires a showing of "proximate cause," with no superseding intervening cause in between. The plaintiff must prove that the defendant's CO2 emissions were a substantial contributing factor to the climate change that dried out my region (9,600 km away), thus leading to the forest fire. Considering that there are millions of CO2 emitters in between Slovakia and California, the California plaintiff would find it nearly impossible to show that the Slovakian emissions (or any individual's emissions for that matter) were a "substantial" contribution to the harm.

153 Kysar, *supra* note 105 at 124.

154 Schwartz, Victor. "Torts Cases and Materials." *Torts Cases and Materials*, 12th ed., 2010, p. 833.

155 Claim by Saúl Ananías Luciano Lliuya from 23 November 2015, 2 (unauthorized translation provided by Germanwatch e.V.), https://germanwatch.org/sites/germanwatch.org/files/announcement/20822.pdf (last updated on Nov. 13, 2018). The German version can be downloaded here: https://germanwatch.org/sites/germanwatch.org/files/static/19019.pdf (last updated on Nov. 13, 2018).

156 *Ibid.*

157 According to § 1004: "[w]ird das Eigentum in anderer Weise als durch die Entziehung oder Vorenthaltung des Besitzes beeinträchtigt, so kann der Eigentümer von dem Störer die Beseitigung der Beeinträchtigung verlangen. Sind weitere Beeinträchtigungen zu besorgen, so kann der Eigentümer auf Unterlassung klagen. (2) Der Anspruch ist ausgeschlossen, wenn der Eigentümer zur Duldung verpflichtet ist."

inter alia that the existence of global climate change is undisputed in Germany and the German *Treibhausgas-Emissionshandelsgesetz* (Greenhouse Gas Emission Allowance Trading Act) is mentioned as an example in this regard.[158] Even though the parent company R.W.E. is based in Germany, it was targeted by the plaintiff on the ground that the main decisions concerning its subsidiaries are taken by the parent company. For this reason it was the plaintiff's view that R.W.E. should have borne the liability.[159] The possibility to try the case before German courts was based mainly upon Article 7 of the Rome II Regulation, which enables a plaintiff to base a claim on the law of origin of the nuisance, i.e. Germany in this case.[160] The relevance of the European Union Environmental Liability Directive[161] and *Umwelthaftungsgesetz* (Environmental Liability Act) in this regard was cited by the plaintiff as well.[162] In the plaintiff's view, the link between the climate change-related emissions and the flood risk in the case could be determined as the result of the cumulative causation, i.e. the accumulation of all contributions that caused the impairment to the use of the property.[163]

The plaintiff's claim was gradually extended to order the defendant to bear the costs for preventative measures to protect the plaintiff's property; or to take adequate measures toward reduction of the water volume of Lake Palcacocha, to contribute to preventative measures by paying monetary contribution to the association of local authorities, or finally to pay a lump sum to the plaintiff.[164] The *Landesgericht* (Regional Court) in Essen dismissed the plaintiff's claims on the one hand for not being sufficiently precise, identifiable, and enforceable[165] and on the other hand dismissed the assertion that the defendant caused the

158 *Ibid.*

159 *Ibid.*

160 According to § 32a of the *Zivilprozessordnung* (Civil Procedure Regulations), jurisdiction in environmental cases is determined by the place where the damage originated, which is applicable in cases with international context as well. See Will Frank, *The Huaraz Case (Lluiya v. RWE) – German Court Opens Recourse to Climate Law Suit Against Big CO2 Emitter,* Climate Law Blog (Dec. 7, 2017), http://blogs.law.columbia.edu/climatechange/2017/12/07/the-huaraz-case-lluiya-v-rwe-german-court-opens-recourse-to-climate-law-suit-against-big-co2-emitter/ (last updated on Nov. 10th, 2018).

161 Directive 2004/35/EC of the European Parliament and of the Council of 21 April 2004 on Environmental Liability with regard to the Prevention and Remedying of Environmental Damage.

162 Claim by Saúl Ananías Luciano Lliuya from 23 November 2015, 22–24, 31 (unauthorized translation provided by Germanwatch e.V.), https://germanwatch.org/sites/germanwatch.org/files/announcement/20822.pdf (last updated on Nov. 13, 2018).

163 *Ibid.*

164 District Court Essen, judgment from 15 December 2016 in the case of Mr. Saúl Ananías Luciano v. RWE AG, No. 2 O 285/15, pp. 3–4 (unauthorized translation provided by Germanwatch e.V.), https://germanwatch.org/sites/germanwatch.org/files/announcement/20823.pdf (last updated on Nov. 13, 2018). The German version can be downloaded here: https://germanwatch.org/sites/germanwatch.org/files/static/19023.pdf (last updated on Nov. 13, 2018).

165 *Ibid.*

disturbance according to § 1004 of the German Civil Code.[166] The evidence in the case was also not sufficient to establish legal causality between the defendant's contribution to the greenhouse gas emissions and flood risk.[167] The concept of cumulative causation[168] relied on the defendant in the case presuming that the theoretical absence of one of the causes would be enough to undo the damage.[169] The court reasoned that because R.W.E.'s emissions represented only a fraction of global greenhouse gas emissions, their absence would not alter the flood risk in the case thereby failing to meet a *condition sine qua non*.[170] Perceiving causation from the other direction, the defendant's emissions alone would also not substantially increase the likelihood of climate change-related events, such as the flood risk in the case.[171]

In contrast to the reasoning of the lower court, the *Oberlandesgericht* (Higher Regional Court) in Hamm did not decline the claim made by applicants. The Higher Regional Court held the claim to be precise enough to be considered by the court.[172] With respect to the causation between climate change and flood risk, the Higher Regional Court held that a decision on this point could be taken only upon the basis of evidence and declined the categorical conclusions made by the court in the first instance.[173] The lawyer representing the plaintiff asserted

166 *Ibid.*

167 *Ibid.*

168 As noted by Will Frank, the lawyer working on behalf of plaintiff in the case, similar to the context of the case *Kivalina*, joint and several liability would be denied before German courts in a similar manner as before the courts of the U.S.A. See Will Frank, *The Huaraz Case (Lluiya v. RWE) – German Court Opens Recourse to Climate Law Suit Against Big CO2 Emitter*, Climate Law Blog (Dec. 7, 2017), http://blogs.law.columbia.edu/climatechange/2017/12/07/the-huaraz-case-lluiya-v-rwe-german-court-opens-recourse-to-climate-law-suit-against-big-co2-emitter/ (last updated on Nov. 10th, 2018).

169 Higher District Court Düsseldorf, NJW 1998, 3720, in District Court Essen, judgment from 15 December 2016 in the case of Mr. Saúl Ananías Luciano v. RWE AG, No. 2 O 285/15, p.6 (unauthorized translation provided by Germanwatch e.V.), https://germanwatch.org/sites/germanwatch.org/files/announcement/20823.pdf (last updated on Nov. 13, 2018).

170 District Court Essen, judgment from 15 December 2016 in the case of Mr. Saúl Ananías Luciano v. RWE AG, No. 2 O 285/15, p. 6 (unauthorized translation provided by Germanwatch e.V.), https://germanwatch.org/sites/germanwatch.org/files/announcement/20823.pdf (last updated on Nov. 13, 2018). *Condition sine qua non is* similar to "but for rule" in common law. The so-called but-for causation assumes that a tortfeasor is liable for an indivisible injury that would not appear without the breach. See Will Frank, *The Huaraz Case (Lluiya v. RWE) – German Court Opens Recourse to Climate Law Suit Against Big CO2 Emitter*, Climate Law Blog (Dec. 7, 2017).

171 District Court Essen, judgment from 15 December 2016 in the case of Mr. Saúl Ananías Luciano v. RWE AG, No. 2 O 285/15, p. 7 (unauthorized translation provided by Germanwatch e.V.), https://germanwatch.org/sites/germanwatch.org/files/announcement/20823.pdf (last updated on Nov. 13, 2018).

172 Order of the Regional Court of Hamm, Mr. Saúl Ananías Luciano v. RWE, p.2, para.1 (unauthorized translation provided by Germanwatch e.V.), https://germanwatch.org/sites/germanwatch.org/files/announcement/21168.pdf (last updated on Nov. 13, 2018).

173 *Ibid.*

that it should be possible to demonstrate causation in the case of "slow onset" events associated with climate change, such as rising sea levels or melting glaciers. Similarly, the defendant's claim that its emissions are only an insignificant fraction of global greenhouse gas emissions was refuted by pointing out that they amount to approximately 0.5 percent of global emissions, which, to get a sense of context, is equivalent to the total emissions of the Netherlands. Seen in this context as well as that of the *Urgenda* case, emissions of this extent are obviously not insignificant.[174] In September 2018, the Higher Regional Court in Hamm appointed experts to determine whether the alleged harms posed a serious threat to the plaintiff's property.[175]

3.3.3 Negligence

Negligence actions can be included in the category of corrective justice elements of tort law.[176] Breach of duty of care is decisive in this regard, irrespective of eventual proprietary interest.[177] Failure to comply with a duty of care can be decisive when deciding the liability of a defendant. To prove negligence, the plaintiff must prove that an established duty of care has been violated. The emphasis put on the duty of care can be characterized "as a judicially imposed regulatory standard."[178] Statutes or regulations can establish a duty of care.[179] For example, the failure to assess and reduce risks of injury, or to remedy a harm to the environment or harm resulting from exposure to hazardous substances released in the environment, can constitute failures to meet duties of care.[180] Still, the application of nuisance and negligence requires the identification of the polluter, proof of causation between the polluting activity and the harm, and the quantification of damage to persons and property, which might be especially difficult in more complex environmental cases.[181] The concept of the duty of care is also evolving. The scientific developments increasing the foreseeability of the negative impact on the environment resulting from car emissions as well as the risks associated with climate change coupled with the progress in technological

174 Frank, Will. "Climate Law Blog." *Climate Law Blog RSS*, blogs.law.columbia.edu/climatechange/2017/12/07/the-huaraz-case-lluiya-v-rwe-german-court-opens-recourse-to-climate-law-suit-against-big-co2-emitter/.

175 Germanwatch, *The "Huaraz Case" at a Glance*, https://germanwatch.org/en/huaraz (last updated Dec. 21, 2018).

176 Shmueli, *supra* note 36 at 751.

177 Dewees, supra note 147 at 428.

178 Abelkop, *supra* note 35 at 395.

179 In this regard, it can be also interesting to consider Kysar's suggestion whether international documents, such as the United Nations Framework Convention on Climate Change, could not stipulate the standards for duty of ordinary care in climate change-related matters. On the evolution of tort law and on what it means to act responsibly in the age of the Anthropocene, see Kysar, *supra* note 105 at 56–57.

180 Latham, *supra* note 32 at 532.

181 *Ibid.* at 753-754; see also Dewees, supra note 147 at 429–430.

and design options can all contribute to the evolution of the duty of care standards that can be expected from and imposed on automobile manufacturers.[182]

3.3.4 Strict liability

Private obligations for environmental harms such as climate change include the concept of strict liability. For example, Indian tort law requires companies to uphold the highest public safety standards when engaged in hazardous and inherently dangerous activities.[183] Companies causing environmental or health-related damage are held to an absolute liability standard without any exceptions, provided that it is proven that these companies are engaged in the activities mentioned above. The damages are determined by considering a company's size and capacity, with higher damages paid by larger and more prosperous companies.[184] As the court held in *M.C. Mehta and Anr v. Union of India and Others*, the concept of liability had to evolve, and since:

> [l]aw cannot afford to remain static … we find that it is necessary to construct a new principle of liability to deal with an unusual situation which has arisen, and which is likely to arise in future on account of hazardous or inherently dangerous industries which are concomitant to an industrial economy.[185]

The internalization of environmental and social costs by the means of absolute liability rules has the potential to significantly enhance and contribute to more effective deterrence of tort law mechanisms.[186]

3.4 Conclusions

In allocating environmental protection among the various state branches, some states, usually with strong civil law foundations, tend to favor administrative law approaches that may feature permits and planning. Other states, usually with strong common law foundations, tend to favor creating statutory norms and administrative details that can be used by public or private litigation, although common law states also have strong permitting systems. In common law states, the historic role of private obligations still plays a role, although most of the harms and norms that could be addressed by private litigation are now subsumed under public law statutes that give implementation authority over the subject matter to the state.

182 Blomquist, *supra* note 17 at 1058.
183 *Ibid.* at 1074–75.
184 *Ibid.* at 1063.
185 M.C. Mehta and Anr v. Union of India and Others, 1987 AIR 1086, 1987 SCR (1) 819.
186 Lin, *supra* note 35 at 1458.

The inception of public environmental law augmented, but did not replace, the law of private obligations. The law of private obligations is still important if the goal of the lawyer is to recover damages for his or her client, and not simply punish or stop the legal person(s) causing the environmental harm. Private remedies can also serve to supplement public actions. Moreover, a private action may serve to demonstrate to the state that a harm has in fact been committed, and thereby spur state action. And finally, it may not be the case that a state has exercised discretion, but rather that it is corrupt and uninterested in protecting the environment. In those cases, private action, individually or collectively, may change the behavior of the legal person(s) causing the harm, not just serve to put money in the pocket of the harmed person. But we must be careful not to go so far as to suggest that environmental law, in the public interest, can successfully meet the needs of the people of that culture through private law alone under a theory of neoliberal economics. Experience and study show that for the protection of public land, water, and air, when the public itself legitimates public law with its support, the protective outcomes of public law are demonstrably better.

Functional comparative law demonstrates that the legal and institutional structures of cultures as diverse as China and Brazil are to a great extent defined by the need to enable the interaction between the state and its citizens and, in the field of environmental law, by the need to address environmental issues. The need to establish such legal and institutional channels, with the underlying objective to address environmental matters, represents to a great extent the shared purpose or *tertium comparationis*.[187]

Literature

Abelkop, Adam D.K. "Tort Law as an Environmental Policy Instrument." *Oregon Law Rev.* 92 (2013), 382–469.

Alexander, Juraj, and Gyárfáš, Juraj. *Legal Research in Slovakia: Including a Brief Description of Slovak Political and Legal System*. Hauser Global Law School Program. February 2013. 5 June 2018 <http://www.nyulawglobal.org/globalex/Slovakia1.html>

Beru, Tsegaye, and Kirk W. Junker. "Constitutional Review and Customary Dispute Resolution by the People in the Ethiopian Legal System." *N.C. J. Int. Law* XLIII (2018), 1–65.

Bignami, Francesca. "From Expert Administration to Accountability Network: A New Paradigm for Comparative Administrative Law." *Am. J. Comp. Law* 59 (2011), 859–908.

Boughey, Janina. "Administrative Law: The Next Frontier for Comparative Law." *Int. Comp. Law Q.* 62 (2013), 55–95.

Canotilho, José Joaquim Gomes, and José Rubens Morato Leite. *Direito Constitutional Ambiental Brasiliero (Constitutional Environmental Law)*, São Paulo: Saraiva, 2015.

187 For more on *tertium comparationis,* see Siems, Matthias. *Comparative Law.* Cambridge University Press, 2014, p. 26.

Changshan, Ma. "Possibility and Limit of New Round Judicial Reform." *Tribune Political Sci. Law* (2015).

Curran, Vivian Grosswald. "Cultural Immersion, Difference and Categories in U.S. Comparative Law." *Am. J. Comp. Law* 46.1 (1998), 42–92.

Dainow, Joseph. "The Civil Law and the Common Law: Some Points of Comparison." *Am. J. Com. Law* 15.3 (1967), 419–435.

Di Pietro, Maria, and Zanella, Sylvia. *Direito Administrativo (Administrative Law).* São Paulo: Forense, 2014.

Elliott, E. Donald et al. "Toward a Theory of Statutory Evolution: The Federalization of Environmental Law." *J. Law, Econ. Organ.* 1.2 (1985), 313–340.

Gadamer, Hans-Georg. *Truth and Method.* 2nd ed. Continuum, 2004.

Green, H. Marlow. "Common Law, Property Rights and the Environment: A Comparative Analysis of Historical Developments in the U.S.A. and England and a Model for the Future." *Cornell Int. Law J.* 30.2 (1997), 541–586.

Greve, Michael S. "The Non-Reformation of Administrative Law: Standing to Sue and Public Interest Litigation in West German Environmental Law." *Cornell Int. Law J.* 22 (1989), 197–244.

Guohua, Jiang, and Xiao Nina. "The Entering of 'Ecological Civilization' into the Constitution and New Development of Environmental Law." *J. Nanjing Univ, Tech. (Social Science Edition)* (2019).

Haisong, Chen. "The Formation and Normalization of Rule of Political Parties in Eco-environmental Law." *Law Sci.* (2019).

Houck, Oliver A. "Save Ourselves: The Environmental Case That Changed Louisiana." *Louisiana Law Rev.* 72.2 (2012), 409–442.

Junker, Kirk W. *Legal Culture in the U.S.A.: An Introduction.* 1st ed. Routledge, 2016.

Klass, Alexandra. "Modern Public Trust Principles: Recognizing Rights and Integrating Standards." *Notre Dame Law Rev.* 82.2 (2006), 699–754.

Kitagawa, Hidetiki. "Environmental Policy Under President Xi Jinping Leadership: The Changing Environmental Norms." *Environmental Policy and Governance in China.* Ed. Hidetiki Kitagawa. Springer, 2017.

Lin, Albert C. "Beyond Tort: Compensating Victims of Environmental Toxic Injury." *Southern California Law Rev.* 78 (2005), 1441–1528.

Machado, Paulo Affonso Leme. *Direito Ambiental Brasiliero (Brazilian Environmental Law).* São Paulo: Malheiros, 2017.

Maohong, Bao. *Environmental Governance in China and Environmental Cooperation in North-East Asia.* Harushobo, 2009, in Kitagawa, Hidetiki. "Environmental Policy Under President Xi Jinping Leadership: The Changing Environmental Norms." *Environmental Policy and Governance in China.* Ed. Hidetiki Kitagawa. Springer, 2017.

McAllister, Lesley. *Making Law Matter: Environmental Protection and Legal Institutions in Brazil.* Stanford University Press, 2008.

Minzer, Carl F. "China's Turn Against Law." *Am. J. Comp. Law* 59 (2011), 935–984.

Nowag, Julian. "Article 11 TFEU and Environmental Rights." *Environmental Rights in Europe and Beyond,* Eds. Sanja Bogojevic and Rosemary Rayfuse. Hart Publishing, 2018.

Peine, Franz-Joseph, and Siegel, Thorsten. *Allgemeines Verwaltungsrecht,* 12th ed. C.F. Müller, 2018.

Prettyman, E. Barret. "The Nature of Administrative Law." *Virginia Law Rev.* 44.5 (1958), 685–701.

Ruskola, Teemu. "Legal Orientalism." *Michigan Law Rev.* 101 (2002), 179–234.

Saunders, Cheryl. "Apples, Oranges and Comparative Administrative Law." *Acta Juridica* 1 (2006), 423–449.

Siems, Matthias. *Comparative Law.* Cambridge University Press, 2014.

Smith, Lionel. "Common Law and Equity in R3RUE." *Washington Lee Law Rev.* 68 (2011), 1185–1202.

Szegedi, Laszlo. "The Eastern Way of Europeanisation in the Light of Environmental Policymaking: Implementation Concerns of the Aarhus Convention-Related EU Law in Central and Eastern Europe." *ELTE Law J.* 1 (2014), 117–134.

Wang, Jin. "Evolution of Environmental Thought and Enforcement of Environmental Protection Legislation in China: The Status Quo." *Environmental Policy and Governance in China*, Eds. Hidetiki Kitagawa. Springer, 2017.

Walch, Angela. "The Path of the Blockchain Lexicon (and the Law)." *Rev. Banking Financial Law* 36 (2017), 713–765.

Yang, Tseming. "Environmental Regulation, Tort Law and Environmental Justice: What Could Have Been." *Washburn Law J.* 41 (2002), 607–628.

Yixiang, Deng et al. "China's Water Management Towards Institutional Integration. A Review of Current Progress and Constraints *vis-à-vis* the European Experience." *J. Cleaner Prod.* 113 (2016), 285–298.

Zhang, Mo. "The Socialist Legal System with Chinese Characteristics: China's Discourse for the Rule of Law and a Bitter Experience." *Temple Int. Comp. Law J.* 24 (2010), 2–64.

4

RESOLUTION OF ENVIRONMENTAL DISPUTES

Vanessa Johnston, Tsegai Berhane Ghebretekle, Ryan Kraski, Jorge Ignacio García Nielsen, Ana Miola, and Mrinalini Shinde[1]

4.1 Introduction: the breadth of environmental dispute resolution institutions

What is the function of resolving a dispute? The answer may appear to be obvious, but, as we saw in Chapter 3, raising the question might lead one to ask why the dispute has arisen and why it is a decision for a court, rather than something for legislators to consider. Shifting environmental matters from legislative bodies, populated with politically elected representatives, to adjudicatory bodies, populated with professional lawyers or scientific experts, is more than just a matter of distinguishing public law from private law or civil law from common law: it can be a legal strategy that increases the likelihood of a favorable outcome for an aggrieved party. Within this context, the question for our study becomes what the function of resolving disputes *through environmental law* is. With world population growth likely to bring the number of humans to nine billion by 2050, it is obvious that competition for the world's limited resources will increase the conflicts over the use of those resources. Environmental conflict is becoming increasingly common. In some cases, the conflict between parties relates to the natural environment directly, regarding access to, or the use of, natural resources such as water, minerals, or timber; protection of flora or fauna; and liability for environmental damage, land use, and planning decisions, to name a few. In other cases, however, conflict only relates to the environment indirectly or incidentally. When those in conflict cannot resolve their differences, they may resort

1 The views and opinions presented in this chapter belong entirely to the authors, and they do not, in any manner or form, represent the views and opinions of the UNU-EHS, the UNFCCC secretariat, or any of their affiliated bodies, organizations, or agents.

to their legal systems. These disputes are brought for resolution at the national or sub-national levels, through administrative or arbitration procedures, before courts and tribunals, before elders who apply customary law, or even as insurance claims.

As we will discuss further in Chapter 5, environmental conflict may also arise as a constitutional matter and, where there is no specialist constitutional court, the conflict will be adjudicated by a court of general jurisdiction. For example, in Germany,[2] Ethiopia,[3] and Chile,[4] citizens have an express constitutional right to a clean or healthy environment. In Germany and Chile, this express right is supported by constitutional obligations imposed on the state to protect or conserve natural resources.[5] Pursuant to Article 20a of the German Constitution, for example, this is expressed as an obligation imposed on the state to "protect the natural foundations of life and animals by legislation and, in accordance with law and justice, by executive and judicial action, all within the framework of the constitutional order."[6] In addition, Germany's Federal Constitutional Court has found on at least one occasion that environmental protection forms part of recognized constitutional rights, such as the right to property and human dignity.[7] In comparison, Ethiopia's Constitution facilitates the implementation of such express rights through constitutional objectives on environmental matters.[8]

The function of resolving disputes[9] can be carried out by a state-authorized court, but that is not the only path. "Think globally; act locally" can include resolving conflicts locally, as with community councils and mediation[10] (U.S.A.),

2 *Grundgesetz für Bundesrepublik Deutschand* (Basic Law for the Federal Republic of Germany) (Germany) BGBL I 1949 1, art 20a; *Basic Law for the Federal Republic of Germany*, http://www.gesetze-im-internet.de/englisch_gg/index.html.

3 FDRE Const. Art. 44.

4 Constitution of the Chilean Republic art 19 N° 8.

5 *Ibid.*

6 *Grundgesetz für Bundesrepublik Deutschand* (Basic Law for the Federal Republic of Germany) (Germany) BGBL I 1949 1, art 20a; *Basic Law for the Federal Republic of Germany*, http://www.gesetze-im-internet.de/englisch_gg/index.html.

7 German Constitution Articles 1, 14. See Volker Oschmann, *Energy in Transition: A Comparative Analysis of the Australian and German Legal Frameworks for the Decarbonization of the Energy Sector in Response to Climate Change* (Master of Laws (Research) Thesis, University of Western Australia, 2011), 163; citing the decision of the Constitutional Court 2 BvL 8/77 8 October 1978. Cf Koch, Luhrs, and Verheyen highlight, however, that it would be extremely difficult to compel the German state to act to protect the climate under these general provisions because of the broad language used in those provisions.

8 FDRE Const. Art. 92.

9 We are using the term "disputes" in the sense of "disagreement over legal duties or rights, or about what extent or kind of compensation may be claimed by the complaining party for breaching said duties or rights." And we are using the term "conflict" in the sense of "friction or opposition rising between parties, based on perceived or actual differences or incompatibilities." Cameron, Steffani. "Conflict vs. Dispute." *Bizfluent*, 29 Apr. 2019, bizfluent.com/about-6610108-conflict-vs--dispute.html.

10 See "Investing in Our Planet." *Global Environment Facility*, www.thegef.org/.

community elders (Ethiopia), and even private arbitration. Compared to dispute resolution through courts and tribunals established by statute, some cultures resolve disputes regarding environmental matters without a formal state forum. In local communities or villages, issues may be handled internally by elders, without state institutional structures or procedural formality. These local methods of conflict resolution are often preferred for various reasons. For the practicing Western lawyer, it might come as a surprise to learn that in some African cultures "the state and state law are alien."[11] Mediation provides an advantage over litigation if all parties can leave the mediation feeling as though they were heard and got some of what they wanted, rather than just winners and losers. On the other hand, it is very questionable whether a state can subject the rights and resources of its citizens to private arbitration, but in the well-known case of *Vatenfall v. Germany*[12] that was precisely what happened. Before a dispute arrives at some formal forum for resolution, the disputing parties have usually tried to resolve the dispute informally with meetings, letters, arguments, threats to sue, politics, and other means. Given the cost of state forum litigation in much of the world, one can realize that going to court is not, for many parties, the likely first method of resolving a dispute. In some states, some types of actions, such as divorce in family law, require that the disputing parties first seek counseling before resorting to state resolution. By comparison, would environmental dispute benefit from such a forced resolution practice, or will it just prolong the problem? The answer in many ways depends upon what the *status quo ante* is. In laissez-faire cultures, where polluters can continue their activities until the state can prove a violation of law, adding a layer of dispute resolution allows the polluter to maintain the polluting *status quo ante.* But if the precautionary principle were in place and institutionalized through something like a mandatory impact assessment (see Chapter 9) or a permit obtained through an administrative process (this chapter, *infra*) then it would be the people who benefit by altering, delaying, or stopping the beginning of the permitted pollution process.

As was discussed in Chapter 3, those harms that are today called "environmental" were at one time solved as violations of the private law of obligations. With the introduction of the state's power and apparatus into the process, the various branches of the state and the tools of the state became the more common way in which one thinks of environmental law. Having discussed the legislative, executive, and administrative institutions and tools in Chapter 3, this chapter will focus only upon those disputes that are resolved through some form of the state system, either in court or through the administrative process.

11 Interview with Professor Hassen Mohammed, Mekelle University School of Law, Ethiopia (Oct.17, 2016) (notes on file with Kirk W. Junker).

12 Deutsche Welle. "US Arbitration Court Rejects Germany's Plea in Nuclear Phaseout Compensation Case: DW: 05.09.2018." *DW.COM*, www.dw.com/en/us-arbitration-court-rejects-germanys-plea-in-nuclear-phaseout-compensation-case/a-45362256.

To better understand the function of environmental dispute resolution, this chapter considers issues of environmental dispute resolution through state bodies in Europe (Germany, Finland, Denmark, and Sweden), Africa (Ethiopia and Kenya), the Americas (Chile, Brazil, and the U.S.A.), Australia, and Asia (China and India). Unsurprisingly, diverse geography, environmental conditions, legal systems, governance structures, economic priorities, the expectations of those in dispute, and other cultural factors will affect the approach taken to resolving environmental disputes. Therefore, in order to usefully compare different approaches taken to the resolution of environmental disputes in these jurisdictions, one must define the scope of comparison carefully.[13]

We begin by "laying out the essentials" according to the functionalist comparative method of Zweigert and Kötz. In doing so, the comparison is limited to the resolution of dispute that relates to the environment directly, through a legal process by which a state arbiter or judge makes a formal decision about the rights or interests of the parties involved (i.e. courts, tribunals, state arbitration panels, review boards, etc.) Some institutions serve judicial and administrative purposes[14] at the national, sub-national (state), or local (county/municipality) levels. Importantly, the definitions given to both concepts contributes to a "neutral syntax"[15] that enables useful comparison between features of these diverse legal systems, minimizing the impact of any special meaning attributed to these terms or concepts in a particular country.[16] In developing this neutral syntax, one must be careful to compare functions and not labels of institutions—that which is called a "tribunal" in some cultures is in fact a constitutionally established state court, and in other cultures a "tribunal" is an administrative body with judges appointed at the pleasure of the executive.

This chapter discusses the function of four types of dispute resolution institutions, namely courts of general jurisdiction, courts of specialized environmental jurisdiction, general administrative law review boards, and specialized environmental law review boards. Comparison among these four types of institutions relates to four topics, including the:

- nature of the institution or forum in which dispute are resolved;
- state jurisdiction over environmental matters;
- trial procedure; and
- reference to environmental law principles and planning objectives.

13 Zweigert, Konrad, and Kötz Hein. *Introduction to Comparative Law*. Translated by Tony Weir, 3rd ed., Clarendon Press, 1998.

14 Meek, Michael K. *Australian Legal System*. Thomson Reuters (Professional) Australia Limited, 2016, p. 26.

15 Zweigert, *supra* note 13 at 44.

16 Regarding translation and understanding of legal terms, see Galdia, Marcus. *Comparative Law and Legal Translation*. European. L., 2003.

Comparing different approaches to resolution of environmental dispute will highlight similarities and differences between jurisdictions, as well as differences in similarity, and similarity in difference.[17] Key findings of this comparison provide a basis on which to make broader observations about the function that the resolution of environmental dispute plays in protecting the natural environment and developing national environmental planning objectives and legal principles.

As an extension of locating environmental law among the functions of the state from Chapter 3, when it comes to dispute resolution, the main difference among the world's cultures is whether to resolve environmental disputes through courts or the administrative system's adjudicatory bodies. Having made that split, we can further divide the former group of the world's cultures according to which of them leave environmental matters before courts of general jurisdiction or special jurisdiction, just as the cultures from the latter group are split on whether to resolve environmental disputes before general administrative tribunals or specialized environmental administrative tribunals. These are functional distinctions, regardless of the names of the bodies in question. The comparativist must be careful not to believe that the title given to a court or tribunal is necessarily indicative of its function. Some courts are in fact named "tribunals," but function as courts.

4.2 Courts of general jurisdiction

Courts of general jurisdiction exist at both national and sub-national levels.[18] They may be created through the legal power of a constitution or statutory law. Courts of general jurisdiction can resolve disputes in civil and criminal matters within the relevant jurisdiction (municipality, state, country, etc.) In order to streamline court functions and improve efficiency, matters may be allocated to "lists" or "divisions" of the court according to subject matter, most often separating civil from criminal matters, but also including the "environmental" in some jurisdictions. Courts of general jurisdiction will have broad power to adjudicate environmental dispute in jurisdictions where there are no specialized environmental courts or tribunals. This occurs in Ethiopia, for example,[19] where gener-

17 Both distinctive and common characteristics of legal systems are relevant to the overall comparison. Reitz, John. "How to do comparative law." *American Journal of Comparative Law*, 46, 1998, p. 624; see, further, Dannemann, Gerhard, and Mathias Reimann, "Comparative Law: Study of Similarities or Differences?" *The Oxford Handbook of Comparative Law*, 2006, p. 383.

18 For example: Australia's courts of general jurisdiction include national (Federal, Federal Court of Appeal, High Court) and sub-national (state/territory) (Magistrate, District, Supreme, Supreme Court of Appeal); in Ethiopia, the Constitution empowers State Supreme Courts to exercise jurisdiction of the Federal Court of First Instance, and also the High Court; see F.D.R.E. Const. Art. 80 (4); D.R.E. Const. Art. 80 (2).

19 F.D.R.E., 2002. Environmental Pollution Control Proclamation, No. 300, Federal Negarit Gazeta, 9th year No. 12, Addis Ababa, Ethiopia, art 2(4).

alist courts adjudicate disputes referred by the Environment, Forest and Climate Change Commission[20] on broad on matters arising under the Environmental Pollution Control Proclamation.[21] While there is no specialized environmental tribunal in Ethiopia, F.D.R.E. Constitution Article 37 provides the legal basis to establish environmental tribunals if required.

4.2.1 Jurisdiction

While courts of general jurisdiction are used to varying degrees in the select jurisdictions to adjudicate environmental disputes, this is not due to any lack of disputes over environmental matters. Courts of general jurisdiction may have original jurisdiction over environmental disputes, appellate jurisdiction over environmental disputes, or both. Original jurisdiction includes disputes arising between private parties, public parties, or a mix of the two, about property or other rights over natural resources, land use, and environmental damage.[22] In Australia for example, the Federal Court adjudicates appeals of a wide range of civil and criminal environmental disputes that arise under the national Environmental Biodiversity and Conservation Act 1999 (Cth),[23] which includes environmental impact assessment and protection, and conservation of world heritage areas. Courts of general jurisdiction often hear appeals by parties of decisions of general administrative law review boards or specialized environmental tribunals or both.

Brazil's Superior Court of Justice (*Superior Tribunal de Justiça* or S.T.J.) provides a key example of the way that specific or localized environmental issues can be resolved by well-designed courts of general jurisdiction. The S.T.J. was established in 1988 pursuant to a constitutional mandate to the High Court of Brazil,[24] under which individuals had the right to an "ecologically balanced environment" and the government a duty to "defend and preserve [the environment] for present and future generations."[25] While constitutional support for environmental tribunals exists in other jurisdictions, the intergenerational aspect of this consti-

20 The Commission of Environment, Forest, and Climate Change was formerly the Ministry of Environment, Forest and Climate Change. At a regional level, some of the environment regulatory bodies are established as authority, agency, and bureau.

21 *Ibid.*

22 Australian examples include *Manderson v. Wright*, VSC 677(2016); *Arnold v. Hickman*, TASSC 55(2016); *O'Brien v. Pittwater Council*, NSWDC 32 (2016); *Spencer v. Commonwealth*, 240 FCR 262 (2015); *Yanner v. Eaton*, 201 CLR 351 (1999); pursuant to the Fauna Conservation Act 1979 (Qld) (Australia); *Cadia Holdings Pty Ltd v. New South Wales* 242 CLR 195, 199–200 (2010).

23 Environmental Biodiversity and Conservation Act 1999, pt 17 (Cth) (Australia).

24 Bryner, Nicholas. "Brazil's Green Court: Environmental Law in the Superior Tribunal De Justica." *Pace Envtl. L. Rev*, vol. 29, 2012, pp. 470, 75 (counting, in the year 2010, that the S.T.J. resolved 330,283 environmental conflicts).

25 Constituiqao Federal [C.F.] art. 225 (Braz.). *Available at* Political Database of the Americas: Federative Republic of Brazil, GEORGETOWN UNIV., http://pdba.georgetown.edu/Constitutions/Brazil/english96.html (last updated Nov. 2008).

tutional support arguably refers to the environmental, cultural, and geographical circumstances of Brazil, which are very different from those in which other environmental courts or tribunals are located. More specifically, as the world's fifth largest country by landmass, and home to about one-fourth of the world's plant species and a proportionately greater amount of animal species, Brazil is one of the most contentious countries in the world in terms of environmental protection and enforcement.[26] Accordingly, rapid economic growth is placing substantial pressure on the environment, and imposing substantial risk of biodiversity loss. This is reflected in the caseload of Brazil's S.T.J., which hears matters primarily relating to deforestation either caused by increasing demand for agricultural land due to rapid growth, urbanization, or industrialization in cities and towns.[27]

The S.T.J. serves as a federal court of last resort for environmental questions referring to federal law,[28] addressing the immense backlog of environmental disputes previously held up within the general listings of the Supreme Federal Court.[29] However, although the S.T.J. hears environmental matters, it is still a court of general jurisdiction.[30] While the S.T.J. has both original and appellate jurisdiction over environmental matters, most environmental disputes it hears are appeals on points of law from the federal appellate courts or state supreme courts.[31] Importantly, the S.T.J. has jurisdiction to hear disputes arising from Brazil's environmental legislation, the Forest Code, which aims to protect and maintain the country's flora (and biodiversity).[32] The S.T.J. has consistently adopted broad interpretations of the code, protecting many more conservation zones, and broadened the definition of illegal burning, protecting many more types of vegetation.[33] Since it first began operating in 1989, the S.T.J. has resolved over 3.7 million cases.[34]

By comparison, to understand jurisdiction in the U.S.A., one must understand the complicated system of federalism in which dual jurisdictions often function simultaneously through a system called "cooperative federalism."[35] The comparativists Zweigert and Kötz have observed that:

> The United States has been said to possess "perhaps the most complicated legal structure that has ever been devised and made effective in man's effort to govern himself." The point of this observation becomes clear when one

26 Bryner, *supra* note 24.
27 Findley, Roger. "Pollution Control in Brazil." *Ecology L.Q.*, 1988, p. 1.
28 Constituiqao Federal [C.F.] art. 129, caput, cl. III (Braz.).
29 Bryner, *supra* note 24 at 82.
30 Constituiqao Federal [C.F.] art. 129, caput, cl. III (Braz.).
31 Bryner, *supra* note 24 at 83.
32 C. Flor [Forest Code] No. 12.651, de 25 de Maio de 2012, Código Florestal (Brazil).
33 *Ibid.* at 86.
34 *Ibid.* at 84–5.
35 See Junker, Kirk W. "Issues of US Environmental Law Encountered from Abroad." *U.S. Law for Civil Lawyers: A Practical Reference Guide*, Nomos, 2020.

> sees the problems which have arisen in the United States from the complexities of the concurrence of federal and state law, and from the fact that both the United States and the several states possess fully equipped courts systems.[36]

That said, there is not one system among the fifty U.S. states that is representative of the others, and any lawyer with environmental legal action in the U.S. must consult the substantive and procedural environmental law of the state in which the problem has arisen in order to begin understanding how a dispute would be resolved and the type of public law-interpreting body before whom one must bring the conflict. Further, one must keep in mind that just one of these states, California, has a larger population than Canada and that another of these states, Alaska, has a greater land mass than all but eighteen sovereign countries in the world. More to the point, the state of Texas would rank seventh in the world in greenhouse gas emissions among the states of the world.

The fifty differing state systems (aside from the country's federal system) include courts of general jurisdiction that adjudicate environmental disputes as quasi-specialist benches, often hearing matters in specialist lists within the court's general jurisdiction. These courts are a compromise between a generalist court that is tasked with adjudicating environmental disputes due to the absence of environmental courts or benches, and a dedicated environmental tribunal that hears only environmental disputes through the administrative process. For example, the Environmental Division of the Vermont Superior Court is a trial court with statewide jurisdiction that adjudicates environmental disputes.[37] For the most part, the Environmental Division of the Supreme Court hears appeals from decisions made by the District Environmental Commission regarding planning and development, environmental permit, or land-use planning;[38] the Agency of Natural Resources regarding enforcement actions against resource use and management;[39] and other local authorities on planning or zoning issues. The majority of the cases are brought *de novo* and heard as a full evidentiary trial in front of two judges. The Environmental Division of the Supreme Court hears more than 300 cases per year.[40]

Environmental dispute may also arise as a constitutional matter and, where there is no specialized constitutional court, the dispute will be adjudicated by a court of general jurisdiction. For example, in Germany,[41] Ethiopia,[42] and Chile,[43] citizens have an express constitutional right to a clean or healthy environment.

36 Zweigert, *supra* note 13 at 249–50.

37 Wright, Merideth. "The Vermont Environmental Court." *J. Ct. Innovation*, vol. 3, 2010, p. 203.

38 Act 250 (Vermont) (1970).

39 Wright, *supra* note 37.

40 *Ibid.*

41 *Grundgesetz für Bundesrepublik Deutschand* (Basic Law for the Federal Republic of Germany) (Germany) B.G.B.L. I 1949 1, art 20a.

42 F.D.R.E. Const. Art. 44.

43 Constitution of the Chilean Republic art 19 N° 8.

While in Germany and Chile, this express right is supported by constitutional obligations imposed on the state to protect or conserve natural resources, the Federal Democratic Republic of Ethiopia's Constitution facilitates the implementation of such an express right through constitutional objectives on environmental matters.[44] In comparison, while Australia's constitution does not expressly address the environment, environmental disputes have been resolved by the High Court of Australia under powers such as "external affairs" by virtue of the disputes involving the protection of world heritage areas under international treaties.[45] In Australia, this is one way that citizens and entities have sought to obtain resolution despite the lack of robust environmental laws to protect aspects of the natural environment at federal or state level from time to time. Consequently, although the environment, as well as the rights and duties of citizens and the states in relation to it, is treated differently in the constitutions of compared jurisdictions, this does not prevent general courts from adjudicating environmental disputes as a constitutional issue.

4.2.2 Trial procedures in general jurisdiction

Disputes relating to environmental matters are challenging for civil and criminal trial procedures in courts of general jurisdiction. (Regarding the norms of criminal environmental law, see Chapter 10, *infra*.) Trial procedures are designed to regulate the way in which substantive rights and obligations are claimed, proved, and enforced, without impacting on the substance of those rights themselves.[46] Whether and how trial procedures accommodate environmental disputes affect both the adequacy of the decision for disputing parties and the integrity of the decision in relation to environmental objectives.

Locus standi, or "standing," is the right to commence legal proceedings before a court or tribunal.[47] In many environmental disputes, standing is governed by rules contained in environmental laws rather than at the discretion of the adjudicating institution according to substantive or procedural law.[48] In Ethiopia and Chile, for example, since environmental cases are fused with civil and criminal cases, the issue of standing in generalist courts is governed by the civil and criminal laws of the country respectively. However, Ethiopia's courts of general jurisdiction hear environmental matters in several ways. For example, persons

44 F.D.R.E. Const. at Art. 92.

45 Commonwealth of Australia Constitution Act 1901 s 51(xxxix) (Cth) (Australia); *Tasmania v. The Commonwealth*, 158 CLR 1 (1983); *Richardson v. Forestry Commission*, 164 CLR 261 (1988); see Preston, Brian. "Benefits of Judicial Specialisation in Environmental Law: The Land and Environment Court of New South Wales as a Case Study." *Pace Envtl. L. Rev*, vol. 29, 2012, p. 396.

46 Colbran, Stephen, et al. *Civil Procedure: Commentary and Materials*. LexisNexis Butterworths Australia, 2015, p. 4.

47 Bates, G. M. *Environmental Law in Australia*. 9th ed., LexisNexis Butterworths, 2016, p. 805.

48 Ryan, Patricia. "Environmental Court System." *UNSW L. Rev.*, vol. 4, 1981, p. 20

with a "vested interest"[49] have standing to make a claim in a court of general jurisdiction relating to the environment. In the U.S.A., only legal persons who can prove that they themselves have been injured typically have standing. The case of *APAP v. FEPA*, which initiated action in a court rather than on appeal from an administrative decision, is the pioneer public interest litigation case in Ethiopia.[50] Article 33 of the Civil Procedure Code requires a party to have such a vested interest in order to have standing and commence a civil action in a court of general jurisdiction.[51] The only exception to the requirement of vested interest as indicated in the Civil Procedure Code is Public Interest litigation as provided in the Environmental Pollution Control Proclamation.[52] The state, of course, has standing to take actions that harm the public generally. The state may have standing to take action when a legal person has harmed the environment as well, under various theories ranging from state ownership of natural resources to the legal doctrine of "public trust," in which the state is only the trustee of the resources for its citizens. Otherwise, matters to resolve environmental disputes as public interest litigation can be initiated by any person without the need to show a vested interest or personal injury.

In Ethiopia, environmental actions can also be initiated by the Office of General Attorney at the federal level or justice bureaus at the state level in line with the Criminal Procedure Code of the country.[53] The Federal Attorney General or regional justice bureaus are able to commence criminal proceedings on environmental matters in courts of general jurisdiction.

In Chile, however, the legitimacy to stand and make claims in a court of general jurisdiction depends on the nature of the case. If it is a civil case of environmental damage, for example, only the affected and municipalities have standing. If it is a criminal case (even though Chile does not have proper environmental crimes), the Public Prosecutor has legitimacy to take an action in a court of general jurisdiction. Nearly anyone has standing in constitutional matters in Chile. The subject matter of the action will determine whether an action is heard by a court of general jurisdiction or an environmental court. Only environmental disputes related to environmental management instruments are recognized by Environmental Courts. If, on the other hand, the dispute is related to some

49 Civil Procedure Code of the Empire of Ethiopia, (1965). Decree No. 52, Negarit Gazeta Extraordinary issue No.3, Addis Ababa, 8th October 1965.

50 A.P.A.P., a social organization, lodged a complaint with the former Federal Environment Protection Authority (F.E.P.A.) about pollution of the Akaki and Mojo Rivers in Addis Ababa caused by the discharge of industrial effluents contrary to the Environmental Pollution Control Proclamation and urged F.E.P.A. to take immediate measures. F.E.P.A. responded that the environmental standards were still in draft form and were awaiting the Environmental Council's approval and therefore they could not be relied upon to decide whether the rivers were being polluted. F.E.P.A. also asserted that it had taken all necessary measures to combat the pollution.

51 Civil Procedure Code, *supra*, note 49.

52 Environmental Pollution Control Proclamation, *supra*, note 19.

53 The Ethiopian Criminal Procedure Code of 1961.

specific natural resources (such as water or forests), the disputes are heard by courts of general jurisdiction.

In Ethiopia, courts of general jurisdiction can impose civil and criminal penalties to remedy environmental disputes. In relation to civil penalties, compensation for environmental damage is often provided through the traditional system of private obligations. However, the traditional private obligations in Ethiopia are fit only to govern traditional liability issues like injury to a person's health or property, not to the environment *per se*. When the Civil Code (where one finds private obligations) was enacted in 1960, the environment was not a major concern that the Code aimed to address. Ethiopia has not yet adjusted its system of private obligations to accommodate environmental issues or introduce a separate comprehensive environmental liability regime in order to ensure environmental justice and promote sustainable development through the application of the international environmental law "polluter pays" principle.

With regard to criminal penalties, the Ethiopian Environment Pollution Control Proclamation, which functions at the highest level of legislation, designated environmental crimes in order to protect the environment from pollution and to subject polluters to punitive sanctions. The crimes created relate to pollution,[54] waste and hazardous materials,[55] environmental inspectors,[56] and records.[57] Overlapping with the Environment Pollution Control Proclamation is the Criminal Code,[58] which also contains environmental crime provisions that prescribe more serious punishment for polluters than the Proclamation. For example, the Criminal Code penalty for intentionally poisoning a well, cistern, spring, water hole, river, or lake is up to fifteen years of rigorous imprisonment.[59] In contrast, the maximum sentence in the Proclamation is ten years.[60]

Notably, in Ethiopia, administrative measures are also employed to protect the environment in addition to civil and criminal penalties. As per the Environmental Pollution Control Proclamation, these administrative remedies include the installation of sound technology, the cleaning up of polluted environment or payment for such an operation, and the closure or relocation of an enterprise causing environmental damage.[61] While, as per the Environmental Impact Assessment Proclamation, the administrative measures are alike: rectifying the damaged environment, and suspension or cancellation of authorization

54 Environmental Pollution Control Proclamation, *supra*, note 19 at Art. 16.
55 *Ibid.* at Art.15.
56 *Ibid.* at Art. 13.
57 *Ibid.* at Art. 14.
58 The Criminal Code of the Federal Democratic Republic of Ethiopia, Addis Ababa, Arts. 514–34.
59 *Ibid.* at Art. 517.
60 Environmental Pollution Control Proclamation, *supra*, note 18 at Arts.15 (2) & 16.
61 *Ibid.*, Art. 3 (3) (4) (5).

or permit.[62] Similarly, other laws also provide different administrative measures that enable them to meet their respective objectives.

In Chile, environmental damage cases are under the jurisdiction of specialized Environmental Courts, but they only address the purely environmental nature of it, and impose restoration plans. The economic aspects of a case, such as compensation, must come under the jurisdiction of general jurisdiction courts. The procedural rules used to prove environmental damage are a mix between Law Number 19.300 and the Civil Code, following a subjective responsibility system.

4.2.3 Reference to environmental law principles and planning objectives

Although courts of general jurisdiction can adjudicate environmental disputes in numerous ways, they generally all do so by reference to legal principles. The range of principles used by courts of general jurisdiction differ based on the legal family, such as civil or common law, and also on whether the culture receives international law through a monist or a dualist approach.

If principles or planning objectives are explicitly part of binding statutes or case law, they may assist in adjudicating environmental disputes. Legislation that includes international environmental principles or reference to environmental treaties is becoming more frequent. For example, in Australia, national climate change laws refer expressly to the United Nations Framework Convention on Climate Change, Kyoto Protocol, and succeeding climate change agreements.[63] Both national and sub-national environmental protection statutes include statements of guiding principles or considerations, such as precaution, intergenerational equity, stewardship, and ecologically sustainable development.[64]

In Ethiopia, Article 85 (1) of the F.D.R.E. Constitution states that government organs (including courts) in the process of implementing the Constitution, other laws (like environmental laws), and public policies (including environmental policy) shall be guided by the Constitutional principles and objectives.[65] From the wording of the Constitution one can see that environmental objectives are elevated to the status of principles. These principles (for external relations and national defense) and objectives (political, economic, social, cultural, and environmental) are indicated in Articles 86–92.[66] Some of the environmental objectives spelled out in the Constitution are that the government shall endeavor to ensure that all Ethiopians live in a clean and healthy environment; that the design

62 Environmental Impact Assessment Proclamation, Art. 12 (2) (3).
63 Carbon Credits (Carbon Farming Initiative) Act 2011s 3 (Cth) (Australia).
64 For example, the Environmental Biodiversity and Conservation Act 1999, ss 3-3A (Cth) (Australia); Environmental Protection Act 1970, 1A, 1C, 1D, 1H (Vic); Environmental Protection Act 1997s 3D (ACT).
65 F.D.R.E. Const. at 10.
66 *Ibid.*

and implementation of programs and projects of development shall not damage or destroy the environment; that people have the right to full consultation and to the expression of views in the planning and implementation of environmental policies and projects that affect them directly; and that government and citizens shall have the duty to protect the environment.[67] Principles of international environmental law are also relevant to the assessment of civil damages in environmental cases; when entertaining environmental cases as civil cases, generalist courts in Ethiopia are supposed to apply the polluter pays principle. However, although the polluter pays principle is recognized in Ethiopia, there is no scheme put in place for complete internalization of costs related to environment.

Courts of general jurisdiction may be guided by a range of legal principles when adjudicating environmental disputes. Based on the experience of the compared jurisdictions, the source and impact of these principles varies, sometimes being referenced specifically in enabling legislation, or otherwise referenced indirectly through the international treaties that embody them. In either case, it is notable that these principles are available and used by courts of general jurisdiction, who otherwise have limited expertise in adjudicating environmental disputes.

Courts of general jurisdiction in many legal cultures around the world are tasked with adjudicating environmental disputes on a range of topics. Despite the difficulty of accommodating technical and complex environmental disputes within the general jurisdiction of these courts, this chapter identifies several tools used by courts in several legal cultures—Australia, Ethiopia, and the U.S.A.—to resolve disputes effectively in relation the relevant environmental objectives. These tools include the ability to deal with environmental matters by reference to constitutional rights and duties which, in some cases, enables the court to apply principles of international environmental law to the dispute at hand, and to impose civil and criminal penalties on polluters or wrongdoers.

4.3 Environmental courts and tribunals

Environmental courts and tribunals are the category of institutions that have competence to adjudicate disputes through the interpretation of environmental constitutional provisions, environmental statutes, or environmental administrative regulations.[68] The adjudicatory institutions may be created by constitutions or statutes for the purpose of resolving disputes. In some cultures, the term "environmental tribunal" includes reference to both tribunals and courts, but other cultures use the terms to denote two different legal forms of institution. When the distinction is made, the tribunal is part of the administrative

67 *Ibid.* at Art. 92.

68 In some cultures, the term "environmental tribunal" includes reference to both tribunals and courts, but other cultures use the terms to denote two different legal forms of institution.

system, while the court is part of the independent adjudicatory branch of the state. Thus a dispute might, under very similar governing law, be brought before an executive-appointed administrative law judge on a tribunal, but in another culture be brought before an appointed or elected member of the independent judiciary.

There are three features of environmental dispute resolution that one should keep in mind: the expertise of the body, the independent legal status of the body, and the centralized or decentralized relationship of the body. In some cultures, such as Australia, the terms "courts" and "tribunals" are used interchangeably, while in other cultures there is a clear difference between a court, which typically enjoys the status of an independent state branch, and a tribunal, which is a dispute-resolving body within the administration, often under the executive branch. Environmental courts are recognized as having some expertise in environmental law matters. But beyond the environmental expertise of the person sitting on such a specialist court or tribunal, it is important to know the legal status of the body, so that one knows the procedure that takes place before it, whether one has a right to appeal, to whom one appeals, and whether the judge or administrative judge are elected or appointed and by whom. Much can be observed about the status and power of environmental adjudication in a culture by learning the status of those empowered to resolve disputes. In the context of achieving ecologically sustainable development, courts and tribunals with specific environmental expertise are increasingly being recognized as the best-placed institutions to interpret, explain, and enforce laws and regulations on environmental matters.[69]

Tribunals may be established as an offshoot of decentralization of government authorities. Decentralization typically results in an increased number of departments, which in turn maximizes the responsibility of the departments. Hence, these departments are given the authority to look after their disputes independently without any interference of courts except when the decisions are challenged in their legality. Any decentralized department may have its own tribunal. In Australia, for example, a tribunal is said to have permanent establishment just as a court would. A bench of adjudicators is responsible for pronouncing a just and fair decision in favor of the aggrieved party. By comparison, the courts normally have a single judge deciding a dispute. The proceedings of a tribunal are often more informal and faster than those of a court. Australian courts are expected to be rigid in their functions because they are directed to do so as per formal rules of procedure and code of conduct. Their performance is reported to the higher courts that initiate misconduct proceedings in absence of obedience of proper conduct. In tribunals, the adjudicators are selected from the organization or the department itself. The department makes its own set of rules

69 Preston, Brian. "Operating an Environmental Court: the Experience of the Land and Environment Court of New South Wales." *EPLJ*, vol. 25, 2008, pp. 385, 386.

and they can be relatively flexible and informal. The court fees and expenses depend upon the subject matter of the case. If the matter in dispute is worth a billion dollars, then the court fee will be charged accordingly, which of course makes courts more expensive than tribunals. Faster decisions, lower costs, and the expertise of the adjudicators all increase the efficiency of Australian tribunals.

However, the jurisdiction of an Australian tribunal is much narrower than that of a court. Since an environmental tribunal is concerned with only environmental matters, its jurisdiction is limited. By comparison, a court hears a variety of disputes including civil, criminal, family, and business matters. Although Australian tribunal decisions are binding upon the parties, they may be appealable in court if the law under which the tribunal is established provides for appeal to higher courts.

Yet, in many countries, for the practicing lawyer, one must distinguish the function of that which is called an environmental "court" from that which is called an environmental "tribunal." In countries that make a distinction, disputes adjudicated by environmental tribunals may include appeals against administrative decisions in planning or land-use approval matters, for example. Accordingly, there is some overlap between the type of matters heard by environmental tribunals and administrative law review boards.

The number of environmental tribunals has grown significantly. In 2009, a study found that more than 350 environmental tribunals existed in forty-one countries worldwide,[70] increasing to approximately 1,200 tribunals in forty-four countries by 2016.[71] Growth of environmental tribunals has been encouraged by constitutional and environmental law reform, and response to civil society demands.[72]

4.3.1 Jurisdiction

The following paragraphs lay out some of the "essentials" as per the functionalist method of Zweigert and Kötz. The essentials are, however, only the beginning of functionalist comparison. Only after laying out these essentials can one make observations that may fulfill the objectives of a comparativist's study, whether it be to borrow ideas, to inform clients in conflict with citizens of another culture, or to construct a plan to relocate to a foreign culture.

The jurisdiction of environmental courts or tribunals depends on their enabling legislation. As a common feature in many countries, environmental tribunals feature within both national and sub-national court structures. Brazil, India, Kenya, and Chile have environmental adjudicatory bodies at a national level, and

70 Pring, George, and Catherine Pring. "Greening Justice: Creating and Improving Environmental Courts and Tribunals." World Resources Institute, Report, 2016, p. 1.

71 Pring, George, and Catherine Pring. "Environmental Courts and Tribunals." United Nations Environment Programme, Report, 2016, p. 1.

72 Pring "Greening," *supra* note 70 at 8–10.

Australia and China have environmental adjudicatory bodies at the state and provincial levels. In most cases jurisdictional choice depends on whether the environment, or environmental protection, is categorized by the state as a national or sub-national responsibility. For example, specialized environmental tribunals operate in four Australian states: New South Wales, Queensland, Tasmania, and South Australia. The New South Wales Land and Environment Court was established in 1980.[73] As the first superior court of record with specific environmental competence worldwide, it is well-regarded for its aims of facilitating rationalization and specialization in environmental matters.[74] Among other features, its success is attributed to a combined and exclusive jurisdiction over eight classes of environmental conflict. Jurisdiction is conferred by numerous environmental laws, relating to merit reviews of administrative decisions, judicial determination of civil and criminal aspects of environmental law, and appeals.[75]

While commentators recognize that this affords comprehensive coverage of planning, environmental protection, land acquisition, mining, and indigenous land claims matters, to name a few, the "jurisdictional maze"[76] of the Land and Environment Court is not as systematic as hoped, and can result in important omissions from the court's jurisdiction over natural resources such as forestry.[77]

In Australia, the Queensland Planning and Environment Court[78] and the South Australian Environment Resources and Development Court[79] are both inferior courts of record, sitting at the same level as District Courts of general jurisdiction. Both established in the early 1990s, they adjudicate between 200–600 matters per annum at either pre-trial mediation or trial proceedings.[80] In each case, jurisdiction is conferred by specific environmental laws across issues including planning and development, environmental protection, coastal management, fisheries, heritage, and vegetation management, within the relevant state.

Lastly, the Tasmanian Resource Management and Planning Appeals Tribunal[81] is part of the Planning System of Tasmania. While appearing to have a limited function on this basis, it contributes to achieving the most environmentally

73 Land and Environment Court Act 1979 (NSW) (Australia).

74 Preston "Benefits," *supra* note 45 at 402–3; Pring "Greening," *supra* note 70 at 21.

75 Land and Environment Court Act 1979, ss 16, 17–21B, 71 (NSW) (Australia); Lyster, Rosemary, et al. *Environmental and Planning Law in New South Wales*. The Federation Press, 2016, pp. 42–45.

76 Ryan, *supra* note 48 at 15.

77 *Ibid.* at 13.

78 Planning and Environment Court Act 2016 (Qld) (Australia); Queensland Courts, *Planning and Environment Court* https://www.courts.qld.gov.au/courts/planning-and-environment-cou rt (accessed 19 October 2018).

79 Environment Resources and Development Court Act 1993s 5 (SA) (Australia); Courts Administration Authority of South Australia, *Environment Resources and Development (ERD) Court* http://www.courts.sa.gov.au/OurCourts/ERDCourt/Pages/default.aspx (accessed 19 October).

80 Rackemann, Michael. "Planning and Environment Court 50th Anniversary Dinner Speech." [9], 2016.

81 Resource Management and Planning Appeal Tribunal Act 1993 (Tas) (Australia).

explicit objectives of any Australian environmental tribunal. Namely, it promotes the sustainable development of natural and physical resources and maintains ecological processes and genetic diversity, in the context of safeguarding the life-supporting capacity of environmental resources for future needs.[82]

In China, up to 134 environmental courts exist as "pilots" at provincial levels.[83] China's first pilot environmental courts were launched in Qingzhen, Guizhou province, in 2007.[84] While attempting to facilitate environmental proceedings, this court was plagued by questions of standing to file public interest lawsuits, payment of court fees, difficulties in obtaining evidence, and major difficulties in enforcing court orders. As of June 2014, Fujian province was the first area to have a pilot tribunal at the provincial high court level.[85] The Fujian court consists of the judicial members along with a dozen technical consultants with technical backgrounds hired to assist the court.[86]

China's Supreme Court has created a special tribunal that will hear only environmental issues.[87] While most environmental disputes are heard by environmental courts at a provincial level, this court works to strengthen the rule of law through summarization, research, and guidance to these lower courts in environmental matters. In this respect, while not being the most powerful tribunal, the Chinese special environmental tribunal assists China's attempts to have more uniformity and accountability throughout the local environmental courts.

The political and social conditions in China have had a significant impact on the development of China's environmental courts, including their structure, jurisdiction, and effectiveness. The Communist Party rules China strictly, permitting limited individual freedoms, and with little or no room for serious political dissent.[88] For example, throughout the 1990s, hundreds of companies contaminated Lake Tai, in the Yangtze Delta of Jiangsu province. A fisherman-turned-environmental activist, Wu Lihong, began monitoring the water quality and organized an informational campaign regarding the contaminants in the water; this was met with harassment, and eventually imprisonment, by local officials and police.[89] In this regard, local courts have been challenged not only by the weak enforcement of environmental laws by local officials, but by the "impossibly high standards" placed on applicants bringing environmental matters

82 Environment Management and Pollution Control Act 1994,sch pt1 (Tas) (Australia).

83 *Ibid.*

84 Alex L. Wang & Jie Gao *Environmental Courts and the Development of Environmental Public Interest Litigation in China,* 3 J. Ct. Innovation 37, 40 (2010).

85 *Ibid.*

86 *Ibid.*

87 *Ibid.* (being admitted as the head of the new tribunal, Zheng Xuelin was formerly a judge in the Supreme People's Court).

88 Nassereith, Jess. "China - Die Welt Des Xi Jinping." *YouTube,* YouTube, 18 Dec. 2018, www.youtube.com/watch?v=ABbacXcp80I.

89 "China's Dirty Secret: The Boom Poisoned Its Soil and Crops." *Yale E360,* e360.yale.edu/features/chinas_dirty_pollution_secret_the_boom_poisoned_its_soil_and_crops.

to the court.[90] Furthermore, once proceedings commence, bribery or undue pressure asserted by local officials onto applicants often results in early settlement or withdrawal of environmental claims.[91] In some respects, this reflects China's priority and focus on economic growth in recent decades.[92] However, in 2015, China enacted an "Environmental Protection Law," which imposes significantly higher penalties on offenders, including short custodial sentences intended to shame offenders for non-compliance, and penalties for local government officials who have failed to enforce local laws.[93] Here we encounter the legal use of a cultural value that may well not be transferable to other cultures—the shaming of an offender. In practice, the Environmental Protection Law affords China's environmental courts much greater jurisdiction and power to resolve environmental dispute involving both private citizens and government authorities. (Compare, for example, citizen suit provisions in Chapter 7, *infra*, on Public Participation.)

The establishment of the National Green Tribunal (N.G.T.) in India was the culmination of decades of political and legislative developments. The Central government in India passed legislation creating the National Environment Tribunal in 1995 and the National Environment Appellate Authority in 1997 for the purpose of revising administrative decisions relating to environmental impact assessments, but the legislation was never implemented and the bodies were never established in actuality.[94] In 2003, the Law Commission of India (L.C.I.) published its 186th report, "Proposal to Constitute Environment Courts," in which the L.C.I. emphasized that while "dealing with the subject of Environmental Courts … [there was a] need to develop a jurisprudence in this branch of law which is also in accord with scientific, technological developments and international treaties, conventions or decisions." In conclusion, the L.C.I. recommended that "a law be made by Parliament under Constitutional Article 253 for constitution of Environmental Courts in each State (or group of States) having original and appellate jurisdiction…"[95] The establishment of the National Green Tribunal in India is particularly significant from the perspective

90 Yee, Wai Hang. "Regulatory Compliance When the Rule of Law Is Weak: Evidence from China's Environmental Reform." *J. Publ. Admin. Res. & Theory*, vol. 14, 2014, p. 9.

91 Lo, Carlos Wing-Hung, and Shui-Yan Tang. *Institutions, Regulatory Styles, Society and Environmental Governance in China*. Routledge, 2014, p. 249.

92 Hart, Melanie, and Jeffrey Cavanagh. "Environmental Standards Give the United States an Edge Over China." *Center for American Progress*, 8 Sept. 2016, www.americanprogress.org/issues/green/news/2012/04/20/11503/environmental-standards-give-the-united-states-an-edge-over-china/.

93 Environmental Protection Law of the People's Republic (adopted at the 8th Meeting of the Standing Comm. Nat'l. People's Cong., Apr. 24, 2014, effective Jan. 1, 2015) 2015 Standing Comm. Nat'l People's Cong., Gaz. 112 (China).

94 Amirante, Domenico. "Environmental Courts in Comparative Perspective: Preliminary Reflections on the National Green Tribunal of India." *Pace Environmental Law Review*, vol. 29, no. 2, 2012, pp. 441–69.

95 *One Hundred Eighty Sixth Report on Proposal to Constitute Environment Courts*. Law Commission of India, Sept. 2003, lawcommissionofindia.nic.in/reports/186th report.pdf.

of comparative environmental law, as, in its report, the L.C.I. used the examples of "special environmental courts in Australia, New Zealand and in some other countries and these are manned by Judges and expert commissioners" in justifying the recommendation to create an environmental court in India.[96]

Following parliamentary debate, the N.G.T. was established via the National Green Tribunal Act, 2010,[97] to determine disputes on matters relating to environmental protection or conservation of resources.[98] Pursuant to Article 14(2)(m) of that Act, the N.G.T. has competence to adjudicate any "substantial question relating to the environment," either as an issue of environmental impact or damage to public health, arising pursuant to breaches of laws listed in Schedule 1.[99] The N.G.T. is a Tribunal created under a statute, as opposed to the High Courts or the Supreme Court which were created under the Constitution of India. The N.G.T. is comprised of a Principal Bench situated in New Delhi and zonal benches in the Eastern, Western, Southern, and Central zones across India. Under the statute, the N.G.T. is not bound by the rules of evidence under the Indian Evidence Act, 1872, unlike the High Courts and the Supreme Court, and relies upon the principles of natural justice while conducting its business. The N.G.T. exercises original jurisdiction in civil cases involving a substantial question relating to the environment, along with certain kinds of appellate and special jurisdiction. One of the most important differences between the N.G.T., and the High Courts and the Supreme Court, is that the N.G.T. bench contains both judicial and expert members, the latter having expertise in environmental topics, enabling the N.G.T. to address technical scientific issues in the settlement of environmental disputes or in the adjudication of questions involving the environment. The statute directs decisions of the N.G.T. to be guided by principles of environmental law such as sustainable development, the precautionary principle, and the polluter pays principle.[100]

The unique nature of the N.G.T. as an environmental adjudication body is summed up by Gitanjali Nain Gill: "[s]cience has a profound effect upon our understanding of and response to environmental issues. Science has a similar role and effect within the N.G.T. Independent, in-house, scientific knowledge

96 *Ibid.*

97 The National Green Tribunal Act, No. 19 of 2010, India Code (2010), vol. 19 (2010).

98 National Green Tribunal Act §14.

99 The National Green Tribunal Act, No. 19 of 2010, *Ibid.* at *Ibid.* at India Code (2010), vol. 19 (listing "Schedule 1" enabling statutes: 1) The Water (Prevention and Control of Pollution) Act, 1974; (2) The Water (Prevention and Control of Pollution) Cess Act, 1977; (3) The Forest (Conservation) Act, 1980; (4) The Air (Prevention and Control of Pollution) Act, 1981; (5) The Environment (Protection) Act, 1986; (6) The Public Liability Insurance Act, 1991; and (7) The Biological Diversity Act, 2002).

100 Brara, Rita. "Courting Resilience: The National Green Tribunal, India." *U.N.R.I.S.D. Working Papers, United Nations Research Institute for Social Development (U.N.R.I.S.D.)*, 2018. *U.N.R.I.S.D. Working Paper 2018-4*, www.econstor.eu/handle/10419/186114.

has become part of the analysis that produces judicially binding decisions."[101] In 2015, merely five years after the establishment of the N.G.T., Gill observed that time would tell if a Thucydides trap had been set, as evidenced by the "growing turf wars between various interests and institutions involved in environmental regulation and environmental jurisprudence" and that the success of the N.G.T. had "resulted in expressions of anxiety focusing on the powerful position achieved by the Tribunal within a relatively short period." Between 2011 and 2016, the N.G.T. issued 2,051 judgments on environmental cases.[102]

Gill's research in 2015 revealed an example of the turf war, which also exemplifies re-centralization of an adjudicatory body, when the Central government, through the Ministry of Environment and Forests, had tried to exercise control through confrontation with the N.G.T. by introducing budgetary restrictions resulting in shortage of infrastructure and amenities.[103] In September 2013, the Central government filed an affidavit before the Supreme Court of India that the N.G.T. was "not functioning as per the provisions of the N.G.T. Act … resulting in embarrassment to the government before Parliament," which resulted in the Supreme Court remarking that the affidavit showed that "the tribunal is an eyesore for the government … first time an affidavit has been filed by government which attacks a quasi-judicial body."[104] The current absence of appointed members at the zonal benches of the N.G.T. is indicative of a possible trend in the dilution of the N.G.T.'s strength in the long term. According to the daily cause lists published by the N.G.T. on its website in July of 2019, hearings at the zonal benches were being conducted through videoconferencing with the Principal Bench in New Delhi, with no presence of appointed judicial or expert members at the zonal benches.

In India, the National Green Tribunal is the primary state forum charged with resolving environmental dispute. While only a single tribunal in a vast continent, the N.G.T. is globally significant as an environmental tribunal for its strong grounding of environmental matters in constitutional law. Since 1976, India's Constitution has included provisions that require the government both to protect the environment, as well as to affirmatively improve the country's forests and wildlife.[105] In addition, Indian citizens have a corresponding constitutional duty to protect and improve the environment.[106] On this premise, India's N.G.T.

101 Gill, Gitanjali Nain. "Environmental Justice in India: The National Green Tribunal and Expert Members." *Transnational Environmental Law*, vol. 5, no. 1, 2015, pp. 175–205.

102 Rengarajan, Sridhar. "National Green Tribunal of India—An Observation From Environmental Judgements." *Env. Sci. & Pollution Research*, vol. 25, 2018, pp. 11313, 15.

103 *Ibid.*

104 Tnn. "Green Tribunal Exceeding Its Brief: Govt to SC: India News - Times of India." *The Times of India*, timesofindia.indiatimes.com/india/Green-tribunal-exceeding-its-brief-Govt-to-SC/articleshow/22527436.cms.

105 India Const. Art. 48 & Art. 51(g), amended by The Constitution (Forty-second Amendment) Act, 2007.

106 *Ibid.* at 48(A).

has the potential to impact and progress environmental protection and conservation in a continent home to more than 1.2 billion citizens.[107]

In 2010, a constitutional amendment resulted in Kenya being the first African state to have a specialized environmental tribunal.[108] Kenya's Environmental and Land Court is a superior court of record, putting it at the same level as the High Court, Supreme Court, and Court of Appeal.[109] Pursuant to Article 162(2)(b) of the Kenyan Constitution, the Environment and Land Court has sole original, supervisory, and appellate jurisdiction over environmental issues in lower environmental court decisions or any other environmental courts or tribunals.[110] Pursuant to Article 165(5)(b) of Kenya's Constitution, the Environmental and Land Court has broad powers to deal with disputes relating to land administration and management, cases relating to public, private, and community land and contracts, and choses in action or other instruments granting any enforceable interests in land, including the environmental impact of same.[111]

One of the largest obstacles facing the Kenyan Environment and Land Court is integration, as a number of generalist tribunals may refer matters to it, pursuant to a variety of enabling statutes. Despite a lack of clarity or coherence about which body may hear the initial environmental dispute,[112] it is clear that the Environment and Land Court has broad remit to hear matters relating to the "environment and the use and occupation of, or title to, land,"[113] as referred by the relevant lower court.

In comparison, the impetus for developing specialized environmental tribunals in the United States has been entirely different. In many ways, the U.S.A. has been a global reform-leader in environmental protection jurisprudence since the 1970s.[114] Since that time, numerous avenues have been developed at federal, state, and local levels to administer and enforce environmental protection and conservation laws, and to resolve disputes relating to environmental matters. One does not find specialized courts dedicated to environmental matters in the U.S.A. Instead, whether in the jurisdiction of one of the fifty states or the federation, most disputes begin before administrative law agencies, such

107 Amirante, *supra* note 94 at 441, 55–58.

108 Const. Art. 162(2) (2010) (Kenya); *see also* Kaniaru, Donald W. "Environmental Courts and Tribunals: The Case of Kenya." *Pace Envtl. L. Rev*, vol. 29, 2012, pp. 566, 71.

109 Const. art. 164 (2010) (Kenya); The Supreme Court Act, (2011) No. 7 (Kenya).

110 Const. art. 162(2)(b) (2010) (Kenya), section 165(5)(b) (having powers to deal with disputes relating to land administration and management, the court is also empowered to hear cases relating to public, private, and community land and contracts, choses in action or other instruments granting any enforceable interests in land).

111 "An Assessment of the Performance of the Environment and Land Court: 12th Scorecard Report." Land Development and Governance Institute, 2013, p. 5.

112 Kaniaru, *supra* note 108 at 575.

113 The Environment and Land Court Act, (2011) No. 19 (Kenya).

114 Wolgast, Anna. "United States' Environmental Adjudication Tribunal." *J. Ct. Innovation*, 2010, pp. 185, 99.

as the Environmental Hearing Board in Pennsylvania or the Environmental Appeals Board at the U.S. Environmental Protection Agency (E.P.A.) The E.P.A.'s Environmental Appeals Board is the final decision-maker on administrative appeals under all major environmental statutes that E.P.A. administers. Located in Washington, D.C., the Board is an impartial tribunal independent for substantive purposes of all Agency components outside the Immediate Office of the Administrator. The Board's caseload consists primarily of appeals from permit decisions and administrative civil penalty decisions, as well as petitions for reimbursement of costs incurred in complying with cleanup orders issued under the Comprehensive Environmental Response, Compensation, and Liability Act of 1980 (C.E.R.C.L.A.) The Board's four Environmental Appeals Judges sit in panels of three and make decisions by majority vote. Although the Board is called an "appeals" Board, the appeals are from E.P.A. administrative decisions, not from adjudicatory body decisions. Eight attorneys and two administrative professionals provide support to the judges. Appeals from the first instance state or federal administrative agency decisions in the U.S.A. are often heard in courts of general state or federal jurisdiction, respectively, at the second instance.

The Environmental Appeals Board offers an Alternative Dispute Resolution (A.D.R.) Program to assist parties in resolving disputes before the Board. The Board's A.D.R. Program offers parties the option of participating in A.D.R. with the assistance of one of the Board's Judges acting as a neutral mediator. As an example of comparative law in practice, according to the U.S. E.P.A.:

> the Board also serves as a repository of expertise within the federal government on environmental administrative adjudication. Judges and attorneys on the Board have provided training and shared their expertise on environmental adjudication, both domestically and internationally, with a significant number of judges, attorneys, environmental specialists, local, state and federal government officials, and civil society organizations. The Board has conducted training for judges from Asia, Central and South America, Eastern Europe, and the Middle East.[115]

A further insight into the status of environmental law in U.S. legal culture stems from the way in which the U.S.A. divides conceptual areas of the law. U.S. legal education and practice do not divide discussions of substantive law into private law and public law and might well write about public and private law issues in the same paragraph[116] while discussing environmental law, or comfortably might insert questions of constitutionality and states' rights into the discussion of a

115 *EPA*, Environmental Protection Agency, yosemite.epa.gov/oa/eab_web_docket.nsf.

116 See Wilder, Keith. "The 'Public Business' Conundrum." *Kölner Schrift Zum Wirtschaftsrecht*, vol. 3, no. 2, 2012, pp. 232–35.

private dispute.[117] Environmental law to a common law lawyer usually involves both public and private interests, as one can observe in the numerous cases brought by administrative agencies of the state against private citizens. During the birth of environmental law as a separate area of law, the court pointed out in the 1970 case of *Boomer v. Atlantic Cement Company* that private law will not solve some problems—in this case, one of the natural environment—and that public law must do so. The court wrote:

> It seems apparent that the amelioration of air pollution will depend on technical research in great depth; on a carefully balanced consideration of the economic impact of close regulation; and of the actual effect on public health. It is likely to require massive public expenditure and to demand more than any local community can accomplish and to depend on regional and interstate controls. A court should not try to do this on its own as a by-product of private litigation and it seems manifest that the judicial establishment is neither equipped in the limited nature of any judgment it can pronounce nor prepared to lay down and implement an effective policy for the elimination of air pollution. This is an area beyond the circumference of one private lawsuit. It is a direct responsibility for government and should not thus be undertaken as an incident to solving a dispute between property owners and a single cement plant...[118]

Several observations can be made here. The judge wrote from the position that, up until that time, one would have expected to solve an air pollution dispute as a private problem. In this vein, Stephen Subrin and Margaret Woo have commented that "it is to civil litigation that Americans frequently turn for redress of social ills; and it is a rights-based culture that has shaped and is reshaped by the process of litigation."[119] But the judge in the *Boomer* opinion tells the audience that private resolution of an air pollution dispute fails, and he concludes therefore that the problem ought to be one of public law and state enforcement. One should note well, however, that, up until that time, the default position was to solve a problem through private action, not through state control.[120]

117 See Barker, Robert. "Constitutional Comparisons between the Law of the U.S.A. and Costa Rica." *American Law Review*, vol. 27, 1992, p. 234.

118 *Boomer v. Atlantic Cement Company*, 26 N.Y.2d 219, 223 (N.Y. 1970).

119 Subrin, Stephen, and Margaret Y. K. Woo. *Litigating in America: Civil Procedure in Context*. Aspen Publishers, 2006, p. 7.

120 Consequently, when the state does decide to take action, the violations are most often the more serious ones, and the state will feel that its legal position must reflect the severity of the situation. This helps to explain the penalty that the U.S. E.P.A. sought in the 2016 litigation against Volkswagen. The German media seemed to believe that, as a foreign corporation, Volkswagen was being penalized outside the normal realm of penalties. That suspicion might have made sense in a civil law planning-and-administration legal culture, as is Germany's, but in a litigation legal culture like the U.S.A.'s, the penalty sought was within litigation

The Chilean Environmental Court, created by Act Number 20.600 of 2012, complements other new Chilean environmental institutions, allowing all citizens to access environmental justice and concerning judicial review of Administrative resolutions and decisions of environmental legal disputes. This court is autonomous and independent from the Administration. It has a mixed composition, because it is comprised of two ministers who are lawyers and one who holds the university degree of Bachelor of Science, integrating legal, scientific, and technical analysis of the issues discussed in front of the Court. However, since its creation, the technical analysis has been deficient, since the focus of the disputes led by private lawyers is almost always based on legal issues, rather than technical.

There are three Environmental Courts. One in the north, one in the center, and one in the south of the country. The final decisions of the Environmental Court, even though it is not part of the general Judicial System, are subject to review only through the "*recurso de casación*" to be viewed at the Supreme Court. Any person or legal institution may bring a lawsuit before the Environmental Court for matters within its competence. The Environmental Court's main jurisdiction is to resolve administrative environmental disputes in order to redress private environmental obligations.

In addition, any person or institution suffering distress or grievance as a result of illegalities of administrative acts or regulations issued by the Ministry of the Environment (M.M.A.), the Environmental Assessment Agency (S.E.A.), the Environmental Superintendence (S.M.A.), and other government organizations with competences over environmental management instruments, such as the E.I.A. system, may be brought before the Court in order to seek redress.

Furthermore, demands to repair environmental damages may be brought before the court by any person or institution who has experienced damage and thus is seeking redress: municipalities (city governments) for claims arising within their own jurisdiction and the state through the *Consejo de Defensa del Estado* (Council of the State Defense). The S.M.A. may also apply for injunctive relief, in response to which the Court can issue a writ of denial or confirmation.

For all countries, the comprehensive jurisdiction to hear, determine, and finalize disputes under all environmental laws in the jurisdiction is a key feature of successful environmental tribunals.[121] The presence of both comprehensive subject matter jurisdiction[122] and an ability to adjudicate various matter types (administrative, judicial, and appellate matters) is key. In combination, this

standards for E.P.A. action, such as the *US EPA and California Air Resources v. Volkswagen*, Case No: MDL No. 2672 CRB (JSC) (2016). Further, the fact that the U.S. does begin with expecting individuals to sort out wrongs before the state gets involved means that in addition to the E.P.A. action against Volkswagen, defrauded car owners took a private, collective "class action" against Volkswagen.

121 Preston, Brian. "Characteristics of Successful Environmental Courts and Tribunals." *Env. L. J.*, vol. 26, 2014, p. 365.

122 *Ibid.* at 373.

affords environmental courts and tribunals with the capacity to adopt a "creative and innovative 'problem-solving' approach to restraining, remediating or compensating for environmental harm"[123] to foster effective determination of legal and non-legal aspects of environmental dispute. It is apparent that broader jurisdiction (for example in India and Brazil) leads to heavy caseloads, and less time to deal with the complex technical issues that may be involved in the cases. In turn, a heavy caseload may have impacts for trial procedural issues, including adjudicator expertise, and the use of experts.

4.3.2 Trial procedure

An environmental court created by constitution or statute often adopts procedural rules or structures to better accommodate environmental dispute, relating to standing, specialist adjudication, evidence (standard of proof), expert material, and remedies, each of which are noticeably different to those in courts of general jurisdiction. As discussed above, traditional standing rules do not align well with environmental dispute, and so are amended in legislation that specifically grants jurisdiction to environmental courts or tribunals to adjudicate dispute. For example, Australia's national and sub-national environment, planning, and development laws that confer jurisdiction on environmental tribunals[124] have broad standing rules that allow any person whose interests are affected by the conduct, or has recently engaged in environmental protection activities,[125] to seek an injunction against conduct contrary to the relevant law.[126] In some cases, the rights of the interested person are not required to have already been breached at the time the claim is made.[127] For example, the case of *Building Owners and Managers Association of Australia v. Sydney City Council* confirmed that a person has standing under environmental laws in Australia despite having no "relevant interest" in the outcome of the proceedings.[128] In many ways, this outcome appears to reflect the growing popularity and acceptance of environmental adjudication bodies being a place in which citizens and interested groups can pursue environmental justice, despite having no formal connection to the

123 *Ibid.* at 376.

124 Environmental Protection and Biodiversity Act 1999 (Cth) (Australia); Environment Planning & Assessment Act 1979(NSW) (Australia). See Bates, *supra* note 47 at 808-9.

125 Environmental Protection and Biodiversity Act 1999s 475(6)-(7) (Cth) (Australia); Environment Planning & Assessment Act 1979 section 123 (NSW) (Australia).

126 Fisher, D. E. *Australian Environmental Law: Norms, Principles and Rules.* Thomson Reuters (Professional) Australia Limited, 2014, pp. 613-4; Stein, Paul. "The Role of the New South Wales Land and Environment Court in the Emergence of Public Interest Environmental Law." *EPLJ*, vol. 13, 1996, p. 179.

127 Environment Planning & Assessment Act 1979 s 123(NSW) (Australia).

128 84 LGRA 54 (1984).

relevant matter.[129] This was upheld on appeal, on the basis that an observer who felt strongly enough about a breach of environmental law to question if the breach should be allowed should be able to direct that question to the appropriate court.[130] In this regard, India takes a similarly broad approach to standing. Pursuant to *National Green Tribunal Act* § 18, claims can be filed by "any person aggrieved, including any representative body or organization," which has been interpreted to encompass claims filed by Non-Governmental Organizations.[131] Such a broad approach has been applauded as developing public interest litigation and enabling civil society to participate in protecting the environment.[132] While this might be an important point of difference between adjudication of environmental disputes in specialized or generalist courts, it may also be a characteristic that distinguishes environmental courts and tribunals from one another in relation to their success in adjudicating environmental disputes in an environmentally progressive, and effective, manner.

4.3.3 Adjudicator expertise

On the premise that environmental dispute demands specialist knowledge and expertise of a range of technical matters, it is important that adjudicators presiding over environmental dispute are "environmentally literate."[133] Among the compared countries, environmental courts and tribunals in Australia, Brazil, Kenya, India, and Chile impose requirements on adjudicators regarding experience and expertise. While the requirements themselves differ in each jurisdiction, the general requirement of an adjudicator to possess some experience and expertise in environmental matters is common to these jurisdictions.

In Australia, the New South Wales Land and Environment Court and the South Australian Environment Resources and Development Court require judges and commissioners to have minimum technical expertise, and undertake continued professional development. For example, judges must have experience at a superior court or be a lawyer with at least seven years' experience,[134] and commissioners must have practical knowledge and experience in areas including ecology, town planning, engineering, native vegetation management, etc.[135] In comparison, while no such standard of knowledge is imposed on judges of the Queensland Planning and Environment Court, judges are called from an

129 For example refer to the public interest in the recent planning decisions for the Adani Coal Mine: *Gloucester Resources Limited v Minister for Planning* [2019] NSWLEC 7.

130 Sydney City Council v Building Owners and Managers Association of Australia388-9(1985) 2 NSWLR 383 (Australia).

131 The National Green Tribunal Act § 4(2); Amirante, *supra* note 94 at 464.

132 Stein, *supra* note 126.

133 Preston "Characteristics," *supra* note 120 at 377.

134 Land and Environment Court Act 1979 s 8 (NSW) (Australia).

135 Land and Environment Court Act 1979 s 12 (NSW); Environment Resources and Development Court Act 1993 s 10 (SA) (Australia).

"informal bar" of District Court (generalist) judges known for their expertise and experience in environmental matters.[136] A similar approach is taken in Kenya, where adjudicators are not only required to have subject matter expertise, but minimum experience in dispute resolution. Pursuant to § 7 of the Environment and Land Court Act 2011 (Kenya), all adjudicators in the Environment and Land Court must have at least ten years' experience as distinguished academic or legal practitioners with knowledge and experience in matters relating to environment or land. Similar requirements are imposed in Denmark's Nature and Environmental Protection Board of Appeal, where a panel of adjudicators can be composed in two ways: by a Chairperson plus two Supreme Court judges or political appointees, or alternatively by a Chairperson plus two to four appointed experts.[137]

In contrast, India addresses adjudicator expertise by imposing strict rules on the membership of the National Green Tribunal, in relation to the number and proportion of judicial and expert members. The *National Green Tribunal Act* provides that, while there may be between twenty-one and forty-one members of the Tribunal, in addition to the single Chairperson, the bench must be composed of between ten to twenty full-time judicial members, and a further ten to twenty expert members.[138] Adjudicators must hold either a Master in Science with a Doctoral Degree or a Master of Engineering or Technology, both with a minimum of fifteen years of experience and five years of technical experience.[139] Brazil also has a large judicial membership for the S.T.J., which consists of thirty-three Justices, selected equally from the pool of Justices serving any of the five Federal Circuit Courts of Appeal, the state Supreme Courts, or more generally from lawyers and prosecutors admitted to practice at state or federal levels.[140]

In Chile, the composition of each environmental court is quite complex. There are conflicts of interest issues, called "disabilities," such that if a person has advised or rendered professional services to individuals or legal entities that have the status of party to the case in question, within two years prior to the date of entry of the former or during the sanctioning procedure by the Superintendence of the Environment that originated it, as well as in the evaluation procedure before the Environmental Assessment Service, that person will be disqualified. In addition, judges are prohibited for a period of two years counted from when they ceased to hold office as minister, advise, or render professional services to natural

136 Walton, Michael. "An Evaluation of the Administrative Law Role of the Queensland Planning and Environment Court Litigation, Mediation & Arbitration - Australia." *An Evaluation of the Administrative Law Role of the Queensland Planning and Environment Court - Litigation, Mediation & Arbitration - Australia*, www.mondaq.com/australia/x/191326/court procedure/An evaluation of the administrative law role of the Queensland Planning and Environment Court..

137 Tegner, Helle. "The Role of Courts in Environmental Law—Nordic Perspectives." *J. Ct. Innovation*, vol. 3, 2010, pp. 111–13.

138 *Ibid.* at § 4.

139 *Ibid.*

140 Constituiqao Federal [C.F.] art. 104, caput, cl. III (Braz.).

or legal persons in any kind of proceedings carried out in the environmental courts located in the region in which they exercised their position. This term is reduced to one year in the case of other environmental courts. Furthermore, a Minister of Environmental Court is an exclusive position that requires dedication and is regarded as being incompatible with any other employment, position, function, or commission, paid or unpaid, in private or public entities, whether as fiscal, municipal, autonomous, or semi-fiscal prosecutors, in companies of the state or in which one has participation by capital contributions. Likewise, the position is held to be incompatible with any position of popular election. And finally, there are expertise requirements. Judges who are lawyers must have a law degree, have practiced the profession for at least eight years, and have excelled in professional or academic activity specialized in administrative or environmental law. The scientific expert must have obtained a Bachelor of Science degree with specialization in environmental matters and have eight years of professional practice.

4.3.4 Rules of evidence and experts

Environmental knowledge and expertise are equally important in the assessment of the technical and expert evidence that is often used in environmental dispute. Unlike in courts of general jurisdiction, environmental tribunal practice rules sometimes allow judges to dispense with standard rules of evidence, in order to inform themselves of technical issues as required.[141] Often this information is provided by independent experts, who are called upon in numerous ways during the adjudication of an environmental dispute.

The role of the judge and independent experts in the context of environmental dispute depends on the nature of the relevant legal system. For example, in the adversarial legal systems of Australia, Ethiopia, the U.S.A., Brazil, or Chile disputing parties make their case (including any expert evidence) to an impartial judge, who makes the final decision. In contrast, in the inquisitorial legal systems of China or Finland, the judge takes a greater role in investigating the arguments and evidence relating to each party's case. In environmental courts and tribunals, however, the role of judges and experts is hybrid, blurring the lines between the standard adversarial or inquisitorial approaches. For example, in all four Australian environmental tribunals, experts have an overriding duty to assist the tribunal, rather than any party.[142] This is in addition to any code of ethics or

141 Land and Environment Court Act 1979 s 38(2) (NSW) (Australia); Planning and Environment Court Practice Rules 2010r 23 (Qld); Environment Resource and Development Court Act 1993s 21 (SA) (Australia).

142 Uniform Civil Procedure Rules 2005 sch 7(2) (NSW) (Australia); Planning and Environment Court Practice Rules 2010 f 26(e) (Qld) (Australia); Environment Resources and Development Court Act 1993s 27(SA) (Australia); South Australia, *Practice Directions of the Resources Management and Planning Appeals Tribunal,* 48 [Practice Direction 12A, r2] (2017).

conduct adopted by experts as part of their professional accreditation.[143] While experts may still be appointed by parties, procedural rules allow an adjudicator to refer any technical question to an expert for investigation,[144] require experts to confer to produce joint reports on agreed technical issues,[145] or provide concurrent evidence in panel sessions to the adjudicator.[146] In a 2015 survey, experts were found to support conferral or concurrent formats because they were able to provide technical information to the tribunal without any pressure on experts to agree with one another.[147] Experts may also be involved in matters adjudicated by India's National Green Tribunal, where the Chairperson has broad discretion to "invite any one or more person having specialized knowledge and experience in a particular case before the [National Green Tribunal] to assist … in that case."[148]

In the compared countries, while experts are a common feature of many environmental courts and tribunals, different requirements are imposed for the expertise and experience of experts. Different rules also apply to how often, and for what purpose, adjudicators may call upon experts in the course of adjudicating environmental disputes. While some commentators consider that the type of legal system plays a part in how actively experts may be used in these matters, more specifically that the "inquisitorial variation on [an adversarial] system" offers advantages in terms of flexibility of process and the handling of expert evidence,[149] improvements are still needed to integrate technical concepts and mechanisms to address uncertainty in both law and scientific disciplines.[150]

4.3.5 Reference to environmental law principles and planning objectives

Due to their specialization and legislative remit, adjudicators are [more likely] to draw on international environmental law principles and planning objectives in resolving environmental disputes, such as the precautionary principle, polluter pays principle, intergenerational equity, and (ecologically) sustainable development. In many cases, these principles form legislative objectives and implementation plans relevant to the environmental tribunal itself,[151] or of the legislation

143 For example, the Environmental Institute of Australia and New Zealand, *Code of Ethics and Professional Conduct* (24 October 2012).

144 Environment Resources and Development Court Act 1993 s 27 (SA) (Australia).

145 Planning and Environment Court Rules 2018 r 27 (Qld).

146 South Australia, n 141, at 49 [Practice Direction 12, r 12.8].

147 De Villiers, Bertus. "Conferral of Experts and Concurrent Evidence: What Do Experts Think." *LS NSW J*, vol. 16, 2015, pp. 92, 93.

148 The National Green Tribunal Act § 4(2).

149 Muirden, April. "Presenting Scientific Evidence in Environmental Court Cases: How Science and Law Meet." *EPLJ*, vol. 25, 2008, pp. 425, 448.

150 *Ibid.* at 439–440.

151 Environment Management and Pollution Control Act 1994 sch 1 (SA) (Australia).

from which jurisdiction is conferred.[152] This has been vital to many cases, including the original hearing of *Gray v. The Minister for Planning*, where the New South Wales Land and Environment Court confirms that intergenerational equity should have been considered when deciding to give planning approval for a new thermal coal mine.[153]

In contrast, adjudicators in Kenya's Environment and Land Court are given express direction to refer to, and apply, principles of international environmental law when adjudicating environmental disputes. In particular, the Environment and Land Court Act 2011 (Kenya) section 18 sets out principles by which the court should guide itself; the legislature created a list: increased public participation, implementation of Kenyan cultural and social principles, principle of international co-operation, polluter-pays principle, and the precautionary principle.[154]

However, where environmental principles are not integrated into legislation, the willingness and approach that environmental courts or tribunals take to applying those principles in environmental dispute differ. In Australia, for example, it has taken some time to develop consistency in applying principles such as precaution to potential environmental harm: while in early cases of *Leatch v. Director-General of National Parks & Wildlife Service* tribunals applied the precautionary principle in a risk-averse approach,[155] a less radical, more cautious manner has been taken in subsequent cases, resulting in orders to modify rather than cease proposed conduct.[156] The cautious approach taken by the Australian judiciary at that time is hardly surprising in the broader context of the conservative position taken by the legislative and executive arms of the Australian government toward climate change. However, this approach was not reflective of the wider view of the Australian public. Nevertheless, the United Kingdom points to a resurgence of the principle in planning dispute, where courts have used precaution to condone planning decisions made on an expectation of adverse consequences of climate change.[157]

Environmental courts or tribunals exist in many of the jurisdictions compared in this chapter. This is significant, because it allows environmental disputes to be adjudicated in a dedicated forum, often brought by a wider range of applicants than in courts of general jurisdiction, and being heard by experienced and

152 For example, the Environment Planning and Assessment Act 1979 s 1.3 (NSW) (Australia); Environmental Protection Act 1994 s 3 (Qld) (Australia).

153 Gray v. The Minister for Planning, Director General of the Department of Planning and Centennial Hunter Pty Ltd, 152 LGERA 258, 291 (2006).

154 *Ibid.* at 576, see also, The Environment and Land Court Act, (2011) No. 19 §18 (Kenya).

155 Leatch v. Director-General of National Parks & Wildlife Service, 81 LGERA 270, 282 (1993).

156 R v Resource Planning &Development Commission; Ex p Aquatus Pty Ltd, 100 LGERA 1 (1998).

157 England, Phillip. "Doing the Groundwork State, Local, and Judicial Contributions to Climate Change Law in Australia." *ELPJ*, vol. 25, 2008, p. 372, *14 Regent Street Pty. Ltd. v. Hobart City Council*, TASRMPAT 26 (2005): *Taralga Landscape Guardians Inc. v. Minister for Planning and RES Southern Cross Pty Ltd.*, 134 LGERA 23 (2004).

knowledgeable adjudicators who have substantial liberty to refer to experts and principles of international environmental law. While there are notable differences in the way each jurisdiction deals with these issues, it is clear that environmental courts and tribunals are becoming more sophisticated, and are used with greater frequency when available, to resolve environmental disputes around the world.

4.4 Administrative law hearing and review boards

Administrative law review boards describe the category of institutions that are charged with resolving disputes arising from administrative decisions made by government ministers or public authorities pursuant to delegated legislation. Administrative boards resolve environmental disputes by reviewing either the *merits* (correctness) or *judicial basis* (legality) of an administrative decision.[158] As a common feature in many countries, administrative law review boards feature within both national and sub-national court structures.

In the U.S.A., the common points in the enforcement process at which a private party finds that he, she, or it must have legal counsel, are the points at which a federal or state environmental agency denies an application for a permit to pollute, issues a notice of violation[159] (N.O.V.), or issues an administrative compliance order. If a party receives an N.O.V., the party should take the matter seriously. Some states treat the N.O.V. as final legal action and it must be appealed to preserve a party's rights to challenge the violation alleged, but other states treat the N.O.V. as a non-final action until a civil penalty is issued to the party for the violation. Finality is an important threshold concept to achieve because if one attempts to appeal an action *before* it is final, the administrative tribunal or court will dismiss the action because it is not yet "ripe." If on the other hand a state action is considered to be final, and a party fails to file an appeal within the required time, the party will have waived his, her, or its right to appeal and the state action will be considered to be final and unassailable. If a party fails to take action within a reasonable period of specified time to the state's satisfaction upon receiving an N.O.V., the state will likely then take the additional step of issuing an administrative penalty. At that point, a party cannot reopen the underlying facts of the violation, but can only argue against the amount of the penalty. Under some federal statutes, such as R.C.R.A.,[160] the U.S. E.P.A. together with the Department of Justice may take a civil penalty action directly in federal court, rather than in the administrative system. This choice may take advantage of the stronger powers of a court over an administrative tribunal, but

158 Lyster, *supra* note 75 at 38–40.

159 The Environmental Council of States (E.C.O.S.) estimates that state and federal environmental agencies issue at least 42,000 notices of violation (N.O.V.) annually, a statistic that shows that the N.O.V. is the most common enforcement tool used.

160 42 U.S.C. § 6992d (d).

the U.S. E.P.A. loses its independent control from the government side over the litigation then.

An administrative penalty may be assessed against a private person who does not need a permit, but is alleged to be in violation of a regulation or statute, or against a private party that does have a permit, but is alleged to be operating in violation of the permit and has failed to sufficiently reply to a compliance order. When an agency denies an application or issues a notice of violation or compliance order, it is regarded as an appealable final action by the government. There is a time limit on when the party may file the appeal. If an appeal is taken from the enforcement action, an administrative hearing will be held. In the U.S. E.P.A., that process begins with an administrative law judge (A.L.J.) If a party is not satisfied with the findings of the A.L.J., the party may appeal to the Environmental Appeals Board.[161]

In summary, administrative law in the U.S.A., both state and federal, typically follows three doctrines—ripeness, finality, and exhaustion. Ripeness and finality were discussed above. The spirit of Article II of the U.S. Constitution, which empowers all federal courts to hear only actual cases or controversies, and not offer advisory opinions, is a spirit that is present in state courts and administrative tribunals as well. One often encounters the doctrine of finality when a party attempts to appeal from an administrative action that is not considered to be final, like issuing a Notice of Violation without assessing a penalty. In most U.S. jurisdictions, a party is not deemed to have suffered damages until an agency follows a notice of violation with an issuance of a civil penalty, and therefore the party may not yet appeal. Once penalized, the individual has suffered damages and may appeal the agency's penalty, based upon the underlying violations.

And last, there is the hurdle of exhausting administrative remedies before one can access the courts. Because the U.S. Congress, like many state legislatures, has delegated its rule-making authority to administrative agencies, which typically include a tribunal and enforcement body, all aggrieved parties must first make use of the administrative system and exhaust the administrative remedies possible before trying to get relief from a court of general jurisdiction. If a party has exhausted all administrative remedies and is still unhappy with the result, then the party can file a complaint with a court of general jurisdiction, state or federal, depending on the agency involved.

But the relationship between the U.S. E.P.A. and state agencies is not one of mutual exclusion. The U.S. E.P.A. can initiate an enforcement action if it determines that the state enforcement action taken is inadequate. Before initiating such an action, however, the U.S. E.P.A. must allow the state an opportunity to correct its deficiency. If the state does not take enforcement to the U.S.

161 The U.S. E.P.A.'s Environmental Appeals Board publishes the *Environmental Appeals Board Practice Manual*, the 2013 edition of which can be found at: https://yosemite.epa.gov/oa/EAB_Web_Docket.nsf/General+Information/Practice+Manual?OpenDocument.

E.P.A.'s satisfaction, the U.S. E.P.A. may file an additional enforcement action of its own, the process that was referred to in the previous section on federalism as overfiling. Overfiling has been upheld by courts as a proper interpretation of federal statutory provisions for R.C.R.A.[162] and the Clean Air Act,[163] but possibly not the Clean Water Act.[164] Because the notion that the U.S. E.P.A. may overfile differs from statute to statute, a German lawyer would be well-advised to check the statute in question and the courts' interpretations of that statute, if a state takes an enforcement action. A second one from the U.S. E.P.A. might still be coming. And finally, if a civil penalty action has been filed by a state or federal authority, it does not prevent another government authority from filing a criminal action.[165]

As with all civil litigation in the U.S.A., the filing of a legal action does not mean the action will result in a trial. Most civil litigation in administrative matters as well as federal and state courts is settled before trial. The document that is typically used is known as a Consent Decree, whereby the state or federal enforcement agency uses its statutory power to agree with a polluter on some combination of civil penalty, remediation work, and a schedule for bringing a site into compliance. If the violator does not comply with a term of the Consent Decree, the agency can go back to court to enforce the Consent Decree, the underlying statutory violations, or both.

Once litigation is commenced, settlements must be approved by the court or tribunal having jurisdiction over the matter. A U.S. Department of Justice report shows that in state trial courts (where most civil trials happen) only about four percent of all tort cases begun were concluded in trials. If, however, a tort case did make it to trial, ninety percent were heard by juries. Still, that means legal disputes are most often disposed of without juries (97.4 percent of the time, to be precise).[166]

A special area of enforcement needs to be separately mentioned here. A "legacy" of industrialism in many places in the U.S.A. are abandoned sites of heavy pollution in need of expensive cleanup. To address these sites, both the U.S. E.P.A.[167] and many states[168] have enacted so-called "superfund" statutes. These cases may have dozens or hundreds of parties and concern themselves with cleanup programs that cost on average twenty-five to thirty million dollars per

162 *United States v. Power Engineering*, 303 F.3d 1232 (10th Cir. 2002), cert. denied, 538 U.S. 1012 (2003).

163 *United States v. LTV Steel Co.*, 118 F.Supp.2d 827 (N.D. Ohio 2000).

164 *United States v. City of Rock Island*, 182 F.Supp.2d 690, 694 (C.D. Ill. 2001).

165 *United States v. Hudson*, 522 U.S. 93 (1997).

166 Cohen, Thomas. "Tort Bench and Jury Trials in State Courts, 2005." *U.S. Department of Justice, Bureau of Justice Statistics*, 2009, p. NCJ 228129.

167 See the federal Comprehensive Environmental Response, Compensation and Liability Act (CERCLA), enacted in 1980 and found at 42 U.S.C. §§ 9601 *et seq.*

168 See for example, the Pennsylvania Land Recycling and Environmental Remediation Standards Act, 35 Pa. Stat. §§ 6026.101 *et seq.*

site. If someone becomes involved in superfund litigation, one will need not only to get a U.S. lawyer, but a U.S. lawyer experienced in superfund litigation. Cases under the federal Comprehensive Environmental Response, Compensation, and Liability Act (C.E.R.C.L.A.) go through eight formalized steps of remediation process to assure that they comply with the National Contingency Plan, among other things.

Potentially responsible parties (P.R.P.s) under the C.E.R.C.L.A. fall into four categories: current owners or operators of the site, under Section 107 (a)(1); owners or operators of the site at the time of disposal of the polluting substances under Section 107(a)(2); those who arranged for the disposal of the hazardous substances, including those who generated the waste, under Section 107(a)(3); and transporters of the waste to the site if they themselves selected the site, under Section 107(a)(4). Any of these parties may be held strictly liable, retroactively liable, and jointly and severally liable. Strict liability is a tort standard that is used outside of a superfund as well, as for instance with manufacturers of hazardous materials. A standard of strict liability means that no recklessness or negligence needs to be proven by the enforcing agency. Superfund liability may also be applied retroactively to various acts such as hazardous waste generation, disposal, and transportation committed before C.E.R.C.L.A. was enacted in 1980. And finally, superfund liability is usually interpreted to be joint and several, meaning one P.R.P. can be liable to pay the costs for all P.R.P.s, especially in situations where the other P.R.P.s are dead, bankrupt, or no longer in business and their assets cannot be found.

In Ethiopia there are no administrative boards to adjudicate environmental disputes by reviewing either the merits (correctness) or judicial basis (legality) of an administrative decision. However, there are administrative agencies[169] (which include environmental institutions) that have been established with different mandates. Some are established with broader mandates while others have specific tasks to implement certain government policies or programs. Almost all administrative institutions in the country are accountable to the executive organ of the government.[170] Very few are accountable to the House of People's Representative (H.P.R.) which is the highest law-making organ in the country.[171] Those administrative institutions which are accountable to the executive organ of the government are known as executive agencies. Those accountable to the H.P.R. are perceived to be autonomous agencies. In the case of environmental institutions, since at the federal level the Commission for Environment, Forest, and Climate Change is accountable to the Prime Minister and the secto-

169 The draft Administrative Procedure Proclamation of 2005, in Art. 2(1), defines an administrative agency as "any ministry, commission, authority of the F.D.R.E., including Addis Ababa and Dire Dawa city administrations, which are competent to render administrative decisions and exercising regulatory or supervisory functions."

170 F.D.R.E. Constitution in Art.77 (2).

171 According to the F.D.R.E. Constitution Art.55 (14) (15), Art.101& Art. 103.

ral and regional environmental organs are also accountable to the relevant competent executive organs, they are all labeled "executive" agencies. In Ethiopia, one who is dissatisfied with the decisions of the administrative agencies has to initiate a new case with a court of law[172] or in some cases the statute that empowers the administrative agency does not say anything about where the dissatisfied party should go.[173]

In the U.S.A., most environmental disputes are resolved by state environmental agencies. Substantive law and procedure of the states vary, so a lawyer must know the substance and procedure of the state in which the environmental problem has arisen. For this reason, administrative boards in states such as Oregon have proven more effective in recent times in resolving environmental disputes for the benefit of the environment. In Oregon, the Department of Environmental Quality was created with the mission of enforcing state and federal environmental laws. Its work consists of providing environmental science and information, regulatory activity, and providing technical assistance to affected individuals.[174] The Department of Environmental Quality receives its direction, in part, through the Oregon Environmental Quality Commission.[175]

Even in the federal system, the federal government's U.S. E.P.A. has a subsidiary, but independent, Environmental Appeals Board serving as a green tribunal.[176] The Environmental Appeals Board uses a four-judge panel to serve as a final decision maker on administrative appeals falling within the E.P.A.'s statutory jurisdiction.[177] The Environmental Appeals Board has original jurisdiction as well as appellate jurisdiction over administrative decisions usually referring to operating permits. In fact, the Environmental Appeals Board is the only federal tribunal devoted exclusively to reviewing appeals on pollution control cases; that includes appellate review over penalties administered by administrative law judges and permits issued by the E.P.A.[178] While the E.P.A. has a strong remit in theory, Dillion observes that the current Trump Administration is making significant attempts to "cripple the [Environmental Protection Agency's] capacity to confront polluting industries and promote public and environmental health."[179]

In other countries, administrative law review boards often adjudicate environmental dispute in jurisdictions where there is no specialist environmental

172 The Environmental Pollution Control Proclamation, supra, note 19.

173 For instance, the Environmental Impact Assessment Proclamation, 2002, No. 299, Federal Negarit Gazeta, 9th year No. 11, Addis Ababa, Ethiopia.

174 *Ibid.*

175 *Ibid.*

176 *EPA*, Environmental Protection Agency, yosemite.epa.gov/oa/EAB_Web_Docket.nsf/General Information/Frequently Asked Questions? OpenDocument.

177 *Ibid*; Wolgast, *supra* note 114 at 185, 99.

178 *Ibid.*

179 Dillion, Lindsey. "The Environmental Protection Agency in the Early Trump Administration: Prelude to Regulatory Capture." *Am. J. Public Health*, vol. 108, 2018, pp. 89, 93.

court or tribunal. Having both a specialist environmental tribunal and an administrative law review board,[180] environmental matters are heard by administrative review boards in Australia at national and sub-national levels, in jurisdictions otherwise without a more specialized environmental tribunal.[181]

4.4.1 Jurisdiction

Unsurprisingly, given their often-complementary relationship with environmental tribunals, the jurisdiction conferred on administrative law review boards to adjudicate environmental disputes mirrors that of environmental tribunals. Namely, jurisdiction is conferred by statute, and covers various matters of planning approvals, land-use approvals, water and mineral licenses, pollution control, fisheries, land acquisition compensation, mining, and indigenous land claims. Notably, in Australia, the Victorian Civil and Administrative Tribunal also adjudicates disputes about forest carbon rights arising from carbon sequestration projects under the state climate change law.[182] In Australia, planning and land-use approvals are the most common environmental disputes adjudicated by administrative law review boards with environmental competence.[183] In 2016, the Victorian Civil and Administrative Tribunal dealt with more than 2,700 disputes in this area alone.[184] In comparison, in the U.S. state of Oregon, the Department of Environmental Quality is the government department responsible for most environmental administrative decisions, and for that reason includes bodies that provide for the adjudication of environmental disputes. The Department receives its direction, in part, through the Oregon Environmental Quality Commission, a lay five-member panel for policy and rulemaking.[185] This Commission, a five-member citizen panel, is appointed by the Governor.

4.4.2 Administrative hearing procedures

Administrative law review boards resolve disputes pursuant to procedural rules. Although the rules adopted by administrative law review boards are not particularly adapted to accommodate the features of environmental dispute, some are nevertheless similar to those which apply in specialist environmental tribunals.

180 South Australian Civil and Administrative Tribunal Act 2013 (SA) (Australia).

181 ACT Civil and Administrative Tribunal Act 2008 (ACT) (Australia); Victorian Civil and Administrative Tribunal Act 1999 (Vic) (Australia); State Administrative Tribunal Act 2004 (WA) (Australia); Administrative Appeals Tribunal Act 1975 (Cth) (Australia); except in South Australia, which has both a Civil and Administrative Tribunal, and specialist environmental tribunal.

182 Climate Change Act 2017 ss 68-8 (Vic).

183 Victorian Civil and Administrative Tribunal, *Annual Report 2016–7*, 54 (Report, 2017).

184 *Ibid*. at 55.

185 *Ibid*.

4.4.3 Standing

In relation to standing, traditional rules generally apply to administrative law review boards to reflect the nature of disputes adjudicated by them; these rules require that an entity has a "sufficient interest" (or similar) connection to the matter before bringing legal proceedings. However, there is still some flexibility, and exceptions, within the compared cultures about standing in relation to environmental disputes. For example, in Australia, standing is approached differently between national and sub-national administrative law review boards depending on the statute relevant to the dispute. Broad rules give standing to "any interested person" in the national Administrative Appeals Tribunal in respect of disputes arising under the Environmental Protection and Biodiversity Conservation Act 1999 (Cth),[186] and to "aggrieved" or "affected" persons in Victorian and Western Australian administrative law review boards in respect of water resource rights disputes.[187] In contrast, entities must have possessed more than strong beliefs or emotions (although less than a "legal interest") in the matter to have standing before sub-national administrative law review boards.[188]

4.4.4 Adjudicator expertise

Similar to environmental tribunals, statutes establishing administrative law review boards require judicial officers to have minimum knowledge and experience.[189] Unlike environmental tribunals, however, these minimum standards relate more generally to legal qualifications, and to knowledge deemed relevant by discretion, rather than prescriptive subject matter knowledge as provided by the statutes which established environmental tribunals.

4.4.5 Rules of evidence and experts

In relation to rules of evidence and experts, administrative law review board legislation is more aligned with that of environmental tribunals than courts of general jurisdiction. In Australia, administrative law review board rules enable adjudicators to seek assistance or inform themselves, as appropriate, from experts who have an overriding duty to the tribunal rather than to their instructing

186 Carrabus, Jason. "Merits Review of Commonwealth Environmental Decision-Making." *EPLJ*, vol. 118–19, 2009, p. 113:

187 Water Rights Act 1989 section 64 (Vic) (Australia); Rights in Water and Irrigation Act 1914 ss 26GG-GI (WA) (Australia); ALEX GARDNER, *Appeals against administrative decisions affecting water resources* 347–8, 17 EPLJ 341 (2000).

188 *ING Development Australia Pty. Ltd. v. Western Australian Planning Commission* [28] WASAT 104 [2008]; *Viento Property Ltd. v. Western Australian Planning Commission* WASAT 229 [2009].

189 For example, the Victorian Civil and Administrative Tribunal Act 1998 sections 10-14 (Vic) (Australia) compared with the State Administrative Tribunal Act 2004 sections 9, 143 (WA) (Australia).

party as would be expected in adversarial systems.[190] The similar approach taken to evidence and experts between administrative law review boards and environmental tribunals is key, enabling administrative law review boards to perform adequately in environmental matters to consider and assess government decisions based on "highly technical evidence incorporating a high degree of uncertainty."[191]

In the U.S.A., the Oregon Department of Environmental Quality not only adopts rules, but the Environmental Quality Commission also establishes policies, issues orders, judges appeals on fines and other actions issued by the D.E.Q., and appoints the D.E.Q. director. In setting policies, other public participation is included; often the legislature will create an environmental program, the D.E.Q. will assemble an advisory committee, a public forum will be left open to reviews, comments, and hearings, and the Environmental Quality Commission will be the body enacting the rules. The Environmental Quality Commission, acting in various capacities, consists of scientific, economic, business, and environmental specialists.

4.4.6 Civil and criminal penalties

Despite their general jurisdiction, administrative law review boards appear to be sensitive to making appropriate determinations in environmental disputes, most commonly to restrict permitted activities, or impose monitoring, auditing, or other conditions under planning permits, land use permits or natural resource licenses. In Australia, it is common for the national Administrative Appeals Tribunal to take a balanced approach to disparate commercial, development, and environmental conservation or protection interests in accordance with the limitations and objectives of enabling environmental statutes.[192]

4.4.7 Reference to international environmental law principles and planning objectives

Like environmental tribunals, administrative law review boards can refer to international environmental law principles and planning objectives to adjudicate disputes, although they are more strictly confined by the substance of the statutes under which the relevant administrative decision was made.

In Australia, sub-national administrative law review boards are referring to international environmental law principles in the course of adjudicating

190 For example, the A.C.T. Civil and Administrative Tribunal Act 2008 section 26 (ACT) (Australia); Victorian Civil and Administrative Tribunal Act 1998 schedule 3 (Vic) (Australia), V.C.A.T. *Practice Note 2* (2012); State Administrative Tribunal Act 2004 section 64 (WA) (Australia).

191 Carrabus, *supra* note 186 at 122–3.

192 *Ibid.* at 123–4, 127.

planning and land-use approval disputes with increasing frequency. Importantly, not only are these principles recognized as relevant considerations,[193] but they are being applied in order to promote environmental policy objectives such as climate change. This is exemplified by cases in which the Victorian Civil and Administrative Tribunal has upheld administrative decisions made by local councils to refuse planning permission for residential developments on coastal land based on the precautionary and intergenerational equity principles.[194] Based on the precautionary principle, the Tribunal considered that there was sufficient risk of sea level rise, inundation, and coastal erosion to preclude development.[195] Importantly, the Tribunal expanded the use of intergenerational equity in the context of development and land-use cases, holding that the local councils were able to refuse development consent to avoid future, long-term liability for supplying infrastructure, emergency services, and a safe community in these coastal areas.[196]

While a number of jurisdictions use administrative review boards as the forum in which to adjudicate environmental disputes, the experience in the jurisdictions compared in this chapter illustrate that the effectiveness of these bodies are limited because they lack the same independence enjoyed by courts from the political process. This remains the main point of distinction and one should not be distracted by discussions of specialization. That said, a lawyer's ability to predict the outcome of a legal dispute is diminished due to these politics. In Australia and the U.S.A. at least, it is notable that the jurisdiction of the review boards is, of course, limited by the nature and scope of decisions made by the relevant administrative authority. In addition, however, the experience in the U.S.A. suggests that despite trial procedure, these boards have far less scope to determine environmental disputes effectively in relation to the environmental outcome; instead, they are constrained by the powers and priorities of the governing authority.

4.5 Conclusions

While environmental courts and tribunals exist in Denmark, Finland, Sweden, and Norway, the approach taken throughout Scandinavia illustrates variations in the design of environmental tribunals, despite other similarities in geography, standard of living, and perceptions about the environment and law. For example, while in Denmark environmental dispute is resolved within the administrative

193 *Wattleup Road Development Co Pty Ltd v. State Administrative Tribunal* (No 2) [53] WASC 279 (2016) (Australia).

194 *Gippsland Coastal Board v. South Gippsland Shire Council* VCAT 1545 (2008) (Australia); *Taip v. East Gippsland Shire Council* VCAT 1222 (2010) (Australia).

195 *Gippsland Coastal Board* [47] VCAT 1545 (2008); *Taip* [96] – [104] VCAT 1222 (2010).

196 *Gippsland Coastal Board* [51] VCAT 1545 (2008); *Taip* [65] – [66] VCAT 1222 (2010).

system,[197] in Finland, these disputes are heard by specialist environmental courts,[198] whereas Norway has no specialized environmental tribunals at all and so channels all of its environmental disputes through its general administrative authorities and courts of general jurisdiction.[199] In Sweden, environmental matters are adjudicated in a range of forums. Notably, being the first European country to enact an Environmental Code, in 1999, Sweden also created five environmental courts of first and second instance granting scientific and technical experts with full judicial powers.[200] These Scandinavian countries demonstrate the diversity in approaches toward environmental problems, even when the problems themselves are relatively similar.

States, members of the public, and environmental interest groups increasingly bring actions before courts of general jurisdiction, environmental courts and tribunals, and administrative review boards. In Australia, for example, this is illustrated by the increasing interest and involvement of community groups in matters that oppose planning approvals for large coal mines in the context of climate change, such as the 2006 decision of *Gray v. Minister for Planning* (Anvil Hill Mine),[201] and the 2019 decision of *Glouchester Resources Limited v. Minister for Planning* (Adani Rocky Hill Mine).[202] Individual and community groups have played substantial roles as claimants, assisted by broad rules on standing, as lobbyists, or as witnesses, contributing in different ways to the overall carriage of cases. In other jurisdictions, state responsibility cases such as *Urgenda Foundation v. Kingdom of the Netherlands*[203] or *United States v. Juliana*[204] made by individuals or activist groups illustrate creative use of judicial and administrative forums for dispute resolution of environmental matters. These trends indicate that highly educated, engaged, and environmentally aware citizens are seeking to hold corporations (and states who approve their activities) accountable at sub-national or national levels for their decisions that harm the environment or the rights of future generations to a healthy and clean environment. While this is a positive development in how environmental courts and tribunals are being used to pursue "environmental justice," it does not necessarily mean that such justice is achieved. In Australia, for example, administrative tribunals in particular have been limited by the statutes and regulations within which they adjudicate disputes. For this reason, environmental dispute resolution is again undergoing development as

197 Tegner, *supra* note 137 at 111, 12-13.
198 *Ibid.* at 114.
199 *Ibid.* at 113-14.
200 *Ibid*; *see also,* Amirante, *supra* note 94 at 441, 99.
201 *Gray v. Minister for Planning* [2006] NSWLEC 720.
202 Quentin Beresford, "The legal challenges from Greenies" in *Adani and the War over Coal* (New South Wales Publishing, 2018) ch 10.
203 *Urgenda Foundation v. Kingdom of the Netherlands,* [2015] HAZA C/09/00456689.
204 *Juliana v. United States,* No. 6:15-CV-01517-TC, 2016 WL 6661146.

new bases for environmental claims emerge from traditionally "non-environmental" sources, namely insurance risk and corporate responsibility for shareholders rights and statutory director's duties.[205]

As disputes about environmental matters become more frequent and more complex, it is important that dispute resolution forums, whether courts, tribunals, or review boards, exist to enable the dispute to be resolved. In the countries compared in this chapter, it is clear that a range of forums are used to adjudicate environmental disputes.

In most cases, the form that this forum takes does not matter; what is important are the substantive characteristics of those forums, which enable the adjudication of environmental disputes in an effective way, and in accordance with environmental objectives. For example, while broad jurisdiction that covers key environmental issues or legislation is important, a broad remit leads to a heavy caseload, making dispute resolution ineffective in terms of time and of the capacity to hear and consider the complex technical matters that come with environmental disputes. Furthermore, in the countries compared in this chapter, experience indicates that expertise is key. Between the compared countries, there is significant inconsistency about educational and experiential standards, including the use of experts. Interestingly, however, between these two roles, most jurisdictions require that one entity, whether adjudicator or expert, contributes to determining the matter at hand. Moreover, it is useful that dispute resolution forums, whether generalist, specialized, or acting to review administrative decisions, can be guided by principles of international environmental law, in order to encourage the relevant body to resolve the dispute as far as possible, in the most environmentally effective way.

As the frequency and complexity of environmental conflicts and disputes increase, it is clear that these issues will come under the microscope, and at that time it is hoped that key practices can be established that allow the judiciary to support and contribute to the protection and conservation of the natural environment.

Literature

Amirante, Domenico. "Environmental Courts in Comparative Perspective: Preliminary Reflections on the National Green Tribunal of India." *Pace. Envtl. L. Rev.* 29 (2012): 441.

Bates, Gerry. *Environmental Law in Australia*. 9th ed, Lexisnexis, 2013.

205 Susan Shearing, 'Raising the boardroom temperature? Climate change and shareholder activism in Australia' (2019) 29 *Environment and Planning Law Journal* 479; *Australasian Centre for Corporate Responsibility v Commonwealth Bank of Australia* (2016) 337 ALR 558; Noel Hutley and Sebastian Hartford-Davis 'Climate change and director's duties: Memorandum of Opinion (Centre for Policy Development and Future Business Council Report, October 2016).

Bryner, Nicholas. "Brazil's Green Court: Environmental Law in the Superior Tribunal De Justica (High Court Of Brazil)." *Pace. Envtl. L. Rev.* 29 (2012): 470.

Carrabus, Jason, "Merits Review of Commonwealth Environmental Decision-Making." *Envtl. Planning L. J.* 26 (2009): 113.

Dillion, Lindsey, et. al. "The Environmental Protection Agency in the Early Trump Administration: Prelude to Regulatory Capture." *Am. J. Public Health* 108 (2018): 89.

England, Phillipa. "Doing the Groundwork State, Local, and Judicial Contributions to Climate Change Law in Australia." *Envtl. Planning L. J.* 25 (2008): 362.

Findley, Roger W. "Pollution Control in Brazil." *Ecology L.Q.* 15 (1988): 1.

Fisher, D. *Australian Environmental Law Norms Principles and Rules.* 3rd ed, Thomson, 2014.

Galdia, Marcus. "Comparative Law and Legal Translation." *European L. F1.* 1 (2003).

Gardner, Alex. "Appeals against Administrative Decisions Affecting Water Resources." *Envtl. Planning L. J.* 17 (2000): 341.

Gui, Tiantian. "China's First Environmental Court: A Seven Year Winning Streak for Public Interest Cases." *China Dialogue* (9 July 2014). https://chinadialogue.net/article/show/single/en/7116-China-s-first-environmental-court-a-seven-year-winning-streak-for-public-interest-cases.

Guangwei, He. "China's Dirty Secret: The Boom Poisoned its Soil and Crops." *Yale Environ.* 360 (30 June 2014). https://e360.yale.edu/features/chinas_dirty_pollution_secret_the_boom_poisoned_its_soil_and_crops.

Hang Yee, Wai, et al. "Regulatory Compliance When the Rule of Law is Weak: Evidence from China's Environmental Reform." *Journal of Public Administration Research and Theory* 26 (1) (2016): 95–112.

Hart, Melanie, and Jeffrey Cavanagh, *Environmental Standards Give the United States the Edge Over China.* Center For American Progress, 2012.

Hollo, Justice Erkki J., et al. "Environmental Law and Administrative Courts in Finland." *J. Ct. Innovation* 3 (2010): 51

Kaniaru, Donald W. "Environmental Courts and Tribunals: The Case of Kenya." *Pace Envtl. L. Rev.* 29 (2012): 566.

Lyster, Rosemary, et al. *Environmental & Planning Law in New South Wales.* 4th ed, Federation Press, 2016.

Madner, Verena. "The Austrian Environmental Senate." *J. Ct. Innovation* 3 (2010): 23, 24.

Mcdermott, Matthew. "Effective Environmental Dispute Resolution in Queensland: Commentary and Critiques from the Planning and Environment Court." *Envtl. Planning L. J.* 30 (2013): 150.

Meek, Michael. *The Australian Legal System.* 5th ed, Thompson, 2016.

Muirden, April, and John Bailey. "Presenting Scientific Evidence in Environmental Court Cases: How Science and Law Meet." *Envtl. Planning L. J.* 25 (2008): 425.

Ohlinger, Theo. "Austria and Article 6 of the European Convention on Human Rights." EJIL 1 (1990): 286.

Preston, Brian. "Operating an Environmental Court: The Experience of the Land and Environment Court of New South Wales." *Envtl. Planning L. J.* 25 (2008): 385.

Preston, Brian. "Benefits of Judicial Specialisation in Environmental Law: The Land and Environment Court of New South Wales as a Case Study." *Pace Env. L. R.* 29 (2012): 396.

Pring, George, and Catherine Pring. *Greening Justice: Creating and Improving Environmental Courts and Tribunals.* World Resources Institute, Report, 2009.

Rackemann, Michael. "Planning and Environment Court 50th Anniversary Dinner Speech." *Qld J Schol* (2016): 16.

Reimann, Mathias, and Reinhard Zimmermann, eds. *The Oxford Handbook of Comparative Law*. Vol. 383. Oxford University, 2006.

Reitz, John. "How to Do Comparative Law." *American J. Comparative Law* 46 (1998): 617.

Rengarajan, Sridhar, et al. "National Green Tribunal of India—an Observation from Environmental Judgements." *Env. Sci. & Pollution Research* 25 (2018): 11313.

Ross, Ashley. *Environmental Conflict Resolution Guidebook*. Texas A & M University, 2016.

Ryan, Patricia. "Environmental Court System." *Unsw L. Rev.* 4 (1981): 11.

Stein, Paul, "The Role of the New South Wales Land and Environment Court in the Emergence of Public Interest Environmental Law." *Envtl. Planning L. J.* 13 (1996); 179.

Tegner Anker, Helle, and Annika Nilsson, "The Role of Courts in Environmental Law—Nordic Perspectives." *J. Ct. Innovation* 3 (2010): 111

De Villiers, Bertus."Conferral of Experts and Concurrent Evidence: What Do Experts Think?." LS NSW J. 16 (2015): 92, 93.

Walton, Michael. "An Evaluation of the Administrative Role of the Queensland Planning and Environment Court." *MONDAQ* (12 August 2012). http://www.mondaq.com/australia/x/191326/court+procedure/An+evaluation+of+the+administrative+law+role+of+the+Queensland+Planning+and+Environment+Court.

Wang, Alex L. and Jie Gao. "Environmental Courts and the Development of Environmental Public Interest Litigation in China." *J. Ct. Innovation* 3 (2010): 37

Wolgast, Anna L., et al. "United States' Environmental Adjudication Tribunal." *J. Ct. Innovation* 3 (2010): 185.

Wright, Merideth. "The Vermont Environmental Court." *J. Ct. Innovation* 3 (2010): 203.

PART III

Norms and tools of environmental legislation, implementation, and dispute resolution

5

CONSTITUTIONAL PROVISIONS

Ryan Kraski, Marek Prityi, and Saskia Münster

5.1 Constitutionalism

What is the function of that which we call a "constitution"? First, constitutions serve to transfer power from the collective individuals of a culture to the state they have created under a political theory of the state. Although we have such things as Aristotle's Constitution for Athens from ancient cultures, the world of states created by constitutions today is largely formed by the Enlightenment's political and legal philosophy. Additionally, in a constitution, citizens usually also reserve their personal rights and freedoms against the collective legal person—the state—that they have just created. Third, once the individuals have created a new public law entity and legal person—the state—and reserved rights for themselves, the constitution often then divides the function of the state among several offices of the state, often in a form of presidential or parliamentary democracy. From these basics, comparative law will show us that the functions of constitutions quickly vary. The Ethiopian constitution, for example, gives the power of constitutional review not to a court, but to the House of the People, which is a politically constituted law-making body.[1] As such, the function of the constitution is to keep control of the constitution away from professional jurists and in the hands of a politically elected body. Other states may in fact find that practice to be precisely what the people do not trust or want.

> Constitutionalism has various advantages, particularly in capturing emerging concepts like environmental constitutionalism. The importance of constitutionalism lies in the supposition that the other constraints on

1 See generally Beru, Tsegaye, and Kirk W. Junker. "Constitutional Review and Customary Dispute Resolution by the People in the Ethiopian Legal System." *NC.J. Int'L 1*, XLIII, 2018.

> power have failed to prevent political overreaching. Constitutionalism also reflects human nature inseparable from human capacities, needs, and deficiencies.[2]

Constitutional provisions are extraordinarily strong mechanisms for reaching legal ends, e.g. environmental protection, for a number of reasons. Constitutional provisions are "entrenched" within legal systems. While statutes and regulations are both subject to legislative politics or executive discretion, it requires much greater exertion and collaboration to change constitutions. Constitutions themselves are the most immutable expression of legal norms. A court cannot strike down a constitutional provision as being itself "unconstitutional."[3] Consequently, all other laws must conform to it. Beyond technical legal requirements, constitutions establish public discourse in a culture. The constitution acts as a way of explicitly stating the nation's values, whereby people will identify more strongly with them.[4] Listening to a political debate in the United States, one will inevitably hear one of the debaters claim that he or she "believes" in something like the First or Second Amendment to the Constitution. Countries such as South Africa have also used their constitutions as mechanisms to form a collective, national identity.[5]

> Why is it important for the right to a healthy environment to be entrenched in a constitution? The logical argument for according constitutional status to the right to a healthy environment is straightforward. Fundamental human rights should enjoy the strongest legal protection available in today's society – constitutional protection – to ensure that they are respected and fulfilled. The right to live in a healthy environment meets the test for recognition as a fundamental human right (significant moral importance, universal, practicable).[6]

2 May, James R., and Erin Daly. *Global Environmental Constitutionalism*. Cambridge University Press, 2016, p. 37.

3 This concept of "judicial review" should be thought of differently in common and civil law jurisdictions. In common law jurisdictions, judges can be thought of as law-makers and law-breakers at the same time. They "make" law through their ability to elaborate, explain, and expand upon existing laws when a legal gray area is contested. They can "break" a law by declaring it unconstitutional, thus making it ineffective, because it does not comply with the land's constitution. Civil law judges, on the other hand, are more restricted in their abilities to do the same actions, but they may make similar decisions as their common law colleagues, when the law proscribes such action and they act within their confined legal permission.

4 May, *supra* note 2 at 32–35; *see also* Tarr, Alan. "Subnational Constitutions and Minority Rights: A Perspective on Canadian Provincal Constitutionalism." *Revue Quebecoise De Droit Constitutionnel*, vol. 2, 2008, pp. 174, 178–85.

5 Tarr, *supra* note 4 at 185.

6 Boyd, David R. *The Right to a Healthy Environment: Revitalizing Canadas Constitution*. UBC Press, 2012, p. 3.

Rather than enumerating what people may do, constitutions such as that of the U.S.A. primarily address what the state may not do in relation to the rights of its citizens (negative rights).[7] In contrast to the U.S.A., many countries' constitutions create enforceable rights of the individual against the state when the state fails to uphold a particular constitutional obligation (positive rights). In contrast, in the German *Grundgesetz* (Basic Law or G.G.), which functions as a constitution, there are not only negative and positive rights guaranteed but also active rights, such as the right to vote.[8] It should be highlighted that, next to the fundamental rights, there are certain *Staatszielbestimmungen* (state objectives) that are identified and can be found in Article 20 of the Basic Law for example providing democracy, sociality, and so forth. Since 1949 various new state objectives have been introduced, such as Article 20a of the Basic Law for the protection of the environment and the animals in 1994. This will be discussed in detail later on.

In the context of environmental law, constitutions granting negative rights, i.e. the rights protecting the individual from government intrusion into their lives, must be distinguished from countries with constitutions granting positive, substantive rights to individuals to take legal action protecting the environment. This fundamental difference, in a comparative context, may be the determining factor in whether an individual may make a constitutional argument to protect the environment, or his or her rights in relation to it, a statutory claim, or no claim at all.[9]

As applied, constitutional provisions have furthered the creation of statutes and regulations pursuant to these rights. Seventy-eight out of ninety-four states with environmental provisions in their constitutions have had subsequent statutes strengthening them.[10] Canada has no constitutional environmental provision and has been criticized for not effectively enforcing what it likes to call the "toughest environmental legislation in the Western world."[11] Compared to the U.S.A., for example, Canada has collected relatively few fines for environmental violations. In Brazil, following its 1988 creation of a right to a healthy environment, the pre-existing criminal enforcement agency was then empowered to

7 See generally Jackson v. City of Joliet, 715 F.2d 1200, 1203 (7th Cir.), *cert. denied*, 465 U.S. 1049 (1983). (Judge Posner describes the U.S. Constitution as "a charter of negative rather than positive liberties ... The men who wrote the Bill of Rights were not concerned that Government might to too little for the people but that it might do too much for them.")

8 *Haller/Kölz/Gächter*, Allgemeines Staatstrecht, S. 334 ff.

9 May, *supra* note 2 at 18 (framing constitutional rights for individuals usually in the context of life, dignity, health, food, housing, education, work, poverty, culture, non-discrimination, peace, children's health, and general well-being and the environment's rights as good quality water, ground, and air).

10 Boyd, *supra* note 6 at 147 (stating further that in Argentina, Brazil, Colombia, Costa Rica, France, the Philippines, Portugal, and South Africa the paradigm has shifted from measures meant to minimize environmental damage to statutory rights securing a healthy environment).

11 *Ibid.* at 148.

conduct investigations and enforce the laws. There was, as a result, a precipitous jump in environmental enforcement.[12] Following these examples, it would seem that, if there were to be an environmental constitutional provision in Canada, more resources would be allocated to environmental protection, citizen participation would increase, and there would be more criminal convictions.[13]

In Chile the constitution guarantees every person the right to develop an economic activity so long as it is not contrary to morality, public order, and national security.[14] Nevertheless, those activities in most of the cases are regulated by statute, which depending on the nature of the activity requires a license to be obtained prior to undertaking the activity. The Chilean Constitution also establishes the right of every person to live in an unpolluted environment, imposing upon the state a duty to protect this right and preserve nature.[15] To implement this, the statute can impose restrictions on the right to develop economic activities.[16]

5.2 State reception of international law through domestic constitutions

Because national environmental law often can be perceived as insufficient to adequately address environmental issues across borders, as for instance with climate change, it is also important to know whether and how a state will receive and implement its obligations made in international law. Constitutions reflect whether the state regards its international agreements as intrinsically part of its domestic law, as in the example of Germany, where Article 25 of the Basic Law provides that "[t]he general rules of international law shall be an integral part of federal law. They shall take precedence over the laws and directly create rights and duties for the inhabitants of the federal territory." This is a clear example of the monist reception of the general rules of international law through which an individual could take an action directly before a German court based upon an international environmental right.

By comparison, dualist states do not immediately commit themselves upon entering into the international obligation, but must take the additional step of national legislation to do so. In those states, the individual can only exercise an international right before a national court when there has been domestic legislation that forms the basis for domestic jurisdiction. The United Kingdom is one such state where entry into an international treaty "has no effect in municipal law until an Act of

12 *Ibid.* at 150.

13 *Ibid.* at 149.

14 Chile, R. D. 2005. Fija el Texto Refundido, Coordinado y Sistematizado de la Constitución Política de la República de Chile. *In:* Presidencia, M. S. G. D. L. (ed.) *Decreto 100.* Available at http://www.leychile.cl/Navegar?idNorma=242302. Article 19 N° 21.

15 *Ibid.* at Article 19 N°8.

16 *Ibid.* at Article 19 N°8 (2).

Parliament is passed to give effect to it."[17] A third, and one might say "hybrid," position is that of the U.S.A., where the Constitution states that the President "shall have power, by and with the advice and consent of the Senate, to make treaties, provided two-thirds of the Senators present concur." Accordingly, as in a dualist state, an international environmental obligation is not actionable in domestic courts until it has been approved by the U.S. Senate, which in theory sounds like a second, domestic step, but in practice is even a bit more restrictive than that of the U.K. because there is not a distinct, first, clear step at which the agreement internationally has any legal meaning. Under such a system, obligation becomes effective in international law and in national law in two conceptually different parts, but at the same time. If the President makes an executive agreement without the advice and consent of the Senate, it is termed only a political executive agreement, and not a legally binding agreement, as for example when President William Clinton signed the Kyoto Protocol to the United Nations Framework Convention on Climate Change, but the Senate never ratified it. That said, since early in the twentieth century, most U.S. international relations agreements are just that—executive agreements, and not full, legally binding treaties with Senate approval.

5.3 Types of constitutional rights

Prior to the 1972 Stockholm Declaration, there really were no constitutional environmental rights anywhere in the world,[18] although there were some provisions such as "safeguard(ing) natural beauty and the historical and artistic heritage of the nation"[19] or the "protection, preservation, and proper use of natural resources"[20] that could be used to protect the environment. Since 1972, there has been a boom in the number of countries that include explicit environmental constitutional provisions. Today, 149 out of the 193 U.N. member states include such provisions.[21]

The spectrum of constitutional environmental rights is generally broad and varied. In many countries, there may be no mention of environmental protection

17 Committee Office, and House of Commons. *House of Commons – The EU Bill and Parliamentary Sovereignty – European Scrutiny Committee*, https://publications.parliament.uk/pa/cm201011/cmselect/cmeuleg/633/63304.htm.

18 Galizzi, Paolo. "From Stockholm to New York, via Rio and Johannesburg: Has the Environment Lost Its Way on the Global Agenda?" *Fordham Int. L. J.*, vol. 29, 2005, pp. 952, 63. The Stockholm Conference was put together by the U.N. in 1972 to confront humankind's impact on the environment. It established an Action Plan, suggesting proactive measures that countries should take, the Declaration on the Human Environment, establishing principles of States' impacts on the environment, and a Resolution on Financial and Institutional Arrangements.

19 *Ibid.*; citing Article 9, Constitution of Italy, and Article 9, Constitution of Malta.

20 *Ibid.*

21 Boyd, *supra* note 6 at 47 (finding a particularly strong correlation with countries belonging to the Organization for Economic Co-operation and Development, the Commonwealth, La Francophonie, and even the Organization of Petroleum Exporting Countries).

in the constitution, but it still may be effectively treated as if it were—often through the protection of human rights. At the other end of the spectrum are countries with explicit duties established in their constitutions, requiring action from the state and perhaps, even, action from the people, as is the case with the Indian Constitution.

Environmental protection may be expressed as rights of the earth (Bolivia), nature and its elements, such as rivers (India), or human access and use of these same things (Pennsylvania). In some cases, it is not clear whether an environmental right exists for the environment itself, or only indirectly, because it supports humans, usually through words like "life," "health," or "dignity." But it is important to note that there are in fact rights in some states that are for nature itself. Human rights lawyers may even find that environmental law ignores the individual human and acts instead as a tool of the state against industry, or worse, as a tool of industry to keep the public pacified.

Generally, human rights regimes are not the most effective means to protect the environment because they are not designed to address environmental rights; they were created before environmental issues were of broad concern, and an environmental right must be tethered to an existing human right.[22] Moreover, as we will consider in Chapter 10, as implemented by states, environmental law can even be antithetical to human rights.

> [i]n the human rights context, environmental concerns are not the cause of action or the injury itself, but the source of the injury: illegal dumping becomes actionable not because it endangers the environment but because it causes cancer in the local population and therefore implicates the right to health; clear-cutting a forest becomes actionable not because of the harm to the trees but insofar as it diminishes the property values of the local residents; using water for irrigation or development is actionable when the diminution in the available supply for personal consumption and sanitation impairs a person's right to live in dignity.[23]

In the alternative, some cultures seek to protect nature solely for the sake of protecting nature itself, without tethering the rights to human interests, but these cultures are still a minority.[24] If, perhaps, a culture regulates environmental degradation through human rights (anthropocentric), it is then likely that this culture's laws were created with the belief that human well-being is the ultimate goal. On the other hand, enacting and enforcing laws protective of sustainable development, biodiversity, and measures against climate change

22 *Ibid.* at 26–27.
23 May, *supra* note 2 at 25.
24 *Ibid.* at 72.

indicate an ecocentric tradition.[25] As a practical difference between the two, regulating from the anthropocentric theory comes with the risk that human interests, perhaps in the form of short-term financial interests, may not coincide with what is generally best for the environment. Also, human standards are susceptible to change; just in terms of the human "dignity" concept, a little more than 150 years ago slavery was a legal institution in the U.S.A. The laws were certainly not shaped to protect all of the people within the country's borders, so why should it be expected that these same laws would incidentally protect the environment?

Some cultures have introduced references to the right to a favorable environment in their catalogues of constitutional rights. Such an anthropocentric right to a favorable environment can be found for example in Article 44 of the *Ústava Slovenskej republiky* (Constitution of the Slovak Republic). The individual right to a favorable environment has been interpreted in the jurisprudence of the Slovak Supreme Court in light of the efforts to preserve a favorable environment for future generations.[26] The right to a favorable environment has been interpreted as a crucial precondition for human existence, which is possible only in an ecologically balanced environment.[27] From the perspective of a citizen searching for his or her place in the sphere of environmental governance, it is of importance to note that the interpretation of the right to a favorable environment in the Supreme Court's jurisprudence explicitly emphasizes the active role of citizens in environmental decision-making as well as their access to environmental information.[28]

The ecocentric tradition of constitutional environmental rights has provided the rights of nature with a form of legal personhood.[29] This radically ecocentric

25 *Ibid.* at 26.

26 "[n]a subjektívne právo jednotlivca na priaznivé životné prostredie nie je možné nazerať inak, ako na snahu ľudstva zachovať priaznivý stav životného prostredia pre nasledujúce pokolenia." See Supreme Court of the Slovak Republic (*Najvyšší súd Slovenskej republiky*), 1 Sž-o-KS 194/2004 of 13 September 2005; see also Supreme Court of the Slovak Republic (*Najvyšší súd Slovenskej republiky*), 1Sžp/1/2010 of 14 May 2013, para.101.

27 "[d]efinícia jednoznačne vyzdvihuje životné prostredie do polohy základnej, neodmysliteľnej a nenahraditeľnej podmienky bytia človeka (conditio sine qua non) a jeho priaznivý stav je zásadne určujúcim faktorom popri ľudskej existencii aj pri reprodukcii človeka, tzn. napomáha [...] naplniť účel objektívneho práva [...] vrátane stability ekologickej rovnováhy planéty." See also Supreme Court of the Slovak Republic (*Najvyšší súd Slovenskej republiky*), 1Sžp/1/2010 of 14 May 2013, para.102.

28 "[a]ko veľmi súvisiace a späté so životným prostredím treba hodnotiť právo verejnosti zapájať sa do procesu hľadania najrozumnejších variantov ľudských činností alebo produktu týchto činností, ktoré z tohto hľadiska nebudú zhoršovať dosiahnutý stav životného prostredia. Aby sa dosiahlo veľmi efektívne zapojenie verejnosti do tohto procesu, musí štát starostlivo zabezpečiť prenos informácií opisujúcich nielen navrhované varianty, ale aj ich dopad na stav životného prostredia zo štátu smerom k verejnosti a naopak." See Supreme Court of the Slovak Republic (*Najvyšší súd Slovenskej republiky*), 1Sžp/1/2010 of 14 May 2013, para.101.

29 Bogojević Sanja, and Rosemary Gail Rayfuse. *Environmental Rights in Europe and Beyond.* Hart Publishing, Bloomsbury Publishing Plc, 2018. pp. 5–6.

vision of constitutional environmental rights is the articulation of a shift in ethical and moral thinking, turning its interest from humanity to the context of the whole planet or parts thereof, such as rivers. These developments are most prominently observable in Bolivia,[30] Ecuador, and India. While Bolivia introduced the rights of nature in a form of statutory legislation, Ecuador and India opted for a constitutional solution. By providing nature with legally enforceable rights, Ecuador made ethical and moral considerations relevant for legal practice.[31]

From the perspective of the interpretation of constitutional provisions, it is important that the reference to *Pacha Mama* (Mother Earth), made in the preamble of the Ecuadorian Constitution and indicating the idea of living in harmony and balance with nature, permeates the rest of the constitutional provisions.[32] The rights of nature are enshrined in Chapter VII of the Ecuadorian Constitution, which provides in Article 71 that:

> Nature, or Pacha Mama, where life is reproduced and occurs, has the right to integral respect for its existence and for the maintenance and regeneration of its life cycles, structure, functions and evolutionary processes. All persons, communities, peoples and nations can call upon public authorities to enforce the rights of nature. To enforce and interpret these rights, the principles set forth in the Constitution shall be observed, as appropriate. The State shall give incentives to natural persons and legal entities and to communities to protect nature and to promote respect for all the elements comprising an ecosystem.[33]

Article 72 of the Ecuadorian Constitution goes on to say that:

> Nature has the right to be restored. This restoration shall be apart from the obligation of the State and natural persons or legal entities to compensate individuals and communities that depend on affected natural systems. In those cases of severe or permanent environmental impact, including those caused by the exploitation of non-renewable natural resources, the State shall establish the most effective mechanisms to achieve the restoration and shall adopt adequate measures to eliminate or mitigate harmful environmental consequences.[34]
>
> The Mother Earth does not have the voice and hands of its own to defend itself. For this reason, the Constitution couples the provisions on the rights

30 Bolivia adopted the Mother Earth Law (*Ley de derechos de la Madre Tierra*) in 2010.

31 Kotzé, Louis. "Somewhere Between Rhetoric and Reality: Environmental Constitutionalism and the Rights of Nature in Ecuador." *T. E. L.*, 6 (3), 2017, pp. 4–5.

32 *Ibid.* at 18.

33 Kotzé, *supra* note 31 at 22-23.

34 *Ibid.*

> of nature with corresponding provisions introducing the responsibilities of citizens. These include the responsibilities to actively protect the environment and act as guardians of nature, while emphasizing the need to exploit resources in a sustainable manner.[35]

From the perspective of enforcing the rights of nature, it is important to emphasize that the wording of Article 71 of the Constitution also provides certain interpretative guidance indicating that the rest of constitutional provisions shall be interpreted in the light thereof. Hence, the emphasis put on the activation of citizens to enforce the rights of nature is translated in the open and liberal understanding of the concept of legal standing in environmental matters. The access to justice provisions enshrined in Article 75 of the Constitution are also to be interpreted in this manner.[36]

However, rights of nature do not constitute a decisive foundational norm in the internal hierarchy of the Ecuadorian Constitution as does, for example, the reference to human dignity in Article 1 of the German G.G.[37] Moreover, experience from the practical application of the constitutional provisions shows mixed results, with the rights of nature not taking clear precedence over other competing interests. Recalling the time before public environmental law gave states sufficient power to control industrial pollution, the Ecuadorian constitution has yet to countervail economic interests.[38]

Other cultures have not only recognized the rights of nature, but also the responsibilities of citizens. In India, for example, the Uttarakhand High Court has attributed legal personality to the river Yamuna as well as to other natural entities, such as glaciers. In its 2017 decision in the case of *Mohd. Salim v. State of Uttarakhand & Others*, the High Court held the rivers "Ganga and Yamuna are juristic/legal persons/living entities [that] have the status of a legal person with all corresponding rights, duties and liabilities."[39] In this regard, the court relied, *inter alia*, on the "*parens patriae*" doctrine, placing the primary responsibility for the protection of natural entities on the state and public authorities.[40] And in the case of *Narayan Dutt Bhattt v. Union of India*, the High Court held that:

> The entire animal kingdom including avian and aquatic are declared as legal entities having a distinct persona with corresponding rights, duties and liabilities of a living person. All the citizens throughout the State of

35 See Article 83 and 399 of the Constitution; Kotzé, *supra* note 31 at 23.
36 Kotzé, *supra* note 31 at 24.
37 *Ibid.* at 26.
38 *Ibid.* at 29-32.
39 O'Donnell, Erin. "At the Intersection of the Sacred and the Legal: Rights of Nature in Uttarakhand, India." *J.E.L.*, vol. 30, 2017, pp. 135–44.
40 *Ibid.* at 139.

Uttarakhand are hereby declared persons *in loco parentis* as the human face for the welfare/protection of animals.[41]

5.4 National constitutions

Importantly, a direct correlation between constitutional environmental provisions and improvements in environmental conditions can be demonstrated. Countries such as Germany, the Netherlands, Norway, Sweden, and Switzerland have made significant progress in improving their air quality following adoption of constitutional provisions. Between 1980 and 2005, emissions of nitrogen oxides and sulphur dioxide fell much faster in countries with explicit constitutional protection for the environment than it did in countries without similar constitutional provisions.[42]

An indirect benefit of constitutional provisions is the mandate given to courts to interpret all statutes so as to protect the environment. The supreme courts of Argentina, Brazil, and India, for example, have proactively called for stronger air quality regulations, based on underlying constitutional rights, and these calls have resulted in stronger standards from the countries' legislatures.[43] There are also plenty of examples of courts directly prohibiting activity, such as a gasoline filling station being built next to a school in Portugal, or requiring that industrial operations reduce pollution, such as has been seen in Colombia, Costa Rica, Greece, Thailand, and other cultures.[44] As mentioned before, the German constitution provides not only fundamental rights but also state objectives. These are meant as a legitimation for the state to thereby serve its citizens. An example is the democracy principle, found in Article 20 I G.G. and described in Article 20 II G.G., and which gives power to the citizens. This equally counts for the states through Article 28 I 1 G.G.[45] Since 1994, the German constitution has had what is known as the state objective of environmental protection, which means that one of the state's express goals is the protection of the environment. It is therefore not a fundamental right but through Article 20a G.G. the state is obliged to protect the environment. It contains the decision of creating something called an "environmental state" which sees environmental protection as a goal and standard of its activities.[46] When comparing to other cultures, it is again important to highlight that Article 20a G.G. does not state a fundamental right. Therefore it does not provide a subjective right for the citizen but is addressed

41 Narayan Dutt Bhatt v. Union of India and others, Writ Petition (PIL) No. 43 of 2014, Judgment of July 4, 2018.
42 Boyd, *supra* note 6 at 154.
43 *Ibid.* at 155.
44 *Ibid.*
45 *Morlok/Michael*, Staatsorganisationsrecht, S. 41 f.
46 *Kloepfer*, Umweltrecht, S. 121.

to the state only.[47] This means that the legislative, judicial, and executive functions of the federation, the states, and the local communities are all bound by Article 20a G.G.[48]

Article I of the U.S. Constitution establishes Congress and enumerates its limited powers in Section 8. Given that the U.S. Constitution was drafted in the eighteenth century, it is not surprising that neither nature nor the environment are included in the catalogue of expressed, enumerated competencies of the U.S. Congress. Several movements have called to amend the U.S. Constitution to include the environment, but none have succeeded in doing so. In its more than two hundred-year history, the U.S. Constitution has only been amended twenty-seven times. Thus one might well ask, if Congress has only limited grants of power, and the protection of the environment is not one of them, under what authority may it enact environmental legislation? One need only read the first clause of the Clean Air Act of 1963 to answer the question: "[t]he Congress finds that the predominant part of the Nation's population is located in its rapidly expanding metropolitan and other urban areas, which generally cross the boundary lines of local jurisdiction and often extend into two or more States..."[49] The operative language in this clause, "extend[ing] into two or more States," implicates the federal Congress' power to legislate under what is known as the Commerce Clause (Article I, Section 8, Clause 3) of the U.S. Constitution which provides that the U.S. Congress shall have power "[t]o regulate Commerce with foreign Nations, and among the several States, and with the Indian Tribes." Were it not for the Commerce Clause, the failure to mention the environment in the federal U.S. Constitution would mean environmental competency remains with the fifty states.

Under the auspices of the Commerce Clause, Congress has similarly enacted the Clean Water Act, Endangered Species Act, and Safe Drinking Water Act along with about a dozen other major pieces of legislation. But legislation alone is inadequate to accomplish any goal, so Congress took further steps to implement these laws with the creation of an administrative body as well as alternative enforcement mechanisms such as "citizen suits" in instances where the Environmental Protection Agency does not follow through with an enforcement of its own. Under federal statutes, the ability to prosecute violators rests with the federal government. However, many federal statutes provide that, when citizens notify the appropriate government entity that a violation is occurring and the government takes no action after a specified period of time (usually ninety days), the ability to prosecute the violation passes to the citizens who may then do so.

47 *Kloepfer*, Umweltrecht, S. 122.
48 *Kloepfer*, Umweltrecht, S. 126 ff.
49 42 U.S.C. § 7401 (a)(1).

5.5 State and other subnational constitutions

To fully understand constitutional law practices, in looking at *national* or *subnational* constitutions, the observer must first identify the nature of the state. Is the state *federal*, *quasi-federal*, or some other form? Federalism in the U.S. is clearly defined by history, being formed by the thirteen original sovereign states. Other countries also formed by once-autonomous states or regions, such as Germany and Italy, are federal states too, but with different characteristics.[50] Therefore, a necessary second step would be to inquire into both the *de jure* and *de facto* interrelationships between the *national* and *subnational* constitutions in the legal culture under examination.

Just like their national counterparts, subnational constitutions come in many different forms. The extent of sovereignty retained in the U.S. state constitutions is unparalleled by any other subnational body. Most countries with federal systems fall somewhere into the middle of the spectrum, with some sort of state, regional, or provincial document guaranteeing rights and, usually, granting very limited governing roles to the subnational governments.[51] Germany's *Länder* (state) constitutions, predating the federal adoption of the Basic Law, contain many provisions, especially with regard to human rights, but do not delegate very much responsibility to the German states.[52] Canada, another country in the middle of the spectrum, has a national constitution, consisting of approximately twenty-six separate documents; thus the provincial constitutions are predictably spread throughout a number of documents as well.[53] Perhaps one of the most notable features of the Canadian subnational constitutions is that rather than being an entrenched document, only changeable through a rigorous amendment process, the provincial documents are statutes created by the Canadian National Assembly.[54] As an obvious consequence, these provincial constitutional documents are susceptible to being altered through other ordinary legislation. Furthermore, in many regards, the provincial governments play a much more restricted role than their U.S. counterparts. This is most apparent in that all Canadian provincial law is subject to interpretation by the Supreme Court of Canada and that, even when it is being interpreted by lower courts, the judges are always federal appointees.[55] Many of the countries in the middle of the spectrum also require conformity between the national and subnational provisions; one could see this in counties like China, Italy, Spain, or Switzerland.[56] This has resulted in a Canadian system in which the national and subnational provisions

50 Ginsburg, Tom. "Subconstitutionalism." *Stan. L. Rev.*, vol. 62, 2010, pp. 1583, 94.
51 See generally Tarr, *supra* note 4 at 176–78.
52 *Ibid.* at 179.
53 *Ibid.* at 176.
54 *Ibid.* at 180.
55 *Ibid.* at 181.
56 *Ibid.* at 187.

have become somewhat indistinguishable.[57] At the other end of the spectrum are countries such as Belgium, Nigeria, and India, which have neither subnational constitutions nor a parallel system of subnational courts.[58]

When European states cede responsibilities to the European Union, it appears to take on many of the functions and features of a state with sub-states.[59] Without a constitution, the European Union is instead a treaty union, bound together by a series of treaties that lay the European Union's foundation and create a series of rights and obligations, similar to the norms in constitutions. The delineation between competencies of the Member State and European Union governments, supremacy of European Union law, binding treaties, hurdles to changing its fundamental principles, and judicial review by European Union courts show many characteristics of a federal system. Furthermore, it has courts, legislative institutions, an executive, and administrative functions; while the European Union and its twenty-seven Member States do not compose a single state, the European Union has certainly transformed in this direction since its much more modest beginning in the 1950s.[60] With regard to environmental protections, the European Union shows a tendency toward federalism; environmental protection is a shared competence between the European Union and the Member States, but any European Union environmental treaty provisions will be supreme law.[61]

By comparison, on its constitutional face, Germany consists of a federation and sixteen sub-national *Länder* (states). Article 30 of the German Basic Law, which functions as a constitution, states: "Except as otherwise provided or permitted by this Basic Law, the exercise of state powers and the discharge of state functions is a matter for the *subnational regions*." Accordingly, the federation as well as each state has its own constitution. The states must follow the basic principles of the federal constitution according to Article 28 I G.G. but are otherwise free to shape their own constitutions.[62] However, one is immediately reminded in Article 31 that "[f]ederal law shall take precedence over state law." In practice, except for some limited areas such as culture (including education) and the administration of the police, the catalogue of law-making competencies is largely national, and the subnational regions administer and enforce national law in addition to the smaller catalogue of state law.[63]

State constitutional provisions are no mere novelty. Over the last decades, litigants have increasingly based environmental actions on subnational, provincial,

57 *Ibid.* at 184.
58 Tarr, *supra* note 4 at 174, 176.
59 Ginsburg, *supra* note 50 at 1583, 84.
60 *Ibid.* at 1583, 95.
61 Treaty of Lisbon (2009) Article 2 C (2)(e).
62 *Michael/Morlok*, Staatsorganisationsrecht, S. 65.
63 More recently, however, the Länder are taking independent actions, as with the Climate Protection Act of North Rhine Westphalia (2013), the first act passed by a federal state in Germany to protect the climate. The Act stipulates a 25 percent reduction in greenhouse gas emissions by 2020 and a reduction of at least 80 percent by 2050.

or state constitutions. Especially in the U.S., this new-found interest in state constitutions has also not limited itself to environmental issues; all areas of the law have experienced a growing interest in their corresponding state constitutional provisions.[64]

Subnational constitutions are foundational documents which may be identical to, similar to, or substantially different than their national counterparts. Using the U.S.A. as an example, while the federal Constitution is characterized by its limited grants of legislative power, by comparison the state governments possess plenary power to legislate.[65] One would notice many textual similarities between the U.S. Constitution and that of any given state, but these must still be understood through the notion of U.S. federalism. Generally speaking, the state constitutions do not grant powers, but, rather, serve to limit the powers of the state governments.[66] In the absence of constitutional limitations, state legislative power still exists; thus, when such limitations are made, they are often narrow in scope and elaborated upon in great detail, which contributes to the tendency of state constitutions, like many other national constitutions in the world, to be lengthy documents.[67] Some commentators have suggested that these lengthy documents have become so detailed that they, at a certain point, cease to be constitutional, and, rather, tend to be legislative outlines.[68] In line with the interpretive principle of *inclusio unius est exclusion alterius* (the inclusion of one thing is to the exclusion of another) this extensive detail regarding the legislatures' duties can also act as a limitation as to how much discretion the legislatures have on many matters.[69]

Borrowing an idea of self-governance from Switzerland, a number of U.S. states have enacted *initiative*, *referendum*, and *recall* provisions.[70] State initiatives allow for citizens to effectively enact their own legislation during general elections, adding particular policy questions for voters at the ballot box. Referenda allow for citizens to effectively repeal laws created by the legislature.[71] In cases where citizens feel that they have not been adequately represented by their elected representatives, or that the representatives have otherwise disqualified themselves

64 Garry, Patrick. "The Rising Role of State Constitutional Law: An Introduction to a Series of Articles on the South Dakota Constitution." *S. D. L. Rev,*, vol. 59, 2014, p. 4.

65 Tarr, Alan. "State Constitutional Design and State Constitutional Interpretation." *Mont. L. Rev.*, vol. 72, 2011, pp. 7, 12.

66 *Ibid.* at 12–13; citing State ex rel. Schneider v. Kennedy, 587 P.2d 844, 850 (Kan. 1978) (stating that "[i]t is fundamental that our state constitution limits rather than confers powers. Where the constitutionality of a statute is involved the question presented is, therefore, not whether the act is authorized by the constitution, but whether it is prohibited thereby").

67 Tarr "State," *supra* note 65 at 7, 13.

68 Williams, Robert Forrest. *The Law of American State Constitutions.* Oxford University Press, 2009, pp. 23–24.

69 *Ibid.* at 24.

70 *Ibid.* at 6.

71 Garry, *supra* note 64 at 5.

to hold office, a recall may be implemented to remove them. These tools, only being available at the state level, are absent from the U.S. Constitution.[72]

In the U.S. it should be recognized that the states' constitutions are not copies of the U.S. Constitution, but rather independent texts with their own unique wording, characters, historical backgrounds, legal histories, and political philosophies.[73] Federalism in the U.S. may also be called *dual constitutionalism*, a term that highlights the essential roles played by constitutions both at the federal and state levels. State constitutions in the U.S. are created on a fundamentally different premise than their federal counterpart. Federalism in the U.S. relies on both the existence of the national and subnational constitutions; the national, i.e. federal, constitution was created by the cessation of state sovereignty in a number of areas. This system of *imperium in imperio* (sovereignty within sovereignty) recognizes that, in relinquishing sovereignty over particular matters to the federal government, the states retain authority on most other matters.[74] One of the main authors of the U.S. Constitution, James Madison, insisted that "[t]he powers delegated by the proposed Constitution to the federal government are few and defined. Those which are to remain in the State governments are numerous and indefinite."[75]

In the field of environmental law, the notion of cooperative federalism means that while the U.S. Constitution has been interpreted to give the power to the federal government to govern environmental matters, in practice a domestic or foreign citizen is far more often going to hear from a state government, not the federal government, if an environmental problem occurs. In fact, the federal government's Environmental Protection Agency (E.P.A.) has too few persons in the field to inspect pollution sites and enforce the law. Instead, the states do most of this work under delegated authority, through such legal instruments as the pre-approved state implementation plans (S.I.P.) in the field of air pollution, submitted by the states to the federal government for agreement and approval.

The Congressionally created E.P.A. is a federal administrative body created for the purpose of enforcing the environmental statutes as well as creating further regulations, which are in line with Congress' statutes. The E.P.A. has received a high degree of deference in its interpretation of the statutes, so long as it "stays within the bounds of its statutory authority."[76] In what has become a well-known case, *Chevron v. Natural Resources Defense Council*, the Supreme Court deferred to the E.P.A.'s judgment on a particular matter, requiring only that the agency act reasonably and that the agency interpret the statute in both the specific context of the case at hand and the broader context of the entire statute as well.[77] The

72 Williams, *supra* note 68 at 32.
73 *Ibid.* at 8.
74 Williams, *supra* note 68 at 17.
75 Hamilton, Alexander, et al. "Federalist No. 45." *The Federalist Papers*, 2009, pp. 97–100.
76 Utility Air Regulatory Group v. E.P.A., 134 S.Ct. 2427, 2439 (U.S., 2014).
77 Chevron U.S.A. Inc. v. Natural Resources Defense Council, Inc., 104 S.Ct. 2778 (1984).

E.P.A.'s interpretation of a statute should, above all, honor Congress' express intent when it is clear; the courts will overrule any impermissible E.P.A. additions that exceed or contradict the intent of Congress.[78] In other words, when Congress has clearly expressed its intention, especially in cases of unambiguous numerical thresholds for contaminants, neither the E.P.A. nor any other executive-branch agency may make an unreasonable interpretation of such thresholds, or edit the provision to fit their unreasonable interpretation.[79]

The practicing lawyer whose client has a problem to be covered by environmental law in the U.S.A. does not usually encounter federal law at the beginning of the action, however. Under the political concept of "cooperative federalism," despite the Commerce Clause giving the federal government competency over the environment, the federal government in turn delegates most law-making and law-enforcing of environmental matters to the states through such mechanisms as State Implementation Plans. For that practitioner, that means that a client must comply first with state law, which must be as strict as or stricter than federal law. Only when the federal government determines that a state's law as made or as enforced is insufficient does it step into a legal matter through a process known as "overfiling." If a state persists in demonstrating that it does not have the wish or ability to make and enforce law as strict as or stricter than the federal law, then the federal government may suspend the state's delegated authority and govern the state's environmental matters from the Commerce Clause power of the federation. In practice, this rarely happens because, as a matter of good governance and pride, elected state officials want to lose neither the power nor the federal support funding that comes along with the state's proper execution of the environmental powers that the federations have delegated to it.

In the U.S.A., state constitutions impose duties on their respective state governments and create rights for citizens of that state, just as the federal Constitution grants power to the federal government as well as imposes limits of such power.[80] In 1970, Illinois was the first state to amend its constitution to include an environmental right, followed quickly one year later by Pennsylvania. Article I, Section 27 of the Pennsylvania Constitution states:

> [t]he people have a right to clean air, pure water, and to the preservation of the natural, scenic, historic and esthetic values of the environment. Pennsylvania's public natural resources are the common property of all the people, including generations yet to come. As trustee of these resources, the Commonwealth shall conserve and maintain them for the benefit of the people.[81]

78 Utility Air Regulatory Group v. E.P.A., 134 S.Ct. 2427, 2445 (U.S., 2014).
79 *Ibid.* at 2446.
80 Tarr "State," *supra* note 65 at 7, 13.
81 Pa. Const, Article I, §27.

Some observers have noted that the text of Section 27 creates two distinct rights with corresponding duties.[82] The rights in the first sentence, such as the right to clean air, pure water, etc., could be referred to as "first sentence rights."[83] All of the other rights arising out of the second and third sentences could be referred to as "public trust rights."[84] On one hand, the first sentence rights do not expressly create a duty on the part of the government, but this does not make them merely aspirational.[85] While it could be said that there is no positive duty, there certainly is a right of the people and corollary obligation for the government to abstain from infringing these rights through legislation or upon executive action.[86] Therefore, the Pennsylvania government is required by Section 27 to make an environmental assessment prior to "any proposed action on the constitutionally protected features."[87]

On the other hand, the public trust rights, arising out of the explicit "conserve and maintain" language, do create a positive obligation for the state to act.[88] The public natural resources are the corpus of the public trust; therefore, they must be protected by the state with the ordinary duties of prudence, loyalty, and impartiality usually associated with fiduciary relationships.[89] Public trust rights are also created implicitly through Section 27's language. First, the state of Pennsylvania "has an obligation to refrain from performing its trustee duties respecting the environment unreasonably."[90] This also entails preventing the depletion of public natural resources both through direct state action or indirectly through its failure to stop harm arising from private parties.[91] Second, the state is, as trustee, required to affirmatively act through legislation to protect the environment.[92] This element has been substantiated by the Pennsylvania General Assembly multiple times, through their passage of the Clean Streams Act,[93] the Air Pollution Control Act,[94] and the Solid Waste Management Act,[95] among others.

Thus, next to a plain-language grant of rights to the people, one finds an implicit obligation that the Pennsylvania General Assembly should create complementary legislation. The separate existence of the two prompted some

82 Kristl, Kenneth. "It Only Hurts When I Use It: The Payne Test and Pennsylvania's Environmental Rights Amendment." *Envtl. L. Rep. News & Analsis*, vol. 46, 2016, pp. 10594, 10595.
83 *Ibid.* at 10596.
84 *Ibid.*
85 Robinson Twp., Washington Cty. v. Com., 623 Pa. 564, 647 (2013).
86 *Ibid.*
87 *Ibid.*
88 Robinson Twp., Washington Cty. v. Com., 623 Pa. 564, 647 (2013).
89 *Ibid.*
90 *Ibid.*
91 *Ibid.*
92 *Ibid.*
93 See 35 P.S. § 691.1 *et seq.*
94 See 35 P.S. § 4001 *et seq.*
95 See 35 P.S. § 6018.101 *et seq.*

interested persons to propose that this amendment is not self-executing in the lawsuit known as the Gettysburg Tower case. That would mean it is aspirational only and a legal action could not rest solely on the amendment without additional implementing legislation, similar to the German *Staatszielbestimmungen* (state objectives). The Pennsylvania Commonwealth Court, an appeals court of second instance, held in the Gettysburg Tower case that Article I, Section 27 of the Pennsylvania Constitution *is* self-executing and therefore could be used by citizens as the legal basis of their lawsuit without any subsequent legislation to further execute the rights created.

In the U.S.A. at present not only Illinois and Pennsylvania have self-executing constitutional environmental rights, but Hawaii does as well. The constitutions of Massachusetts and Rhode Island explicitly include the right in their constitution, but they are not self-executing and require further legislation to enable citizens to act upon their rights. In that sense, they are comparable to the German Basic Law's state objectives. Montana's constitutional provision is neither clearly self-executing nor clearly dependent—further judicial interpretation is necessary.[96] In exercising comparison, we can raise the question of what the function of a constitutional provision is, and when it is dependent upon legislation to be exercised (as with a state objective), when compared to the function of a self-executing environmental right in practice.

Even with express constitutional rights and obligations in place, counterbalancing factors will put the courts to task; they will interpret the provisions and apply them to cases as they arise. Returning to the example of Pennsylvania's Article 1, Section 27, expressly laying out what is to be protected does not define the exact parameters of environmental protection afforded. Shall the environment be completely untouched? In the 1973 case *Payne v. Kassab*, the Pennsylvania Commonwealth Court held that there are limits to the extent of protection afforded under Article I, Section 27, allowing for the "normal development of property."[97] In permitting a street-widening project, the *Payne* court distinguished this as a case of "controlled development" as opposed to the other possible alternative "no development."[98] In order to allow for controlled development, in the sense of Article I, Section 27, the court laid out a three-part test for subsequent courts to follow: they should ask 1) if the developers complied with all applicable statutes and regulations relevant to the Commonwealth's mandate to protect the environment, 2) whether there is proof that potential disturbance to the environment was minimalized, and 3) whether the environmental harm from the challenged action clearly outweighs the benefits of the project.[99]

96 English, Article "State Constitutions and Enviromental Bills of Rights." *Council of State Governments*, 2015.

97 Payne v. Kassab, 312 A.2d 86, 94, 11 Pa.Cmwlth. 14, 29 (Pa.Cmwlth. 1973).

98 *Ibid.*

99 *Ibid.* at 29–30.

5.6 Conclusions

In comparing any culture's constitutional environmental rights and liabilities, we must begin with understanding a state's federal structure. Constitutional organization of the twenty-first century sovereign state spans a spectrum from exceptional unitarianism (France) to a complete dual federal–state system (U.S.A.) Most states of the world (China, Italy, Spain, Switzerland, et cetera) organize their national–subnational relations somewhere in the middle of this spectrum. After one has understood a state's constitutional federalism, one must of course remain mindful of the allocation of environmental rights and liabilities, not just between the federal units, but among the branches of the units, as we discussed in Chapter 4. With that two-dimensional matrix in hand, one can then turn to the nuances of politics in any culture and how they interface with legal institutions, bodies, and sources of law in life and legal practice as they serve to protect the citizens' air, water, land, health, and enjoyment of life.

Literature

Boyd, David R. *The Right to a Healthy Environment: Revitalizing Canadas Constitution*. UBC Press, 2012.

Brennan, William. "State Constitutions and the Protection of Individual Rights." *Harvard Law Rev.* 90 (1977), 489–504.

Fernandez, Jose. "State Constitutions, Environmental Rights Provisions, and the Doctrine of Self-Execution: A Political Question?" *Harv. Envtl. L. Rev.* 17 (1993), 333–388.

Garry, Patrick. "The Rising Role of State Constitutional Law: An Introduction to a Series of Articles on the South Dakota Constitution." *S. D. L. Rev.* 59 (2014), 4–13.

Ginsburg, Tom. "Subconstitutionalism." *Stan. L. Rev.* 62 (2010), 1583–1594.

Haller, Walter, et al. *Allgemeines Staatsrecht*. 5th ed., Nomos, 2013.

Kotzé, Louis. "Somewhere Between Rhetoric and Reality: Environmental Constitutionalism and the Rights of Nature in Ecuador." *T. E. L.* 6.3 (2017), 1–33.

Kristl, Kenneth. "It Only Hurts When I Use It: The Payne Test and Pennsylvania's Environmental Rights Amendment." *Envtl. L. Rep. News & Analsis* 46 (2016), 10594–10607.

May, James R., and Erin Daly. *Global Environmental Constitutionalism*. Cambridge University Press, 2016.

O'Donnell, Erin. "At the Intersection of the Sacred and the Legal: Rights of Nature in Uttarakhand, India." *J.E.L.* 30 (2017), 135–144.

Tarr, Alan. "Subnational Constitutions and Minority Rights: A Perspective on Canadian Provincal Constitutionalism." *Revue Quebecoise De Droit Constitutionnel* 2 (2008), 174–193.

Tarr, Alan. "State Constitutional Design and State Constitutional Interpretation." *Mont. L. Rev.* 72 (2011), 7–26.

Williams, Robert Forrest. *The Law of American State Constitutions*. Oxford University Press, 2009.

6

INTERFACES BETWEEN LAW AND POLITICS

Tsegai Berhane Ghebretekle and Marek Prityi

6.1 Introduction

Never departing far from the basic question that enables comparative law—what is the function of the topic?—we address in this chapter the function of the interfaces of law and politics for environmental law across cultures. There are various approaches to the relationship between the realms of law[1] and politics.[2] Politics is not law and law is not politics. The interfaces between the two are most often exemplified in the political process of choosing law-makers and the indirect result those political choices have on the law that will be made by them.

We have already seen that the decision to establish a green tribunal, and not a green court, in India was political. As of the writing of this book, it was the agents of those politics who have effected their desire to minimize the accessibility and impact of the N.G.T. by eliminating all N.G.T. judges except the central ones and consequently eliminating all regional N.G.T. access except by videolink. A constitutionally established court, on the other hand, could not be

1 Miro Cerar defines *law* as: the binding value-normative system established and carried out by the state in national law and carried out by international organizations and institutions in international law, which are intended for the establishment and maintenance of a balance between justice and order and solving and preventing pressing societal and international conflicts. Cerar, Miro. "The Relationship Between Law and Politics." *Annual Survey of International & Comparative Law*, 15 (1), 2009, pp. 19, 21.

2 Politics is a result of a collective choice made by a society. It deals with the way states organize and govern themselves. Thus, one could argue that rule-making and decision-making (be it economic, social, or environmental) are inherently political, in the meaning that they seek to resolve inherent conflicts and to reconcile environmental goals with other competing values, such as economic well-being, individual rights, and social justice. See Kraft, Michael E. *Environmental Policy and Politics*. Pearson, 2016, pp. 7, 10.

politically suspended from operating. Politics can also be responsible for creating and reflecting citizen attitudes toward state legal actions. When the state fails to respond to the political expression of citizens' attitudes, the law may provide for direct citizen environmental legal action, as will be discussed in the next chapter.

Thus, this chapter aims to look at the disciplinary context of environmental law rather than turn inward on itself. The chapter will primarily focus on historical and political contexts shaping the character of environmental law. Staying true to its comparative nature, the points made in the chapter will be demonstrated in examples from various regions divided for the sake of clarity in the global North and global South. Like law itself, politics is part of the culture[3] in which we conduct the present comparisons. The wider context shaping the development of environmental law is illustrated by examples highlighting the active role of citizens in addressing environmental issues through the use of their civil and political rights.

6.2 Relevance of the intersection between law and politics for legal comparison

Even though the intersection between law and politics might at the first glance appear a bit far away from the concept of comparative law (at least in its simplified meaning), the postmodern and critical approaches to comparative law see it as a valuable contribution to comparative legal research.[4] The argument that the fields of law and politics should not be seen as strictly divided was vocalized already by the critical studies movement in the United States in the 1970s.[5] If one adopts the immersion approach discussed in Chapter 2, foreign law can be best understood by immersing oneself in all of the foreign culture, including politics, history, economics, and especially language.[6]

The political identity of cultures might also be seen as an articulation of their historical development and cultural influences.[7] Based upon his empiricist philosophy of law, U.S. Supreme Court Justice Oliver W. Holmes famously stated "[t]he life of the law has not been logic but experience,"[8] and, further, that that experience has been shaped by the necessities of the time, by moral and political theories, and by the public policy in which it is gained.[9] Hence, comparative environmental law research will benefit from an insight into the cultural,

3 Luhmann, Niklas, et al. *Law as a Social System*. Oxford University Press, 2009, pp. 371–74.

4 Siems, Mathias M. *Comparative Law*. Cambridge University Press, 2014, p. 114.

5 *Ibid*.

6 Grosswald Curran, Vivian. "Cultural Immersion, Difference and Categories in U.S. Comparative Law." *American Journal of Comparative Law*, vol. 46, 1998, pp. 43, 59.

7 Siems, *supra* note 4 at 101-02.

8 Holmes, Oliver Wendell. *The Common Law*. Dover, 2014.

9 Holmes, Oliver Wendell. "The Path of the Law." *Harvard Law Review*, vol. 10, 1897, p. 457.

historical, and political developments shaping legal systems, which form the object of comparison.[10]

As comparative law should not be an exercise that begins and ends with textual comparison, it is necessary to pay attention to and acknowledge the *context* of institutions and sources of law involving national customs and cultural research.[11] The same ideological and political factors that were visibly traceable in constitutional environmental law in Chapter 5 are also traceable at the interface between politics and other sources of law. The role of political parties in the legislative process is crucial, as they are fundamental in aggregating and communicating political preferences and priorities and translating them into binding legislation.[12] Politics might be influenced also by environmental and social movements which do not necessarily transform into political parties. Nevertheless, political parties represent a power with which lawyers must reckon, with a potential to influence the legislature, executive, and judiciary.[13]

6.3 Emergence and evolution of green politics in the global North

The emergence of green politics is inextricably linked to social movements and activism which appeared in Western countries in the 1960s and 1970s.[14] Characteristic for these movements was not only a new approach to environmental issues, but also a new perspective on the broad set of social and collective values, articulated in the form of human rights, women's rights, and civil rights.[15] New thinking questioned the ideology of capitalist culture based on unlimited growth and instead emphasized the natural limits associated with economic and population growth.[16] The ideas and principles fought for by activists in this period formed the main ideological foundations of green politics. Green politicians began to institutionalize their politics in the form of Green Parties in the 1980s.[17]

The German Green Party (*Bündnis 90/Die Grünen*), might be seen as an example of a political party built upon the idea of giving political representation not only to environmental activists, but also to peace and feminist activists

10 Junker, Kirk W. "A Focus on Comparison in Comparative Law." *Duquesne Law Review*, vol. 52, 2014, pp. 69, 81; see also Grosswald, *supra* note 6 at 43, 51.

11 Junker, *supra* note 10 at 88–90.

12 Lindberg, Björn, et al. "Party Politics as Usual? The Role of Political Parties in EU Legislative Decision-Making." *Journal of European Public Policy*, vol. 15, 2008, pp. 1107–1108.

13 Chopra, Kanchan. *Development and Environmental Policy in India the Last Few Decades*. Springer, 2017, p. 47.

14 Burchell, Jon. *The Evolution of Green Politics: Development and Change within European Green Parties*. Earthscan Publications, 2002, p. 9.

15 *Ibid.* at 11.

16 *Ibid.* at 12.

17 *Ibid.* at 9.

opposing a large variety of topics, ranging from military actions and nuclear energy to industrialization and authoritarianism. At the beginning, the German Green Party was confronted with the challenge of balancing its dual role as a movement for social change and as an aspiring party within an established political system. However, it gradually evolved to its current form of a relatively mainstream, institutionalized, and professional party with regular representation in the parliament at federal level as well as at the level of respective federal states.[18] The transformation from a movement to a political party—and in this way becoming associated with legislature—is a crucial step for an environmental movement to have a long-standing impact on law.[19]

The power of processes initiated by environmental movements in the 1960s and 1970s might be demonstrated also by the fact that the ministries of the environment nowadays form a relatively standard part of governments not only in Western countries, but also worldwide.[20] Such a milestone might, however, also mark the end of the life-cycle of the environmental movement, as it modified the original social order by becoming a part of it.[21]

The field of environmental law was also influenced by developments in the academic, political, and activist fields. For instance, the discussion revolving around Christopher D. Stone's famous article, "Should trees have a standing?"[22] in which he argued for the recognition of legal personality and the "rights" of nature, significantly shaped the issue of access to justice in environmental law not only in the U.S.A., but in other countries as well. Stone's article influenced the discussion about access to justice in environmental matters in Germany, where a certain form of action in public interest (involving environmental matters) was demanded already in 1967 by the German Council for Landscape Management.[23] Even though it was for a long time resisted by the legal community in Germany, Stone's provocation served as a powerful incentive for the discussion in German legal, environmental, and academic circles.[24]

The countries behind the Iron Curtain have experienced a different kind of development. Under the dictatorial rule of the communist parties and under the influence of the Soviet Union, the room for civic activities and engagement was

18 *Ibid.* at 52-58.

19 Chopra, *supra* note 13 at 47.

20 Rootes, Christopher. "Environmental Movements: From the Local to the Global." *Environmental Politics*, vol. 8, no. 1, 1999, p. 1. However, it must be noted that, in the U.S.A., on multiple occasions attempts have been made to change the status of the U.S. Environmental Protection Agency from an administrative agency to an Executive Cabinet level position, and each time the attempt has failed.

21 Chopra, *supra* note 13 at 38.

22 Stone, Christopher D. "Should Trees Have Standing?–Towards Legal Rights for Natural Objects." *Southern California Law Review*, vol. 45, 1972, pp. 450–501.

23 Schmidt, Alexander, et al. *Die Verbandsklage Im Umwelt- Und Naturschutzrecht UmwRG – BNatSchG U.a.* Beck, C H, 2014, p. 3.

24 *Ibid.* at 4.

considerably smaller. However, the situation began to change toward the end of the Cold War. In its last years, the Soviet Union gradually eased its involvement in the internal politics of its satellite states, which created room for engaged individuals and groups to test the limits of the regime.[25] In countries such as the German Democratic Republic or Czechoslovakia, groups focusing on environmental and cultural heritage protection activities began to form.[26] Initially, their activities did not cause major confrontation with the regime. However, they created room for the emergence of a new social and environmental ethos that gradually transformed these occasional activities and discussions into a movement embracing civil and political rights and public advocacy. This is the reason that the people organizing and participating in these activities were monitored by the intelligence services and were soon branded in the intelligence and governmental circles as internal enemies.[27] In 1977, a number of Czech and Slovak citizens signed a "charter" that criticized the Czechoslovak government for failing to implement the human rights provisions of a number of applicable international and domestic laws, including their own constitution. Known thereafter as Charter 77,[28] from at least 1987 onwards it became the focal point for a number of newly founded groups in Czechoslovakia.[29] Thanks to its initial activation during the reign of the Communist regime, the environmental movement was one of the main forces shaping the revolutionary events of 1989 and entered the era of the newly discovered democracy as one of the most well-prepared elements of the civil society.[30]

In the first years after the revolution in Czechoslovakia, an unprecedentedly open system of political representation, coupled with the public's support, served as a catalyst for the adoption of basic legislation in the field of environmental protection and an institutional framework implementing the legislation. The *Zákon o životnom prostredí* (Act on the Environment) No.17/1992 set the basic

25 Šuška, Pavel. *Politické a diskurzívne príležitosti lokálneho environmentálneho aktivizmu: prípad premien bratislavského ochranárstva [Political and Discursive Opportunities of Local Environmental Activism: The Case of Environmental Protectionism Transformation in Bratislava]*, Sociológia-Slovak Sociological Review, vol. 46, 2014, pp. 60, 65.

26 In the German Democratic Republic (colloquially known as Eastern Germany), peace and environmental groups had been active under the shelter of Lutheran churches since the end of 1970s. In the 1980s, they formed an integral part of a decentralized network structure which enabled the communication between the various groups through workshops, seminars, or lectures. See Wielgohs, Jan. "Comparative Perspectives on Dissent and Opposition to Communist Rule, Dissent and Opposition in Communist Eastern Europe: Origins of Civil Society and Democratic Transition." Edited by Detlef Pollack, pp. 19, 38.

27 Šuška, *supra* note 25 at 66.

28 The Charter 77 was a document which served as an incentive for the establishment of an informal civic initiative in the former Czechoslovakia operating under the same name.

29 Wielgohs, *supra* note 26 at 19, 38.

30 Šuška, *supra* note 25 at 68–69.

principles of environmental law, and was adopted relatively shortly after the fall of the totalitarian regime, and is an example of such development.[31]

Similar dynamics and developments were present also in the newly unified Germany. The revolutionary spirit and zeal of East Germans peacefully conquering the dictatorial Communist regime was translated for example in the push for having environmental protection interests enshrined in the Basic Law (*Grundgesetz*). The unification presented also the opportunity to improve the Basic Law, which was demonstrated for instance by integrating environmental protection as the constitutional goal in Article 20a.[32] Even though the discussion about the role of environmental protection in the constitutional framework had been going on also in the former West Germany before unification, there are valid arguments for stipulating that it was the push by East German revolutionaries that enabled the turn from theoretical discussions into reality.[33]

Paradoxically, the post-revolutionary period is also associated with a gradual weakening of the environmental movement in Czechoslovakia. Interestingly, the former unity of the movement, strengthened through the opposition against the Communist regime, was gradually disrupted in the democratic society for at least two different fundamental reasons. First, environmental activism lost its strength because it lacked its former autocratic opponent when the newly founded Green Party had to compete with a more mainstream party at the time, the *Verejnosť proti násiliu* (Public Against Violence), while both parties provided shelter to the former colleagues and activists of the pre-revolutionary environmental movement.[34] The party then degenerated into a marginal organization, regularly not meeting the threshold for entering the parliament. In addition, the established political process disrupted the focus of former activists, who were now politicians and occupied with other matters that were not related to environmental protection. In the long-term perspective, this led to the gradual weakening of the environmental movement.[35]

Moreover, the topics of environmental protection were in the general discourse soon superseded by other issues such as the economic and political transformation of the country that were at the time perceived as more imminent.[36] Green Parties also disappeared in other countries of Central and Eastern Europe shortly after the initial period of transformation. In Slovakia, the Green

31 Katrlík, Radovan. Zákon o posudzovaní vplyvov na životné prostredie: Komentár (Environmental Impact Assessment Act: Commentary) Wolters Kluwer 2016, p. 16.

32 Jaggi, Stephen. "Revolutionary Reform in German Constitutional Law." *Hastings International & Comparative Law Review*, vol. 41, no. 3, 2018, pp. 173, 179.

33 As noted by Jaggi, some even argue that Art. 20a was not going far enough, since it is formulated "as an objective constitutional goal, not as a subjective individual right." See *Ibid.* at 179–83.

34 Šuška, *supra* note 25 at 60, 70.

35 *Ibid.* at 68–69.

36 *Ibid.* at 71.

Party is marginal.[37] In the Czech Republic or Hungary, the Green Parties are relatively livelier and more capable of meeting the threshold minimum number of votes for entering the legislative bodies. However, their influence remains marginal, especially in the nationalistic climate currently prevalent in the region.[38]

The different historical context explains why so many Eastern European countries relied on the Western market economy to solve environmental problems. In Czechoslovakia, the narrative linked the environmental problems with the critique and failures of the socialist system, while the Western market economy had generally been seen as superior. The environmental activists were lured by the promise that, in a democracy, a market economy would automatically protect the environment.[39] Eastern European countries viewed Western economies as superior and thought they would solve their environmental problems, and therefore adopted them under false impressions. The assimilation of the former German Democratic Republic (East Germany) into the Federal Republic of Germany (West Germany) created such pressure in Germany. In addition, the historically less economically fortunate situation of Eastern European states pressured them into economic structural changes that resulted in damage to the environment. Eastern Germany was economically weaker than its Western counterpart and the pressure to achieve the same standard of living as quickly as possible was clearly present. The transformation processes associated with infrastructure and industrial projects were often detrimental to the environment.[40]

The introduction of market logic into all spheres of society and skepticism toward regularization became the defining traits of the progressive and democratic political spectrum also in the Slovak Republic. Some scholars describe the environmental activists' situation at the time as a kind of schizophrenia: on the one hand they were essentially aligned with the progressive political spectrum, but on the other hand they were in a profound disagreement with respect to certain political–economic questions.[41]

37 The result of the Slovak Green Party (*Strana zelených Slovenska*) in the last elections to the European Parliament was under one percent. See Euro voľby 2019 na Slovensku (*European election 2019 in Slovakia*), *Strana zelených Slovenska*, https://www.vysledkyvolieb.sk/euro-volby/strana/218-strana-zelenych-slovenska.However, Green candidates can be found on the lists of more relevant parties.

38 When we take a look at the broader post-socialist region, covering also Baltic countries, some exceptions from this trend might be mentioned. For instance, the Latvian Green Party in 2004 was even able to install its representative in the office of Prime Minister. See Horvat, Vedran, and Tomislav Tomašević. Green Parties in Europe: Potential Routes of Future Development." *Croatian International Relations Review*, vol. 17, no. 64/65, 2011, pp. 15–16.

39 Šuška, *supra* note 25 at 69–70.

40 Zschiesche, Michael. *Öffentlichkeitsbeteiligung in umweltrelevanten Zulassungsverfahren: Status Quo und Perspektiven*. BWV, Berliner Wiss.-Verl., 2015, p. 117.

41 Šuška, *supra* note 25 at 72 (2014).

The peculiar nature of environmental (and civil) activism formed during the totalitarian rule of the Communist Party, coupled with the rapid post-revolutionary political and economic transformation emphasizing the narrative of free market economy, led also to the blurred understanding of the concept of public interest. While the notion of public interest was originally associated with the interests of the state, gradually it has become associated with the activities of the private sector. The defense of development activities in Bratislava, capital of the Slovak Republic, by the former Minister of Culture in 2005–2006, who argued for the curtailment of the cultural heritage zone in the city center by referring to the public interest, is an example. In his view, public interest was represented by the development activities of the private sector contributing to the development of the city infrastructure.[42]

However, the transformation of former activists into politicians also created permanent or temporary bonds between civil society and political parties or state institutions. For instance, the progressive provisions on access to justice in environmental matters reveal the implementation of proposals made by civil society representatives participating in the working groups preparing the law.[43] The Pezinok landfill case is a genuine example of civic activism in the Slovak Republic and a situation when an environmental issue led to a spontaneous emergence of civic engagement using public pressure and legal tools. Despite the protests by the municipality and the public, the authorities approved the building of the controversial landfill even though it did not meet formal requirements.

Since 1997, data has indicated that residents of Pezinok were at higher risk of cancer in comparison to areas in the rest of the country.[44] The case presented a battle between a municipality and the public, on the one hand, and the state authorities favoring businessmen with ties to political circles standing behind the landfill project, on the other. The legal steps undertaken to stop the landfill were accompanied by a grassroots campaign that attracted significant media coverage.[45] The case came before the Constitutional Court and the Supreme Court. The Supreme Court referred the case to the European Court of Justice for guidance, which the E.C.J. provided. The Supreme Court followed the guidance and ruled in favor of the public and municipality.[46] This outcome is the result of a collaborative effort between the interested public, represented by lawyers and

42 *Ibid.*

43 Vozár, Imrich. "Na Čo Sa v Zozname Úspechov Smeru Zabudlo." *Denník N*, Denník N, 20 June 2015, https://dennikn.sk/163676/na-co-sa-v-zozname-uspechov-smeru-zabudlo/.

44 *Pezinok Case One of the Most Important Cases of Civic Engagement in Slovakia since the Velvet Revolution.* Pezinok Case one of the most important Cases of Civic Engagement in Slovakia since the Velvet Revolution.

45 *Ibid.*

46 Judgment of 15 January 2013, *Križan et al. v. Slovenská inšpekcia životného prostredia*, C-416/10, EU:C:2013:8.

N.G.O.s who fought the case before the courts, and the citizens supporting the cause politically at the grassroots level.[47]

Even though most of the countries of Western, Central, and Eastern Europe are now members of the European Union, the different role and influence of the Green Parties still run along the symbolic Iron Curtain. The countries of Central and Eastern Europe are still in the process of searching for their political identities. The emergence of equivalents of political forces that had formed in the old democracies of Western Europe is met with region-specific obstacles in Central and Eastern Europe.[48] The Green Party in the Czech Republic is an exception. In the years 2006–2010, the Czech Green Party not only entered the parliament, but also became a member of the coalition government.

6.4 The Europeanization of Green Parties

The emergence of the environmental movement in the 1970s, accompanied by the developments at the international level, influenced also the role of the European Union in the environmental field. The increasing power and competences of the European Union in the field of the environment have gone hand in hand with the growing realization of its ability to exert influence not only at the national, but also at the international, level.[49] The Green Parties have been represented in the European Parliament since 1984. In these early elections, the dominance of the Green Party in Germany was translated also in the results at the European Parliament level when seven out of eleven Green M.P.s came from Germany. Hence, it is possible to say that the transformation of the green movement into a political force happened not only at the national, but also at the supranational, level. In the period of 1989–1994, i.e. the years immediately after the fall of the Iron Curtain, Green politicians actively promoted the deepening of ties with the new democracies in Central Europe, which was in line with their policy of supporting greater European integration. As was mentioned above, the different historical and socio–economic developments of the countries of Central Europe in comparison to their Western European counterparts meant that the Green politics faced the challenge of being established and gaining support in the region. This trait might be observed also in the 2004 and 2009 elections of the European Parliament, in which the Green Parties did not win any seats in the new Member States.

Given the fact that, according to estimates, approximately 50–80 percent of Member States' national legislations are made by the European Union, its impact

47 *Pezinok Case One of the Most Important Cases of Civic Engagement in Slovakia since the Velvet Revolution*. Pezinok Case one of the most important Cases of Civic Engagement in Slovakia since the Velvet Revolution.

48 Horvat, *supra* note 38 at 15–16.

49 See Van Zeben, Josephine. "Subsidiarity in European Environmental Law: A Competence Allocation Approach." *Harvard Environmental Law Review*, vol. 38, 2014.

might be even more significant in practice than that of the national parliaments.[50] European legislation in environmental matters is based on the idea of minimum harmonization, which leaves the Member States freedom to adopt more stringent laws going beyond the protection provided by the European Union legislation.[51] The European Commission has a strong mandate in proposing environmental legislation.[52]

Nevertheless, even though the European Parliament is not comparable with legislative bodies at the national level, which usually have exclusive legislative competences, the European Parliament plays an important role in E.U. legislation. Its role has gradually been strengthened in the course of the last thirty years, culminating in the Lisbon Treaty in 2009.[53] As the body representing the interests of the citizens of the European Union, it shares its legislative competences with the Council of the European Union, representing the interests of the E.U. Member States.[54] In fact, the European Parliament was the only body in the institutional framework of the European Union which was able to enhance its impact on environmental planning after the changes introduced by the Lisbon Treaty.[55] The strengthened position of the European Parliament goes hand in hand with the enhanced role of political parties in the law-making process within the European Union. The relationship between the European Parliament and the European Commission resembles the relationship between the government and legislative body at the national level, with the government depending on the support of the legislature and the legislature promoting the government's agenda.[56] Hence, when there is enough support for the green agenda in the European Parliament, there is a good chance that the legislation proposed by the European Commission will be passed. Even though this is often the case, the consensus is the result of negotiating and bargaining between the main political groups in the European Parliament, none of which has a dominant role.[57]

The process of European integration has also influenced the national Green Parties in European Member States, which compete in the European election not only on their own behalf but are also united under the banner of the European Green Party. When elected to the European Parliament, the members of the European Green Party are active in the Greens/EFA political group in the

50 Miller, Vaughne. "How Much Legislation Comes from Europe?" *Commons Library Briefing - UK Parliament*, 13 Oct. 2010, http://researchbriefings.parliament.uk/ResearchBriefing/Summary/RP10-62.

51 Van Zeben, *supra* note 49 at 428–29.

52 "Environment: Basics - Welcome," http://ec.europa.eu/environment/basics/home_en.htm.

53 Shackleton, Michael. "Transforming Representative Democracy in the EU? The Role of the European Parliament." *Journal of European Integration*, vol. 39, no. 2, 2017, pp. 191–92.

54 *Ibid.* at 195.

55 Van Zeben, *supra* note 49 at 430.

56 Shackleton, *supra* note 53 at 198.

57 *Ibid.* at 191.

European Parliament.[58] In this way, the European Greens have a strong opportunity to influence E.U. legislation.[59]

It might be concluded that the environmental movement especially in the Western European countries emerged among the middle class in relatively affluent societies, which had already reached a certain level of economic development. Even though the countries of Central and Eastern Europe were considerably poorer under the leadership of the Communist Parties and in the first years of democratic transition in comparison to their Western European counterparts, the standard of living, for example in Czechoslovakia, was relatively decent.[60]

6.5 Green politics in global South: environmental justice in Ethiopia and India

In the countries of the global South, such as Ethiopia, it is possible to identify a tension between development-related objectives and environmental protection, which manifests itself in the field of law as well as politics. In Ethiopia, it is possible to observe the efforts to reconcile these objectives under the scope of the principle of sustainable development in Article 43 of the Constitution,[61] which states that "[t]he Peoples of Ethiopia as a whole, and each Nation, Nationality and People in Ethiopia in particular have the right to improved living standards and to sustainable development." Further, it states *inter alia* that "Nationals have the right to participate in national development and, in particular, to be consulted with respect to policies and projects affecting their community."

It is also possible to identify the tension between human rights politics and those of environmental and wildlife protection. In Ethiopia, as will be discussed in Chapter 7 on the topic public participation, environmental activism is at its infant stage. The poverty-stricken public is mostly concerned with satisfying its economic needs. Even if there are some environmental NGOs, they are mostly concerned with awareness-creation issues. The existing media are also preoccupied with promoting development issues. Environmental activism for a long time had been considered a prerogative of those whose standard of living enables them to care about the issues beyond the scope of satisfying their economic and social needs. The environmentalism of the poor, coupled with the unique Ghandian

58 The Greens together with members of the European Free Alliance, uniting regionalists and democratic nationalists, formed the Greens/EFA political group in 1999. See "About Greens/EFA." *Greens/EFA*, https://www.greens-efa.eu/en/our-group/about-greens-efa/.

59 Horvat, *supra* note 38 at 15, 17.

60 "The Czechs and Their Communism, Past and Present." *IWM*, 14 Nov. 2013, http://www.iwm.at/publications/5-junior-visiting-fellows-conferences/vol-xvii/muriel-blaive/.

61 Art. 43 of the Constitution of the Federal Democratic Republic of Ethiopia See Constitution of the Federal Democratic Republic of Ethiopia, https://www.wipo.int/edocs/lexdocs/laws/en/et/et007en.pdf (last accessed May 27, 2019). However, it is appropriate to note that sustainable development has been criticized as ambiguous and inadequate in a manner it attempts to reconcile at times often conflicting interests.

legacy of *satyagraha* and developments in the field in international environmental law following the adoption of the Stockholm Declaration in 1972, significantly shaped Indian environmental law and policy, although the poor people were considered "too poor to be green."[62] Practicing lawyer and human rights advocate Kranti L.C., who is the Executive Director of the Human Rights Law Network and Advocate at the Bombay High Court, maintains that the interests of public environmental law generally remain at odds with human rights.[63] At least historically, there have been exceptions to this position, most prominent among which was the Chipko[64] movement, called by some an example of the environmentalism of the poor,[65] and an exception to the tension.[66]

The "*Chipko*" (tree-hugging) movement emerged in 1973 when a forest department in Garhwal Hymalayas allotted ash trees, allegedly for felling, to a private company. As a reaction to that, the local cooperative organization *Dasholi Gram Swarajya Sangha* (Dasholi Society for Village Self-Rule) burned resin and timber depots. When these steps failed to stop the tree-felling, villagers on the initiative of their leader, Chandi Prasad Bhatt, started to embrace or hug the trees in the forest to prevent them from being felled.[67] The Chipko movement quickly spread to various parts across India, from Uttar Pradesh to Himachal Pradesh in the north, Karnataka in the south, Rajasthan in the west, Bihar in the east, and to central India.[68] Gradually, it became a representative movement for a wide range of environmental conflicts in the 1970s and 1980s, ranging from conflicts over forests, fishes, pastures, and water dams to the issues associated with unregulated mining.[69] In 1987, the Chipko movement was awarded the Right Livelihood Award.[70] The foundation awarding the prize noted that the movement contributed to the decision of then-Prime Minister Indira Gandhi[71] to order a fifteen-

62 Chopra, *supra* note 13 at 40.

63 Interview with Kranti LC, Advocate, Bombay High Court, Executive Director, Human Rights Law Network, March 14, 2019. Notes on file with Kirk W. Junker.

64 The word *chipko* means "to embrace" in Hindi. One popular folk song by Raturi referred allegedly also to the embracing the trees as a method for protecting them from felling. See Shiva, Vandana, and J. Bandyopadhyay. "The Evolution, Structure, and Impact of the Chipko Movement." *Mountain Research and Development*, vol. 6, no. 2, 1986, pp. 133, 137.

65 The environmentalism of the poor might be understood as a movement linking the social justice concerns with sustainability. Environmental issues are perceived through the lens of disadvantaged local communities and the impact of the resource use on natural environment and their livelihoods. See Poonia, Malvika. "Reinvestigating the Chipko Movement of 1973." *International Journal of Environmental Sciences*, vol. 6, no. 5, 2016, pp. 837, 838.

66 Chopra, *supra* note 13 at 40.

67 *Ibid.* at 39–40.

68 Shiva, Vandana, and J. Bandyopadhyay. "The Evolution, Structure, and Impact of the Chipko Movement." *Mountain Research and Development*, vol. 6, no. 2, 1986, pp. 133, 136.

69 Poonia, *supra* note 65 at 840.

70 *Ibid.* at 838.

71 In 1980, Indira Gandhi referred to the Chipko movement in the following manner: "Well, frankly, I don't know all the aims of the movement. But if it is that trees should not be cut, I'm all for it ... the cutting of trees has immediately brought havoc because it has increased our

year ban on green felling of the Himalayan forests in 1980, which was later implemented also in the states Himachal Pradesh and Uttaranchal.[72] Even though the Chipko movement did not transform into a political party, it was nevertheless able to shape politics and make a lasting impact thanks to the influence on politicians receptive to environmental and social issues and willing to adopt some of the objectives of the movement as their own.[73]

In 1989, the Chipko movement focused its attention also on the dangers posed by dams in the Himalaya region with a series of hunger strikes, which led to the Save Himalaya movement. This tactic was applied by Sunderlal Bahuguna, one of the leaders of the Chipko movement and Ghandian philosopher and activist, again in 1995 against the Tehri dam project.[74] As a result, the Prime Minister personally committed to conducting a review of the dam. In the years afterwards, Bahuguna continued to lobby for environmental issues, ranging from the receding of Himalayan glaciers, to water scarcity and the enhanced protection of forests.

As mentioned above, similar to the environmental movements in the West, the Chipko movement was not only about environmental problems but tied together a bundle of interrelated issues. Apart from the general social justice perspective, seeing the environment and forest resources as a prerequisite for human survival, the Chipko movement might also be seen in the context of women's empowerment and eco-feminism in India. Interestingly, the authors writing about the Chipko movement largely[75] conclude that village women in the hills and forests

---

drought, it has increased our floods and it has made vast areas much more difficult to live in ... naturally, anybody who lives in a backward country has to be concerned with that [poverty], too." See Poonia, *supra* note 65 at 841.

72 In 2004, the ban was lifted in Himachal Pradesh despite the protests initiated by the Chipko movement. See "The Chipko Movement." *The Right Livelihood Award*, http://www.rightlivelihoodaward.org/laureates/the-chipko-movement/.

73 Indira Gandhi, who sympathized with the causes fought for by the Chipko movement, might be rightly seen as a political leader who ingrained the fundamental environmental awareness in Indian laws and policies. See Rangarajan, Mahesh. "Ideology, the Environment and Policy: Indira Gandhi 33." *India International Centre Quarterly*, vol. 50, 2006, pp. 50–64.

74 Shiva, Vandana, and J. Bandyopadhyay. "The Evolution, Structure, and Impact of the Chipko Movement." *Mountain Research and Development*, vol. 6, no. 2, 1986, p. 133–142. It is not coincidental that Ghandi's politics were interwoven with law, given that he had been a practicing lawyer himself. See DiSalvo, Charles R. *M.K. Gandhi, Attorney at Law the Man before the Mahatma*. University of California Press, 2013.

75 Some authors, such as Jayanta Bandyopadhyay, emphasize that the Chipko movement was not based on gender conflicts, but might be rather seen as a movement, in which the notion of equality prevailed, and women played roles as significant in the movement as their male counterparts did. It cannot be unequivocally concluded that the women acted as opponents of change and bearers of tradition, while the men acted as agents of economic change and development. Moreover, Bandyopadhyay also mentions that, contrary to other sources, there was actually only one reported instance of the use of the method of embracing the trees by a male activist in 1977 and, in most cases, other forms of non-violent resistance were applied. See Bandyopadhyay, Jayanta. "Chipko Movement: Of Floated Myths and Flouted Realities." *Economic and Political Weekly*, vol. 34, no. 15, 1999, pp. 880–82.

were one of the main driving forces of the movement when they hugged and protected the trees with their own bodies.[76] The Chipko movement originally consisted mostly of men, but it was naturally attractive for women. Many of them were the victims of landslides and floods in the Garhwal region and had suffered most from environmental disasters or were responsible for cultivation, livestock, and children. Because of various factors, including male migration for work, women in these areas were aware of the importance of self-reliance. Based on their experiences, they were able to link environmental problems with social and economic issues. As reported, in certain instances a disagreement about the priorities and the course of development in the village between the village men and women stirred. While women argued for the protection of trees, for instance, village men would sacrifice the trees for infrastructure.[77]

The Chipko movement was not a solitary episode in the history of environmental struggles in India. The first example concerns a project to build a dam in the Silent Valley area in the 1970s. The area designated for the project, covering approximately 89.52 km^2 with rich flora and fauna in the remote region of Kerala state, triggered a vocal opposition from ecological activists. While the Chipko movement might be seen both as an environmental and a social movement, the Silent Valley case drew attention almost exclusively from the environmentalists' community. The area in question was unique for its allegedly "uninterrupted evolutionary history of almost fifty million years."[78] The scientific arguments presented by Indian naturalists with the support of European scientists in the *Journal of the Bombay Natural History Society*, coupled with the grassroots action taken by N.G.O.s focusing on raising environmental awareness, such as Kerala Sastra Sahitya Parishad, all led to a state reaction. The taskforce under the auspices of the National Council for Environmental Planning and Conservation conducted a study on ecological and environmental impacts in the pertinent area, which recommended not to commence the dam project. Consequently, this view was adopted also by the government of Kerala. Despite the initial reluctance (spanning over a period of several years) to abandon the dam project by the central government, the project was finally stopped in 1983. The Silent Valley case represents the most prominent example of a significant project sanctioned and promoted by the government that was thereafter abandoned for ecological reasons.[79] Within party political systems, citizens can indirectly interface with the law by voting for law-makers. But politics can also mean disagreeing with, and even fighting, the extant political system already in place. In this second category, it is difficult to evaluate whether citizen politics have a legal impact,

76 Poonia, *supra* note 65 at 838.

77 Jain, Shobita. "Standing up for Trees: Women's Role in the Chipko Movement." *Unasylva – No. 146 Women in Forestry Standing up for Trees: Women's Role in the Chipko Movement*, http://www.fao.org/docrep/r0465e/r0465e03.htm.

78 Chopra, *supra* note 13 at 38.

79 *Ibid.* at 38–39.

but in both of the Chipko examples presented above, Prime Ministers explicitly stated that they were responding to the citizens by implementing a legal tool on the citizens' behalf.

Another example of the politicization of environmental law issues in India is the Vedanta case from Orissa, in which a mine project threatened to endanger the life of local communities and destroy an important wildlife habitat in the area. There were concerns about the non-compliance of the project with environmental laws and the Ministry of the Environment and Forests had denied the project approval. A senior leader of the ruling government party in the state assured local tribal people of his support, since he saw himself as their representative (*sepoy*), but the government elected in the next election, in 2014, took a more pro-industry stance. Contrary to the previous environmental movements, which originated mainly in the 1970s, local people and activists could rely also on the international legal framework applied by the Indian judiciary.[80] The Supreme Court ordered an environmental referendum on the issue in the area on the question of obtaining prior informed consent to the project from local communities, as is required also in the Convention on Biological Diversity. The project was rejected in the referendum.

6.6 Conclusions

The nuances of the interplay between the law and politics are as diverse as the legal cultures across the world. It is possible to observe the differences even between the countries in the global North, which were shaped by profoundly different historical and political developments after the end of World War II. After the fall of the Iron Curtain and subsequent accession of a large part of the Central and Eastern European countries to the European Union, it has been possible to observe a process of convergence between Western European countries and their Central and Eastern European counterparts. Nevertheless, the political identities of these countries still differ to a certain extent. This might be possible to observe on the example of Green Parties and the relevance of green topics for mainstream political parties in the Western European countries on the one hand and Central and Eastern European countries on the other. Still, thanks to their membership in the European Union, the European Union legislation in the field of the environment significantly influences also the legislation—and, as a side-effect, the political discourse—at national levels. With respect to environmental movements and civic engagement in Western European countries and Central and Eastern European countries, the common denominator might be seen in the fact that it is primarily driven by the middle class. This distinguishes it from the situation in the countries of the global South. As shown on the example of India

80 Chopra, *supra* note 13 at 41–42.

earlier in the text, the emergence of environmental movements was very often inextricably linked to grave human rights violations.

These various stories of the interplay between green politics represented either by political parties or environmental and social movements on the one hand and environmental law on the other support Curran's claim that law is not an isolated matter, but is entangled in the web of complex and interrelated societal, historical, and political relations. Hence, when reading the text of environmental laws or judicial decisions, it is wise to be aware of the fact that there might be a story behind them. The laws are often a reaction to certain societal developments. The stories of political, historical, and often personal struggles provide a valuable context for the understanding of the laws and the culture in which they are embedded.

Literature

Amos, Robert. "Reassessing the Role of Plants in Society." *Int. J. Law Context* 13.3 (2017), 295–315.

Baker, Susan, et al. *The Politics of Sustainable Development: Theory, Policy and Practice within the European Union*. Routledge, 1997.

Bandyopadhyay, Jayanta. "Chipko Movement: Of Floated Myths and Flouted Realities." *Econ. Political Weekly* 34.15 (1999), 880–882.

Blaive, Muriel. "The Czechs and Their Communism, Past and Present." *Inst. Hum. Sci.* 25 May 2018. http://www.iwm.at/publications/5-junior-visiting-fellows-conferences/vol-xvii/muriel-blaive/.

Boyle, Alan. "Human Rights and the Environment: Where Next?" *Eur. J. Int. Law* 23.3 (2012), 613–642.

Burchel, John. *The Evolution of Green Politics: Development and Change within European Green Parties*. Earthscan, 2002.

Cerar, Miro. "The Relationship between Law and Politics." *Annu. Survey Int. Comp. Law* 15.1 (2009). 27 May 2019. http://digitalcommons.law.ggu.edu/annlsurvey/vol15/iss1/3.

Chopra, Kanchan. *Development and Environmental Policy in India the Last Few Decades*. Springer, 2017.

Curran, Vivian Grosswald. "Cultural Immersion, Difference and Categories in U.S. Comparative Law." *Am. J. Comp. Law* 46.1 (1998), 43–92.

Horvat, Vedran and Tomašević, Tomislav. "Green Parties in Europe: Potential Routes of Future Development." *Croatian Int. Relat. Rev.* 17.64/65 (2011), 15–29.

Jaggi, Stephen. "Revolutionary Reform in German Constitutional Law." *Hastings Int. Comp. Law Rev.* 41.3 (2018), 173–243.

Jehlička, Peter, et al. "Czech Green Politics After Two Decades: The May 2010 General Election." *Environ. Politics* 20.3 (2011), 418–425.

Junker, Kirk W. "A Focus on Comparison in Comparative Law." *Duquesne Law Rev.* 52 (2014).

Junker, Kirk W. *Legal Culture in the United States: An Introduction*. Routledge, 2016.

Katrlík, Radovan. *Zákon o posudzovaní vplyvov na životné prostredie: Komentár (Environmental Impact Assessment Act: Commentary)*. 1st ed., Wolters Kluwer, 2016.

Kraft, Michael E. *Environmental Policy & Politics*. 7th ed. University of Wisconsin, 2018.

Lawrence, Jennifer. "Managing the Environment: Neoliberal Governmentality in the Anthropocene." *Sustainability and Peaceful Coexistence for the Anthropocene*, Ed. Pasi Heikkurinen. Routledge, 2016, E-Book.

Lele, Sharachandra. "Sustainable Development: A Critical Review." *World Dev.* 19.6 (1997), 607–621.

Lindberg, Björn, et al. "Party Politics as Usual? The Role of Political Parties in EU Legislative Decision-Making." *J. Eur. Public Policy* 15.8 (2008).

Luhmann, Niklas. *Law as a Social System.* Oxford University Press, 2009.

Miller, Vaughne. "How Much Legislation Comes from Europe?" House of Commons Briefing Papers. 2010. 21 May 2018. http://researchbriefings.parliament.uk/ResearchBriefing/Summary/RP10-62.

na Slovensku (*European election 2019 in Slovakia*), *Strana zelených Slovenska*, 2019. 27 May 2019. https://www.vysledkyvolieb.sk/euro-volby/strana/218-strana-zelenych-slovenska

Poonia, Malvika. "Reinvestigating the Chipko Movement of 1973." *International Journal of Environmental Sciences* 6.5 (2016).

Rangarajan, Mahesh. "Ideology, the Environment and Policy: Indira Gandhi." *India Int. Centre Q.* 33.1 (2006).

Rootes, Christopher. "Environmental Movements: From the Local to the Global." *Environ. Politics* 8.1 (1999).

Schmidt, Alexander, et al. *Die Verbandsklage im Umwelt- und Naturschutzrecht.* C. H. Beck, 2014.

Shackleton, Michael. "Transforming Representative Democracy in the EU? The Role of the European Parliament." *J. Eur. Integr* 39.2 (2017).

Siems, Mathias. *Comparative Law.* Cambridge University Press, 2014.

Shiva, Vandana, and J. Bandyopadhyay. "The Evolution, Structure, and Impact of the Chipko Movement." *Mountain Res. Dev.* 6.2 (1986).

Šuška, Pavel. "Politické a diskurzívne príležitosti lokálneho environmentálneho aktivizmu: prípad premien bratislavského ochranárstva [Political and Discursive Opportunities of Local Environmental Activism: The Case of Environmental Protectionism Transformation in Bratislava]." *Sociológia-Slovak Sociological Rev.* 46 (2014).

van Zeben, Josephine. "Subsidiarity in European Environmental Law: A Competence Allocation Approach." *Harvard Environ. Law Rev.* 38 (2014).

Whittington, Keith E., et al. *The Oxford Handbook of Law & Politics.* Oxford Handbooks Online, 2009.

Wielgohs, Jan, and Pollack, Detlef. "Comparative Perspectives on Dissent and Opposition to Communist Rule." *Dissent and Opposition in Communist Eastern Europe*: Origins of Civil Society and Democratic Transition. Eds. Detlef Pollack and Jan Wielgohs. Ashgate Publishing Limited, 2004. 25 May 2018. https://www.kuwi.europa-uni.de/de/lehrstuhl/vs/politik3/Lehre_WS_10_11/Wielgohs-Pollack-Dissent.pdf.

Zschiesche, Michael. *Öffentlichkeitsbeteiligung im umweltrelevanten Zulassungsverfahren.* BWV-Berliner Wissenschaftsverlag, 2015.

PART IV

Persons subject to environmental legislation, implementation, and dispute resolution

7

PUBLIC PARTICIPATION

Mrinalini Shinde, Tsegai Berhane Ghebretekle, and Kirk W. Junker

7.1 Public participation in environmental law

Lawyers can become so entwined in law as their area of special concern that we lose sight of the fact that, unlike perhaps other specialist areas of human study and practice, the law is for all citizens. In this chapter, we compare how, when, and why citizens can directly participate in various legal cultures when it comes to their environmental concerns. What is the function of the citizen in public environmental law? As discussed in Chapter 4, litigation is one of the foremost avenues through which public participation in environmental law is enabled. Seen optimistically, effective public participation functions to interject the non-jurist into the legal process. Too often, when asked what the legal role of the public is, the answer is limited to voting for political representatives, as was discussed in Chapter 6. Legal roles are more direct, and include commenting on proposed legislation, administrative regulation, or permit applications; public hearings on legislation, administrative regulation, or permit applications; and access to government information on permit applications and their review (see Chapter 3). Public participation can also include reviewing the government compliance records for various environmental actors, both individual and corporate, to establish evidence of compliance and the effectiveness of government actions to bring about the same. And finally, if a dispute arises, does the public have the opportunity to sit in judgment of the case, to act as advisors or witnesses to the parties, or to bring actions on their own collective behalf, as "the public?" All of these, and not political voting, are roles for public participation in environmental law. More subtle social scientific investigations can also include determining the existence and level of the public's trust in its government to enact and enforce environmental statutes and regulations as well as issue permits and enforce any of these. Social science may also investigate the extent to which the public accepts the norms announced in various legal instruments. Many citizens of various cultures can point to nicely worded statutes and regulations in their cultures that are never enforced. The violators know that and the environmental victims know that. What then becomes the function of public participation in environmental law?

To focus these potential issues, this chapter compares the interaction between the public in the environmental litigation process across three cultures, through citizen suits in the United States and public interest litigation in India and Ethiopia. These jurisdictions make for notable comparisons given that, with respect to the U.S.A., the legal system is rooted in the British common law tradition, with binding judicial precedent, while India and Ethiopia have mixed systems that include common law, civil law, and local customary law. Moreover, all of these countries have been privy to historical and persistent injustice meted out to indigenous peoples and lower socio–income groups concerning drinking water, making citizen-driven litigation important from environmental, health, and social perspectives. The differences in the size of their economies and their geographic locations in North America, Africa, and Asia allow one to examine how differently or similarly economic factors and motivations affect environmental preservation.

The comparison between citizen suits (C.S.) and public interest litigation (P.I.L.) is of consequence because they are both contained squarely within the judicial sphere, providing an opportunity for the judiciary to assess a situation on merits and relevant statutory provisions, theoretically independent of political influence. In the United States[1] and in India,[2] industrial polluters have been significant contributors to political election campaigns, blurring the line between industry and state, and consequently between regulatory bodies and polluters. The independence of the judiciary is vital in clarifying these blurred lines to enable environmental justice for the public.

As was discussed in Chapter 3, actions for environmental harms have shifted to the state to rebalance legal power between those who caused harm and those who suffered harm. Once that shift was made, although many cultures maintain the legal ability of harmed individuals to take actions against large industries or the state, individuals most often rely upon the state to take action on their behalf.

1 Goldenberg, Suzanne, and Helena Bengtsson. "Oil and Gas Industry Has Pumped Millions into Republican Campaigns." *The Guardian*, Guardian News and Media, 3 Mar. 2016, https://www.theguardian.com/us-news/2016/mar/03/oil-and-gas-industry-has-pumped-millions-into-republican-campaigns.

2 "Sterlite: Vedanta Smelter in India Fined 1bn Rupees for Pollution." *BBC News*, BBC, 2 Apr. 2013, http://www.bbc.com/news/world-asia-india-21999901; "BJP, Congress Accepted Foreign Funding Too, Vedanta Top Donor." *Firstpost*, 20 Dec. 2014, http://www.firstpost.com/business/bjp-congress-accepted-foreign-funding-too-vedanta-top-donor-1330877.html. The news reports reveal how the largest donor to the election campaign for the current Central government in India have also been inducted in water pollution cases, and have had to pay significant damages.

But what if the state fails to do so? In these instances the state has failed to carry out its enforcement function in public law. In these cases, even within the realm of public law, the individual must have some recourse directly against the persons causing harm, the state official who failed to enforce the law, or both. In Chapter 4, we mentioned that, in China, citizens may take legal action against the state official who fails to enforce the law. By comparison, in the U.S.A., several states give the power back to the citizen to directly prosecute the harmful actor under the public law statute when the state fails to do so.

7.2 Public interest litigation in India

In India, Public Interest Litigation (P.I.L.) was a litigation mechanism introduced by the judiciary, and not the legislature. A P.I.L. can be heard by the Supreme Court[3] or the High Courts[4] as a writ petition, since, under the Constitution of India, High Courts and the Supreme Court can exercise original jurisdiction, and hear cases in the first instance.[5] The Supreme Court has issued guidelines[6] under which "petitions pertaining to environmental pollution" are included among the petitions eligible to be heard as P.I.L.[7] P.I.L. is combined with epistolary jurisdiction, wherein judges can take *suo moto* actions in the event of a cause of action being brought to their attention through a simple letter, in order to minimize the inherent procedural and financial challenges faced by the poorest public when approaching the courts.[8]

The scope of P.I.L. was built on the expansion of *locus standi*, with the Supreme Court (in a matter relating to judicial transfers)[9] deciding that lawyers who demonstrated an interest in the maintenance of an independent judiciary were interested parties who had *locus standi* to bring the petition, irrespective of whether they were directly affected parties.[10] The Court went on to establish that a:

> member of the public having sufficient interest can maintain an action for judicial redress for public injury arising from breach of public duty or from violation of some provision of the Constitution or the law and seek enforcement of such public duty and observance of such constitutional or legal provision.

3 Art. 32, Constitution of India.
4 Art. 226, Constitution of India.
5 Shinde, Mrinalini. "Environmental Law of India." *Comparative Environmental Law and Regulation*, 2018.
6 Supreme Court of India, *Compilation of Guidelines to be Followed for Entertaining Letters/Petitions Received in this Court as Public Interest Litigation*, http://supremecourtofindia.nic.in/circular/guidelines/pilguidelines.pdf.
7 Shinde, *supra* note 5.
8 Bandhua Mukti Morcha v. Union of India, 1984 AIR 802.
9 S.P. Gupta v. Union of India (1982) 2 S.C.R. 365.
10 Shinde, *supra* note 5.

In subsequent decisions, the Supreme Court held that P.I.L. played an important role in ensuring social audit.[11] In *L.K. Koolwal v. State of Rajasthan,*[12] the High Court of Rajasthan, while giving directions to improve sanitation in the city of Jaipur, explained that Article 51A(g), under which protection of the environment is a fundamental duty of citizens, also consequently gives citizens the right to approach the court in carrying out that duty. In *Subhash Kumar v. State of Bihar,*[13] the Supreme Court went on to include the "right to enjoyment of pollution free water and air," as part of the right to life under Article 21. Subsequently, the right to clean drinking water was explicitly included as part of Article 21 in *A.P. Pollution Control Board II v. M V Nayudu.*[14]

The evolving nature of the P.I.L. has resulted in a phenomenon one could describe as "living goals." The boundaries and extent of what it seeks to achieve are determined by the facts before the Court, all the while upholding constitutional and fundamental rights, and ensuring access to justice for the all citizens irrespective of their financial ability to bring suits. The right to clean water and the right to clean drinking water have become law through precedent and, therefore, the goals sought by the Water Act are indirectly also goals sought to be achieved by P.I.L.

The consequence of vesting the function of environmental protection in the judiciary through P.I.L. can also be problematic. When the Court functions as a quasi-administrative agency, there is an inability to ensure direct enforcement on the ground, after the judgment and directions have been given. After a judgment is passed, the responsibility shifts to the administrative agencies such as the Central Pollution Control Board for implementation. On several occasions, there seems to have been a disparity between what the judgment seeks to achieve and the motivations of agencies in enforcing the same. There is in fact the possibility for the agencies to file a contempt petition against defaulters for non-compliance, but this is rarely done.

For example, the data from the registers of the Andhra Pradesh High Court between 1990 and 1999[15] reveal that in all environmental P.I.L.s (196 in total) disposed of by the Andhra Pradesh High Court, forty-three were cases relating to water (a sum of cases limited to water and those including both air and water) in which the decision was in favor of the plaintiff; four were cases relating to water in which the complaint was withdrawn; four were cases relating to water in which the complaint was made in favor of the defendant; and eleven were cases dismissed on technical grounds. It is understood that a P.I.L. decision in favor of a plaintiff who has filed the case in the public interest is consequently also in the

11 Fertilizer Corporation Kamgar Union and Ors. v. Union of India and Ors. 1981 AIR 344.

12 AIR 1988 Raj 2.

13 AIR 1991 SC 420.

14 (2001) 2 S.C.C. 62.

15 Prasad, P.M. "Environmental Protection: The Role of Liability System in India." *Eco. & Pol. Wkly.*, vol. 39, 2004, pp. 257, 263.

public interest. However, the opposite must also be true, because precedent and the legislative interpretation dictate that a decision in favor of the defendant is also in public interest, as the Court has been convinced that the public interest was better served by dismissing the petition. Similar to the calculation made for the U.S.A. as a Type I comparison, it is appropriate to judge the ability of the P.I.L. device to enforce the law through the number of decision outcomes, rather than just outcomes in favor of the plaintiff. Therefore, out of the sixty-two water-related environmental P.I.L.s, forty-seven were decided in favor of the plaintiff or defendant, without being withdrawn or dismissed on technical grounds, making it 75.80 percent successful in its ability to enforce water quality standards.

In response to a media report regarding pollution in the river Yamuna, a P.I.L. was filed in 1994 entitled *In Re: News Item Published in Hindustan Times Titled 'And Quiet Flows the Maily Yamuna.*[16] The claim involved the issuance of several interim orders giving directions to responsible authorities to monitor and prevent the pollution.[17] The standards for water quality in a drinking source set by the Central Pollution Control Board (C.P.C.B.), the highest central authority responsible for monitoring and ensuring water quality standards, include several criteria. One standard requires that drinking water sources without conventional treatment but with disinfection should have a maximum B.O.D. level of 2 mg/liter. B.O.D. is the biological oxygen demand: the oxygen used by living organisms inside the water; a high demand indicates the presence of contaminants such as nitrates and organic matter.[18] Research conducted by Misra[19] revealed that B.O.D. levels across the River Yamuna in 2009 had a maximum B.O.D. level of 51 mg/liter in the Palla area in New Delhi. Research conducted by Parmar and Singh showed that B.O.D. levels in New Delhi reached a maximum of 58.2 mg/liter by 2014.[20] Malik et al. claim[21] that between 1980 (before the institution of the P.I.L. in 1994) and 2005 (over a decade after the case was first heard) there was a 2.5 times increase in B.O.D. levels. A 250 percent increase in B.O.D. levels across a time period in approximately the middle of which the case was instituted is evidence of an alarmingly negative trend in the enforcement of standards by the P.I.L. as a tool.

7.3 Public interest litigation in Ethiopia

The fundamental environmental right included in the Federal Democratic Republic of Ethiopia (F.D.R.E.) Constitution is the right to a healthy and

16 W.P. (Civil) No. 725 OF 1994, Supreme Court of India.

17 Shinde, *supra* note 5.

18 Misra, Anil Kumar. "A River about to Die: Yamuna." *J. of Water Resource and Protection*, vol. 02, no. 05, 2010, pp. 489, 492.

19 *Ibid.*

20 Parmar, Shobhika. "Water Quality Parameters of River Yamuna in Delhi after 20 Years of the Yamuna Action Plan." *Int'l J. Sci Nature*, vol. 6, 2015, pp. 662, 665.

21 Malik, Darshan. "Heavy Metal Pollution of the Yamuna River: An Introspection." *Int'l J. Current Microbiology Applied Sci.*, vol. 3, 2014, pp. 856, 858.

clean environment.[22] The right to a healthy and clean environment is part of the democratic rights enshrined in the Constitution.[23] This right is expected to be enforced by persons bringing legal actions pursuant to Article 37 of the Constitution, which deals with access to justice, and through other subsidiary legislation like the Pollution Control Proclamation,[24] the Environment Impact Assessment Proclamation,[25] and the Solid Waste Management Proclamation.[26] Proclamations in Ethiopian law function as legislation.

However, the Pollution Control Proclamation, which has the power of legislation, is one of the core laws that aim to safeguard the right to a clean and healthy environment. Similar to the Indian Constitution, the Proclamation in its preamble declares "the protection of the environment, in general, and the safeguarding of human health and well-being, as well as the maintaining of the biota and aesthetic value of nature, to be the duty and responsibility of all."[27] It is in this Proclamation that Public Interest Litigation is expressly allowed.

In Ethiopia, the concept of P.I.L. was inferred from the cumulative reading of Articles 37 and 9 (2) of the Constitution. Article 37 of the constitution states that, "Everyone has the right to bring a justiciable matter to, and to obtain a decision or judgment by, a court of law or any other competent body with judicial power;" and:

> the decision or judgment referred to under sub-article 1 of this Article may also be sought by: (a) [a]ny association representing the collective or individual interest of its members; or (b) [a]ny group or person who is a member of, or represents a group with similar interests.[28]

Article 9 (2) of the Constitution reads as: "All citizens, organs of state, political organizations, other associations as well their officials have the duty to ensure observance of the Constitution and to obey it."[29] Article 37 (1) of the F.D.R.E. Constitution establishes the right of access to justice to everyone whereby each person can bring a justiciable matter to a court of law and obtain a decision or judgment from the court of law. Specially, Article 37 (2) (b) enables any person (be it a natural or juridical person) to bring an action by representing a group with

22 FDRE, 1995.The Constitution of the Federal Democratic Republic of Ethiopia, Proclamation No. 1/1987, Federal Negarit Gazeta, 1st year No. 1, Addis Ababa, Art. 44.

23 *Ibid.*

24 FDRE, 2002. Environmental Pollution Control Proclamation (EPCP), No. 300/2002, Federal Negarit Gazeta, 9th year No. 12, Addis Ababa.

25 Environmental Impact Assessment Proclamation(EIAP), No. 299/2002, Federal Negarit Gazeta, 9th year No. 11, Addis Ababa.

26 FDRE, 2007.Solid Waste Management Proclamation No. 513/2007, Federal Negarit Gazeta, 13th Year No.13, Addis Ababa.

27 EPCP.

28 FDRE Constitution.

29 *Ibid.*

similar interest. In this provision, the groups to be represented could be residents of a particular area that are affected by environmental disaster. The rationale is that the affected group might not be able to act in its own interest due to lack of financial ability or other constraints. Hence, the Constitutional provision enables the affected person or group of persons to be represented in the courts of law and accordingly seek remedies.

As indicated above, in the Environmental Pollution Control Proclamation, unlike the Civil Procedure Code,[30] there is no requirement for vested interest to initiate a case. It clearly states that "any person shall have, without the need to show any vested interest, the right to lodge a complaint at the Authority or the relevant regional environmental agency against any person allegedly causing actual or potential damage to the environment," and:

> when the Authority or regional environmental agency fails to give a decision within thirty days or when the person who has lodged the complaint is dissatisfied with the decision, he may institute a court case within sixty days from the date the decision was given or the deadline for decision has elapsed.[31]

From the reading of the provision one can infer that any person, natural or juridical, can lodge a complaint at the Authority[32] or the relevant regional environmental body[33] before instituting a court case. Thus, an exhaustion of remedy is part of the very provision. The concept of the exhaustion of remedy was introduced with the thinking that, unlike ordinary courts of law, environment regulatory bodies at the federal or regional level are technically capable of assessing damage and taking immediate measure. However, to make this provision truly effective, there is a need to further codify its procedure. Moreover, the provision in the pollution control proclamation must also apply to other serious environmental matters. P.I.L. needs to be accompanied by integration of environmental protection in judicial interpretation of procedural and substantive laws, in line with the five-year Growth and Transformation Plan.[34] The Plan also promises to promote awareness among the public about constitutional rights.[35]

30 Civil Procedure code of the Empire of Ethiopia, (1965). Decree No. 52, Negarit Gazeta Extraordinary issue No.3, Addis Ababa, 8th October 1965, Art.33.

31 EPCP Art. 11.

32 Recently, the environmental regulatory body at the federal level has been established as a Commission for Environment, Forest, and Climate Change. However, when we trace its genesis, it was first established as a Ministry, then degraded into an Authority, then again upgraded into a Ministry, and currently it is also degraded into a Commission.

33 However, it is good to note that currently in Ethiopia there is no uniform establishment of regional environmental bodies. Rather, some are established as Bureaus, some as Authorities, and others as Agencies.

34 FDRE Ministry of Finance and Economic Development (2010) "Growth and Transformation Plan (GTP) 2010/11-2014/15," vol. 1, p. 27.

35 *Ibid.*

The case of *APAP v. FEPA*[36] is the pioneer P.I.L. in Ethiopia. A.P.A.P., a social organization, lodged a complaint with F.E.P.A. about the pollution of the Akaki and Mojo Rivers in Addis Ababa caused by the discharge of industrial effluents contrary to the Environmental Pollution Control Proclamation, and urged F.E.P.A. to take immediate measures. F.E.P.A. responded that the environmental standards were still in draft form and awaiting the Environmental Council's approval and could therefore not be relied upon to decide whether the rivers were being polluted. F.E.P.A. also asserted that it had taken all necessary measures to combat the pollution.

A.P.A.P. was dissatisfied with this and initiated a case in the Federal First Instance Court against F.E.P.A. It requested the court to order F.E.P.A. to take administrative and legal measures to stop the alleged pollution, to direct F.E.P.A. to clean the rivers, and to employ environmental inspectors to ensure that F.E.P.A. takes the necessary measures. The court ruled that Article 11 of the Environmental Pollution Control Proclamation does not empower A.P.A.P. to sue F.E.P.A. but does empower it to sue the polluter. The F.E.P.A. had filed a preliminary objection, raising the issue of *locus standi*, and the court sustained the objection, thereby dismissing F.E.P.A. from the action without hearing arguments on the merits concerning F.E.P.A.'s liability. The Federal Higher Court including the Cassation Bench of the Federal Supreme Court (which is the highest court in the country) subsequently endorsed the Federal First Instance court's ruling.

The A.P.A.P. case illustrates the fact that public interest litigation as envisaged in Article 11 of the Pollution Control Proclamation can only be initiated where the specific claim benefits a public interest purpose. It is nearly impossible to quantify the success of P.I.L. versus citizen suits in terms of environmental effects, unlike the straightforward enactment of a statute and observed results of its implementation. This is due to the inherent nature of citizens' litigation wherein citizens or persons are driven to the judiciary to ensure that federal statutes are safeguarded. The separation of powers doctrine would dictate that the goals of the legislature do not necessarily dictate the goals of the judiciary, and, more often than not, the job of the judiciary is to keep a check on the government in environmental disputes. Statutes and constitutional provisions are some of the several legal mechanisms to support environmental safety and quality. Therefore, it would be presumptuous to attribute any rise or fall in water quality to citizens' litigation alone while making such a comparison. However, since these litigation tools are significant components of the environmental law enforcement stage, their direct effect on giving teeth to legislation cannot be ignored, and their correlation to the success of that legislation must be evaluated. No legislation is introduced with the goal of it failing to be implemented by administrative or

36 *Action for Professionals' Association for the People (APAP) v. Federal Environmental Protection Authority (FEPA),* (October 31, 2006).Civil File No. 64902, Federal First Instance Court.

judicial channels. Litigation can act as a viable option for citizens to pursue when the ideal process of implementation fails in practice. The comparison to make asks how well they fulfill that role of secondary implementation.

7.4 Citizen suits in the United States

In the U.S.A., a person with concerns regarding lawmaking, even from abroad, can keep an eye on new federal and state regulations online as they are announced. The environmental law that lawyers use most in the federal system is the administrative *Code of Federal Regulations* (C.F.R.), the electronic version of which is available at http://www.ecfr.gov/cgi-bin/ECFR?page=browse. All E.P.A. regulations are found in Title 40, Chapter I of the C.F.R. The Administrative Procedure Act (A.P.A.), a federal statute, requires that when an agency such as the E.P.A. develops a regulation, it must publish two documents in the *Federal Register*.[37] The *Federal Register* is the daily journal of the U.S. government, which anyone can read at https://www.federalregister.gov/. It includes legal notices, proposed rules, and many other documents.

First, the E.P.A. (or any other agency that proposes a regulation) must publish a proposed new regulation. Other agencies who could propose relevant regulations, in addition to the E.P.A., are the U.S. Fish and Wildlife Service, the Department of the Interior, and even the U.S. Army Corps of Engineers. But more importantly, a lawyer must remember that most areas of environmental compliance are delegated to the governments[38] of states, not the federation.

It is, however, through these notices in the *Federal Register* (or its state equivalent) that environmental lawyers from anywhere in the world can be on notice that their clients in the U.S. might have new obligations or liabilities coming. After the publication of the proposed new regulation occurs, the public then has sixty days to comment in writing on the proposed regulation. The public often includes environmental organizations, non-governmental organizations, industries, trade groups, and citizen groups. Although it is possible, it is less often the case that an individual citizen would know, or spend the time or energy, to monitor proposed new regulations. Moreover, it is usually only affected industry groups or organized citizen groups who are aware of the regulatory comment process just described and can make the types of comments that can change regulators' positions between proposed and final regulations. But a citizen's chance to comment on new regulations is an extremely important opportunity in the legal system.

37 Administrative Procedure Act, 5 U.S.C. § 553 (2006).

38 When the word "government" is used in the U.S.A., it does not mean what "*Regierung*" typically means in Germany, nor even what "government" does in the U.K. For the U.S.A., due in part to the fifty states being called "states," the federation is usually called the "U.S. Government," in reference to all parts of federal administration, not just the political party in control of the legislature, as it would be in the U.K.

Second, the E.P.A. (or any other agency that proposes a regulation) must publish the final regulation in the *Federal Register*, and with the final regulation the agency must address, in writing, each of the comments that it received in writing to its proposed regulation. If the final regulations on their face do not provide clear guidance for the public, then one must return to the *Federal Register* and examine the explanations for the proposed rule and the responses to the public comments.

Consistent with the description of dual federalism in Chapters 3 and 4, all that has been outlined above for the federal system is in place for the U.S. states as well. Pennsylvania, for example, has the Pennsylvania Code, containing 246 chapters or "titles" of administrative regulations. Environmental protection is found in Title 25.[39] As a state equivalent to the *Federal Register*, the *Pennsylvania Bulletin* publishes proposed legislation, consent decrees of case settlement, and other administrative actions in Pennsylvania.[40]

In addition to the public having a role to play in law-making, citizen suits are a legal tool through which the public has a role to play in implementation and adjudication of environmental law. The following two statutes with a mandate for maintaining water quality provide examples: the U.S. Federal Water Pollution Control Act (C.W.A.)[41] and the U.S. Safe Drinking Water Act (S.D.W.A.)[42] The C.W.A. is intended *inter alia* to prohibit water pollution, for human life and biodiversity, while the goal of the S.D.W.A. is to protect public health by creating regulations and standards for drinking water supply, including the protection of water sources such as lakes, rivers, and wells.[43]

Sections 101 and 102 of the Clean Water Act (C.W.A.)[44] state the goals and policy that the statute seeks to achieve. Apart from the multiple administrative and organizational goals, a quantifiable goal of the C.W.A. was eliminating the discharge of all pollutants into navigable waters by 1985.[45] The provision for C.S. under Section 505[46] provides for private enforcement of violations of the C.W.A. The pollution control and effluent standards enacted by Congress under the C.W.A. are primarily enforced by the states, but final authority lies with the E.P.A. administrator.[47] C.S. under Section 505 of the C.W.A. grants citizens the ability to sue for public rights and not private injury, against a party violating effluent standards or administrative orders, or the states. The citizen is required

39 See "Title 25." *Pennsylvania Code*, http://www.pacode.com/secure/data/025/025toc.html.
40 See *Pennsylvania Bulletin Online*, http://www.pabulletin.com/index.asp.
41 33 U.S.C. 1251.
42 42 U.S.C.A. §§ 300f et seq.
43 United States, Congress. "Understanding the Safe Drinking Water Act." *Understanding the Safe Drinking Water Act*, EPA 816-F-04-03, 2004.
44 33 U.S.C. 1251.
45 33 U.S.C. 1251 § 101(a)(i) et seq.
46 33 U.S.C. 1251 § 505.
47 McKinney, Carie. "Statute of Limitation for Citizen Suits Under the Clean Water Act." *Cornell L. Rev.*, vol. 72, 1987, pp. 195, 197.

to give sixty days advance notice to the violating party, the E.P.A., and the State before filing the suit. This period allows the government to take enforcement actions, thus preventing immediate involvement of the judiciary.[48]

The Safe Drinking Water Act[49] (S.D.W.A.) provides for the setting of minimum standards for the management of all public water systems within the United States, and providing for the creation of regulations under it. Under the S.D.W.A., the provision for C.S. is referred to as "citizens' civil actions."[50] The S.W.D.A. provision is similar to the C.S. provision under the C.W.A. in allowing the civil action only if the state is not "diligently prosecuting" a civil action, and, if such an action has already commenced, granting citizens the right to intervene in said action.[51]

In 2009, federal data revealed that, of the 54,700 water systems in the U.S. at that time, over 20 percent were in violation of key provisions of the S.D.W.A.[52] In 2015, according to a report by the Natural Resource Defense Council, 1,110 community water utilities providing water to around 3.9 million people in the U.S.A. contained lead exceeding the prescribed levels. The same report states that, in 2015, there were 8,000 violations with respect to water suppliers, and the E.P.A. took enforcement actions against 11%, with penalties sought for 3%.[53]

The C.S. provisions have been successful in the C.W.A. and S.D.W.A. due to the ability of citizens to collect samples from water sources and to have them tested without requiring them to trespass on private property to make a strong case under the C.S. provisions.[54] By comparison, and through the U.S. system of cooperative federalism, it is even simpler for citizens in Pennsylvania, for example, to monitor whether a permitted facility is in compliance with the Pennsylvania Clean Streams Law by checking online or going to a local state office of the Pennsylvania Department of Environmental Protection, and, under a citizen's Right to Know statutory request, reviewing the Discharge Monitoring Reports (D.M.R.s) filed by the National Pollution Discharge Elimination System (N.P.D.E.S.) permit holder. Those D.M.R.s indicate the permitted limits of certain pollutants; the actual values measured at the permitted outfalls; and whether the value indicates a violation of the permit, the Safe Drinking Water Act, or both. In this example, a citizen need not even take his or her own samples.

48 *Ibid.* at 198-201.

49 42 U.S.C.A. §§ 300f et seq.

50 42 U.S.C.A. § 300j-8.

51 42 U.S.C.A. § 300j-8(b)(1)(B).

52 Duhigg, Charles. "Millions in U.S. Drink Contaminated Water, Records Show." *The New York Times*, The New York Times, 7 Dec. 2009, http://www.nytimes.com/2009/12/08/business/energy-environment/08water.html.

53 United States, Congress, Olson, Erik, and Kristi Fedinick. "What's in your water? Flint and Beyond." R: 16-06-A, 2016.

54 Coplan, Karl. "Citizen Litigants Citizen Regulators: Four Cases Where Citizen Suits Drove Development of Clean Water Law." *Colo. Nat. Resources Energy & Envtl. L. Rev.*, vol. 25, 2014, pp. 61, 70.

The legislative intent behind introducing C.S. in the U.S. state and federal systems was to provide private citizens with the ability to bring actions against the government or private parties, in cases of environmental pollution, similar to being a private prosecutor rather than just a victim. However, conservative interpretation of the separation of powers resulted in the Supreme Court significantly narrowing the scope of standing in C.S., limiting it to parties whose interests are affected.[55] During the floor debates preceding the Clean Air Amendments in the 1970s, there was considerable opposition to C.S. provisions on the grounds that it would open the floodgates for frivolous litigation, and there was also considerable support for them on the grounds that enforcement agencies lacked resources and motivation to successfully enforce the statute without citizen intervention.[56]

The legislative history of the C.W.A. reveals that, along with protecting the quality of the environment, a goal of the Act was to ensure public participation in the enforcement of its provisions[57] and to expand the access of citizens to the courts while enforcing the statute.[58] The C.S. provision ensures a check on enforcement agencies,[59] with the possibility of litigation if there are inadequacies or discrepancies in their non-discretionary functions—to "stir slumbering agencies."[60]

Citizen suits are intended to bring enforcement rather than be seen as compensatory mechanisms based on damages as would a private tort case, although civil penalties are possible under the C.W.A. A hybrid type of action exists in which citizens assert their rights as citizens to personal protection from pollution, while primarily seeking remedies akin to the government.[61] In addition, since the C.W.A. authorizes the imposition of civil penalties up to $25,000 per day of violation, the action can be both for enforcement and deterrence.[62] From an intended safety valve for the efficacy of the C.W.A, Section 505 has progressed to often being used as a judicial enforcement mechanism.[63]

55 Kellner, Martin. "Citizen Participation in Environmental Law Enforcement in Nicaragua: A Comparative Study of Nicaraguan, United States and German Environmental Law." *L. And Pol. in Afr. Asia and Latin Am.*, vol. 42, 2009, pp. 385, 391–93.

56 Dunn, Daniel. "Environmental Citizen Suits Against Natural Resource Companies." *Nat. Resources & Env't.*, vol. 17, 2003, p. 161.

57 S.REP. No. 414, 92d Cong., 1st Sess. 80 (1971).

58 Natural Resources Defense Council, Inc. v. Train, 510 F.2d 692, 700 (D.C. Cir. 1975).

59 S. Rep. No. 1196, 91st Cong., 2d Sess. 35-36 (1970).

60 Friends of the Earth v. Carey, 535 F.2d 165, 173 (2d Cir. 1976).

61 Hecker, James. "The Citizen's Role in Environmental Enforcement: Private Attorney General, Private Citizen, or Both?" *Nat. Resources & Env't.*, vol. 8, 1994, p. 31.

62 Thompson, James. "Citizen Suits and Civil Penalties under the Clean Water Act." *Michigan L. Rev.*, vol. 85, 1987, pp. 1656, 57.

63 Abell, Charles. "Ignoring the Trees for the Forests: How the Citizen Suit Provision of the Clean Water Act Violates the Constitution's Separation of Powers Principle." *Va. L. Rev.*, vol. 81, 1995, pp. 1957, 77.

Another goal of C.S. is to ensure that "public interest" and not just private interests are litigated. A federal appeals court[64] further expanded the notion of public benefit in a C.S., stating that instead of focusing on the outcome of the litigation alone it was crucial to consider whether the suit was of the type that Congress had "intended to encourage," which was "a prudent and desirable effort to achieve an unfulfilled objective of the Act."[65] From judicial interpretation and legislative history, it appears that the core goals of the C.S. provisions in the C.W.A. and S.D.W.A. are essentially enforcement of water quality standards and prevention of pollution, which includes deterrence of future violations and safeguarding public interest.

The contamination of drinking water in the city of Flint, Michigan in the United States has been central to the discussion on safe drinking water in the United States in recent years, garnering global media attention and generating renewed efforts by civil society and the administrative agencies to reinforce the goals of the C.W.A. and the S.D.W.A. The city water supply had suffered from instances of bacterial contamination in 2014, and the consequent chlorine treatment resulted in the water becoming increasingly corrosive. However, the high level of lead in the drinking water was at the center of the case, resulting in citizens filing a federal citizen suit, joined by the Natural Resources Defense Council and the American Civil Liberties Union. The C.S. was a concrete attempt by civil society to enforce standards of water quality, especially relating to lead. In the case, *Concerned Pastors for Social Action et al. v. Khouri et al.*,[66] the plaintiffs alleged past and continuing harm from contaminated drinking water containing high levels of lead, including violation of the Lead and Copper Rule[67] (L.C.R.) pursuant to the S.D.W.A., and related obligations and implementing regulations. The Court denied the motion to dismiss by the defendants, maintaining that the plaintiffs had satisfied the requirement for standing under the S.D.W.A.

The L.C.R. is a federal regulation pursuant to the S.D.W.A. which relates particularly to lead and copper levels through water plumbing systems and not pollution at the source of water itself, but nevertheless is an integral component of contamination in drinking water. The L.C.R. requires water management authorities to take additional actions to combat lead contamination through plumbing corrosion, and the levels at which action must be taken (called the "action level") for lead is 15 p.p.b in more than ten percent of sampled households.

The citizen suit seeking enforcement of S.D.W.A. obligations was filed in January, 2016, and the judgment denying the defendants' motion to dismiss was

64 *Metropolitan Washington Coalition for Clean Air v. District of Columbia*, 639 F.2d 802 (D.C. Cir. 1981). In *Sierra Club v. Gorsuch*, 672 F.2d 33, 36 (D.C. Cir.), the Court stated that it was important to assess whether the plaintiffs in the C.S. had served the goals of the Act.

65 Schwartz, Richard. "Citizen Suits Against Private Industry Under the Clean Water Act." *Nat. Res. Lawyer*, 17(3), 1984, pp. 327, 369–70.

66 E.D. Michi., Case No. 16-10277.

67 40 CFR Part 141 Subpart I.

made by the Court in July, 2016. The E.P.A. data showed a 65 percent decrease in action levels (15 ppb) of lead in drinking water between March and November 2016. Using this data, when one evaluates the ability of the C.S. provision to enforce the goals of the L.C.R. pursuant to the S.W.D.A., although there might not be a direct causal correlation between the C.S. decision in favor of the plaintiffs and the decrease in action levels of lead contamination in the drinking water, one sees that they did occur consecutively and the causal effect cannot be ruled out, considering how the language of the judgment reiterated the obligation on behalf of the state and water utility systems management to enforce S.W.D.A. standards. The 65 percent decrease after a few months following the judgment therefore strongly suggests the enforcement efficiency of the C.S. provision.

The C.W.A. mandates that the filing of a citizen suit be preceded by the dispatch of a notice of intent to sue to the enforcement agencies and to the polluter, sixty days prior to the filing. A report by the Office of Enforcement in California revealed that sixty notices to sue were received under the Clean Water Act in California between March 2009 and June 2010, with information on 70 percent of the cases, making it a total of 42 percent. Of those 42, 18 were settled, 21 were involved in litigation or negotiation and 3 were not pursued.[68] This translates to a minimum of 92.85 percent of total notices in that period being successful enforcement devices by initiating legal action, irrespective of the outcome of those cases.

7.5 Comparisons across cultures

A perfunctory statistical comparison between C.S. and P.I.L. in the U.S.A. and India, respectively, using available data leads to potential conclusions about how these tools drive enforcement of environmental standards. The goal examined for Flint was action levels of lead, while the goal examined for the Yamuna River in India was the B.O.D. level in the river water. In Flint, because of the 65 percent decrease in action levels for lead in the months following the C.S. judgment, we can say that there is positive trend in action. On the other hand, there being a nearly 250 times increase in B.O.D. levels in the Yamuna from 1980 to 2005, with maximum B.O.D. levels recorded being twenty-five times the prescribed water quality standard, is testament to the fact that the case was not able to steer the course of the river as it sought to do by bringing the *suo moto* action in 1994. The judges faced conflicting data on pollution from enforcement agencies like the State Pollution Control Boards (S.P.C.B.s), who as the enforcement agencies are the defendants in these cases, while simultaneously serving the

68 Sato, Reed. "Citizen Suit Enforcement Under the Federal Clean Water Act A Snapshot of the California Experience Based on Notices of Intent to Sue March 2009 through June 2010." *Citizensuit Enforcement Under the Federal Clean Water Act A Snapshot of the California Experience Based on Notices of Intent to Sue March 2009 through June 2010*, 2011.

role of scientific experts since they are the government sources of data driving the judgment. The fact that the matter was taken up as an application before the N.G.T.,[69] and that its judgment in 2015 required all enforcement agencies and concerned government bodies to coordinate and work on achieving the "*Maily Se Nirmal Yamuna*" ("from a filthy to a clean Yamuna") by 2017, bodes well that the data after 2017 will reveal positive information about the ability of P.I.L. to enforce water quality standards. The Yamuna case illustrates the involvement of the news media in aiding public interest litigation in India, a phenomenon that can be observed in Ethiopia as well.

Due to the imbalance of powers between the public and the state or private companies, there might be a possibility that litigants are being compelled to withdraw complaints under peer pressure in India, while the dismissal on technical grounds could be an indication of inadequate and low-quality legal representation or awareness; but one can only make conjecture at this point. However, it would be safe to say that both C.S. and P.I.L. have expanded the ability of citizens to keep a check on government agencies, through the development of standing, which was the original intention with which they were introduced in both countries.

As seen in India, we can observe that, in Ethiopia, P.I.L.'s success also depends on the level of public awareness of the issue brought before the court. If environmental issues are to be protected through the means of P.I.L., the public must have awareness. The Ethiopian society is mainly unaware of many of the rights stated in the Constitution or international instruments.[70] Lack of education is one reason. The limited number of civil societies working on environmental issues is another reason. Many industries in Ethiopia are found in the peripheries and the immediate victims live in the suburbs or rural areas possibly with no knowledge of their rights, let alone how they may exercise them. Furthermore, in Ethiopia, there is no systemic dissemination of environmental information to the public.[71] Access to legal information is limited even for the judges. There are no adequate libraries; decisions are not published, except those of the Federal Supreme Court Cassation Division; legal documents are not disseminated and when they are, it is not done fast enough.[72] Judges handle more court cases than required, which challenges the quality of decisions.[73] The media also plays a major role in disseminating information about environmental concerns. It can influence the choices private institutions make. More importantly, it helps consolidate the position

69 *Manoj Mishra v. Union of India and Ors*, National Green Tribunal, (Principal Bench) Original Application No. 6 of 2012.

70 FDRE Ministry of Finance and Economic Development (2010) "Growth and Transformation Plan (GTP) 2010/11-2014/15" vol. 1, p. 27, 101.

71 FDRE Ministry of Capacity Building, 2005, Comprehensive Justice System Reform Program (2005) baseline study Report." p. 175.

72 *Ibid.* p. 14.

73 *Ibid.*

of public interest litigants by providing resources that would help their cases. However, there are problems related to media and free press in Ethiopia.[74]

The notion of sustainable development, were it to be applied as a Type II standard domestically, would require the state to balance the social, not just the economic and environmental, interests of the citizens. The social is often the opportunity to bring concerns of justice, not simply compliance, into state action. The German Basic Law (*Grundgesetz* or G.G.) states for example that the country's courts are to apply "law and justice" ("*Recht und Gerechtigkeit*").[75] Environmental justice means more than just working within the law, and, in many cultures, it is possible for the public, not just a state agency, to bring that justice about, either through direct action or the indirect action of monitoring the state. While state control of environmental matters through legislation enables the state to regulate certain matters in greater detail, addressing environmental problems through litigation on an *ad hoc* basis enables the state to focus on specific problems individually.

7.6 Conclusions

Both C.S. and P.I.L. can benefit from the knowledge and study of each other, in achieving internal goals. For example, a major problem of the P.I.L. system is that enforcement agencies are characterized as defaulters by virtue of their lack of enforcement action. To rectify this situation they could be statutorily required to benefit from a notice of intent to sue. The P.I.L.'s goal of making justice accessible to the poorest sections of society can also be achieved by allowing contingency fees[76] for lawyers, and improving the quality of legal representation. Private prosecution of crimes is available for example in the U.K. and although the barrister should be paid by his or her private client, experience shows that the barristers very often can and do recover their fees, or substantial portions of them, from the state. Criminal charges,[77] such as those filed against government officials for concealing relevant information in Flint, could have contributed to the positive turnaround in lead levels in 2016.

A crucial question one must ask is how one essentially evaluates the success of litigation, because considering a decision against a plaintiff as a failed suit in terms of environmental interests stems from the assumption that all plaintiffs are owed a victory by the justice system, and the lack thereof is a reflection on the

74 ADB and African Development Fund, Country Governance profile-Ethiopia, (March 2009), p. 10.

75 Article 20, § 3 of the German Basic Law.

76 The charging of contingency fees by lawyers is prohibited under Rule 20 of the Bar Council of India Rules.

77 Haimerl, Amy, and Abby Goodnough. "6 More State Workers Charged in Flint Water Crisis." *The New York Times*, 29 July 2016, https://www.nytimes.com/2016/07/30/us/flint-michigan-water-crisis.html.

inefficiency of the legal device, an assumption which is antithetical to the idea of the judiciary acting as an independent and impartial interpreter of the law. However, at the same time, the Courts in the U.S.A., India, and Ethiopia have consistently been an avenue for the protection of the rights of the most vulnerable and marginalized, and citizen participation in the judicial process through these suits performs not just a legal function, but a social one.

Employing a Type I comparison, one may make a useful and intelligent comparison of *locus standi*. In neither Ethiopia nor India does the public need a vested private and personal interest in order to commence P.I.L, nor does a person need one to commence a C.S. in the U.S.A. This legal recognition of using litigation as a tool for the common good, plus the rights extended to other entities discussed in Chapter 5, contribute to the recognition of a norm that values air, water, and land as a shared public benefit, rather than requiring private ownership and private injury in order to use litigation.

For these tools to work effectively, especially in countries where poverty and illiteracy are high, the success of a suit is unthinkable without the support of strong civil society organizations and freedoms. Even if technically available to them, non-profit and civil society organizations or individuals are not generally able to file claims against private investors who pollute the environment.

Literature

Abell, Charles S. "Ignoring the Trees for the Forests: How the Citizen Suit Provision of the Clean Water Act Violates the Constitution's Separation of Powers Principle." *Va. L. Rev.* 81 (1995): 1957, 1977.

Coplan, Karl S. "Citizen Litigants Citizen Regulators: Four Cases Where Citizen Suits Drove Development of Clean Water Law." *Colo. Nat. Resources Energy & Envtl. L. Rev.* 25 (2014): 61, 70.

Dunn, Daniel J. "Environmental Citizen Suits against Natural Resource Companies." *Nat. Resources & Env't.* 17 (2003): 161, 161.

Elliott, E. Donald, Bruce A. Ackerman, and John C. Millian. "Toward a Theory of Statutory Evolution: The Federalization of Environmental Law." *J. L. Econ. & Org.* 1 (1985): 313, 316.

Haimerl, Amy and Abby Goodnough. "6 More State Workers Charged In Flint Water Crisis." *N. Y. Times* 29 Jul. 2016. 2 Apr 2019. Https://Www.Nytimes.Com/2016/07/30/Us/Flint-Michigan-Water-Crisis.Html.

Hecker, James M. "The Citizen's Role in Environmental Enforcement: Private Attorney General, Private Citizen, or Both?" *Nat. Resources & Env't.* 8 (19931994): 31, 31.

Kellner, Martin. "Citizen Participation in Environmental Law Enforcement in Nicaragua: A Comparative Study of Nicaraguan, United States and German Environmental Law." *L. and Pol. In Afr., Asia and Latin Am.* 42 (2009): 385, 391–393.

Machado, Paulo Affonso Leme. *Direito Ambiental Brasiliero (Brazilian Environmental Law).* São Paulo: Malheiros, 2017.

Mckinney, Carie Goodman, "Statute of Limitation for Citizen Suits under the Clean Water Act." Cornell L. Rev. 72 (19861987): 195, 197.

Office of Water, Environment Protection Agency, Understanding the Safe Drinking Water Act 1. 2004. Https://Www.Epa.Gov/Sites/Production/Files/2015-04/Documents/Epa816f04030.Pdf (Last Visited Apr. 2, 2019).

Parmar, Shobhika and Vir Singh. "Water Quality Parameters of River Yamuna in Delhi after 20 Years of the Yamuna Action Plan." *Int'l J. Sci Nature* 6 (2015): 662, 665.

Prasad, P. M. "Environmental Protection: The Role of Liability System in India." *Eco. & Pol. Wkly* 39 (2004): 257, 263.

Robinson, Gail J. "Interpreting the Citizen Suit Provision of the Clean Water Act." *Cas. W. Res. L. Rev.* 37 (1987): 515, 519–520.

Schwartz, Richard E. and David P. Hackett. "Citizen Suits against Private Industry Under the Clean Water Act." *Nat. Res. Lawyer* 17.3 (1984): 327, 369–370.

Shinde, Mrinalini. "Environmental Law of India." *Comparative Environmental Law and Regulation*. Ed. Elizabeth Burleson, Lin-Heng Lye, and Nicholas A. Robinson. Toronto: Thomson Reuters, 2018.

Thompson, James L.. "Citizen Suits and Civil Penalties under the Clean Water Act." Mich. L. Rev. 85 (1987): 1656, 1657.

Timms, Gwynn Goodson, "Statutorily Awarding Attorneys' Fees in Environmental Nuisance Suits: Jump Starting the Public Watchdog." S. Cal. L. Rev. 65 (1991-1992): 1733, 1734.

Yang, Tseming. "Environmental Regulation, Tort Law and Environmental Justice: What Could Have Been." Washburn L. J. 41 (2002): 607, 616–167.

8

ECONOMIC CHOICES ENABLED BY ENVIRONMENTAL LAW

Sacha Kathuria

8.1 A case study of integrating clean energy in the electric grid

In Chapter 3 we raised the question of whether sufficient environmental planning could eliminate the need for litigation. In this chapter, we are looking at energy as an example of an area in which economic instruments function as a key feature of planning. Economic instruments can be used in both planning and litigation. In legal planning, tariffs, license fees, permit fees, and taxes may all be used to protect the environment. In litigation, state penalties for public law litigants can be criminal or civil. Civil penalties, with limits set by legislation and known by polluters in advance, can be sufficiently large so as to deter some activities.[1] In this chapter, we will look to just one example of economic instruments being used to regulate and promote electric energy production and distribution.

When it comes to electricity, technological changes mean opportunities for societies to integrate renewable energy resources into their portfolios to create an electric grid that is more reliable and has less impact on the environment. Among its many advantages, integrating renewable energy can help societies reduce greenhouse gas emissions (G.H.G.s) in an effort to address climate change and provide cost savings and reliability for consumers, and doing so can also serve as means to enable individual states to become more energy independent.

1 Relying upon this sort of deterrence requires that legislation give great flexibility to judges and administrators in assessing penalties, however, so as to assure that the state can assess a penalty large enough to deter highly profitable businesses that can incorporate "compliance costs" (penalties) into their business model and operating system and remain undeterred from polluting activities. At the same time, small operators can be deterred with much smaller penalties. Therefore the degree of flexibility required to make assessed penalties effective deterrents might not be concrete enough to provide clear notice to would-be polluters in advance.

Nevertheless, that which motivates states to create and utilize renewable energy varies. The rate at which renewables are supported economically, either by market forces or through government incentives, is dependent upon whether a culture supports it and whether its regulatory regime reflects the culture's expressed values. We will compare the planning and performance of the Netherlands and U.S.A., as two states underway in the process of changing their approaches to renewable energy. Part of this evolution includes compensating customers for adding renewable energy, such as solar, to the electric mix with tariff schemes. Ultimately, the decision of whether renewables are valued by a jurisdiction rests upon whether they function to serve the needs of the society in which they are deployed. Given that the author practices in the U.S.A., we will use the Mansel principle and present considerations from the Netherlands first.

8.2 Electricity transmission and distribution in the Netherlands

The Netherlands' abundant natural resources for electricity generation include some of the world's largest gas reserves in the province of Groningen.[2] Consequently, the Netherlands is the second largest exporter of natural gas in Europe and a significant producer of liquid fuels, placing it as an important transportation and processing hub.[3] In 2015, Dutch electricity generation was 80 percent driven by fossil-fuel power plants.[4] Although famous for its windmills, that year wind made up only seven percent of electricity generation, of the twelve percent generated from renewable resources.[5] Despite its progressive reputation for being "clean and green," the Netherlands is a huge polluter in Europe.[6] It ranks near the bottom of European Union (E.U.) countries for renewable energy use and is failing to meet its E.U. target goals for green energy use. The Netherlands is facing increased pressure from the E.U. and the public to change.[7] As we saw in Chapter 5, Dutch citizens have found Dutch state planning and legislation to be insufficient to address their needs for clean energy that meets climate crisis demands, so they filed a suit, along with the Urgenda Foundation, to achieve their goals instead through litigation.

2 *Netherlands – Energy,* U.S. Department of Commerce, International Trade Administration, (October 18, 2018), https://www.export.gov/article?id=Netherlands-Energy.

3 "U.S. Energy Information Administration - EIA - Independent Statistics and Analysis." *Netherlands - International - Analysis - U.S. Energy Information Administration (EIA),* https://www.eia.gov/beta/international/analysis.php?iso=NLD.

4 *Ibid.*

5 *Ibid.*

6 Kakissis, Joanna. "The Coal Dependent Netherlands Searches for Alternatives." *NPR,* 1 Jan. 2018, https://www.npr.org/2018/01/01/574985944/the-coal-dependent-netherlands-searches-for-alternatives.

7 *Ibid.*

Since natural gas is clean-burning, its increased use helped the Dutch economy decarbonize already in the 1970s, as coal production waned in the coal region of Limburg.[8] In 2015, the Dutch parliament voted to phase out the country's existing eleven coal-fired power plants, some of which were scheduled to be closed permanently while others will convert to biomass co-firing.[9] However, gas production is widely believed to have caused earthquakes in the region. Public pressure is increasing to cease production immediately for safety concerns.[10] The Dutch government compromised and decided to phase out the Groningen gas field by 2030.[11] In addition, Groningen is running low on reserves.[12] Although the goal is to replace natural gas with renewables, this will be expensive and take time.[13] In the meantime, it is likely that the country will rely on natural gas imports instead, which may increase costs, be time-consuming, and negate any environmental benefits of natural gas.[14] Like their American counterparts, Dutch politicians are developing plans, laws, and regulations that use new technology to achieve public demand for renewable energy.

8.2.1 The Electricity Act of 1998

The Dutch electric grid is connected to Belgium, Germany, Norway, and the United Kingdom via high voltage lines that run along the North Sea bed.[15] Its most common forms of energy for electricity generation have been natural gas and coal.[16] TenneT, owned by the Dutch government, is the sole grid operator for transmission in the Netherlands.[17] In 2010, it acquired German E.ON's extra-high voltage grid, becoming the first cross-border grid administrator in Europe.[18] Dutch law requires unbundling the transmission and distribution.[19] As a result, the Netherlands has eight distribution service

8 Van de Graff, Evert. "The Termination of Groningen Gas Production – Background and Next Steps." *Columbia SIPA Center on Global Energy Policy*, 2018.

9 U.S. Energy, *supra* note 3.

10 Meijer, Bart H. "Dutch Citizens Demand End to Quake-Hit Groningen Gas Production." *Reuters*, Thomson Reuters, 17 Jan. 2019, https://www.reuters.com/article/us-netherlands-gas-court/dutch-citizens-demand-end-to-quake-hit-groningen-gas-production-idUSKCN1PB0TD.

11 Van de Graff, *supra* note 8.

12 Verdonck, Rob. *Europe's Biggest Natural Gas Producer Is Running Out of Fuel*. Bloomberg, 16 Sept. 2016, https://www.bloomberg.com/news/articles/2016-09-16/europe-s-biggest-natural-gas-producer-is-running-out-of-fuel.

13 Van de Graff, *supra* note 8.

14 *Ibid.*

15 U.S. Energy, *supra* note 3.

16 *Ibid.*

17 "Print TSO Focus." *Global Transmission Report: Dutch-German TenneT: Playing a Central Role in Creation of NWE*, https://www.globaltransmission.info/archive.php?id=26063.

18 *Ibid.*

19 U.S. Energy, *supra* note 3.

operators, over twenty-five producers, and thirty-five electricity retailers.[20] TenneT is part of the European Network of Transmission System Operators (E.N.T.S.O.-E.), which represents forty-three electric transmission operators from thirty-six countries in Europe.[21] E.N.T.S.O.-E. stems from E.U. regulation EC714/2009, which is meant to enhance cooperation between transmission system operators to achieve a more secure and reliable network, facilitate cross-border development, and enhance the creation of an internal electricity market.[22]

The Electricity Act (E-Act) of 1998 forms the main basis for electricity law in the Netherlands, allowing producers and suppliers of electricity to offer it on a market.[23] The E-Act allowed distribution companies and large consumers to buy electricity from generators other than local ones and allowed self-generation by industry.[24] The E-Act forced distribution companies to take in all electricity that was locally generated and supplied to them, and to pay a feed-in tariff essentially equal to the avoided costs.[25] The law also established an independent regulator for the energy sector, the *Directie Toezicht Energie* (DTe), the Office for Energy Regulation, as a part of the *Nederlandse Mededingingsautoriteit* (NMa), and the Netherlands Competition Authority, to supervise the sector "to guarantee nondiscriminatory access to the grid for generators and to regulate the prices of the captive end users."[26]

In 2009, the E.U. issued the Renewable Energy Directive (2009/28/EC), which set binding targets for all Member States to obtain 20 percent of their final energy consumption from renewable resources by 2020.[27] In response, E.U. Member States adopted national renewable energy action plans outlining their targets for various energy usage such as for electricity, transport, and heating.[28] In its initial national renewable energy action plan, issued in 2010, the Netherlands cited energy savings, climate change, and the need for energy diversification as drivers for innovation and economic activity.[29] The plan laid out a number of central instruments to achieve this goal. These instruments contain efforts to include *Verplichting Biobrandstoffen* (biofuels at the mandatory

20 *Ibid.*
21 "The Third Energy Package." *ENTSO*, https://www.entsoe.eu/about/inside-entsoe/official-mandates/.
22 *Ibid.*
23 "Electricity Act." *Grantham Research Institute on Climate Change and the Environment*, http://www.lse.ac.uk/GranthamInstitute/law/electricity-act/.
24 *Ibid.*
25 *Ibid.*
26 *Ibid.*
27 "Renewable Energy." *Renewable Energy | Energy*, https://ec.europa.eu/energy/en/topics/renewable-energy.
28 *Ibid.*
29 *The Netherlands National renewable energy action plan*, Directive 2009/28/EC, EUROPEAN COMMISSION, 10 (September 7, 2010).

minimum share) for transport; the *Rijkscoördinatieregeling* (R.C.R.), a government coordination scheme which is an effort to speed up licenses for energy infrastructure projects; the *Wet algemene bepalingen omgevingsrecht* (Wabo), a general provision of the Environment Act to increase transparency and speed up small scale energy installations; and the *Voorrang voor duurzaam* (Priority for Sustainability Bill), which gives priority to sustainable energy in the event of network congestion.[30] It also included the *Stimuleringsregeling Duurzame Energieproductie* (S.D.E.), an incentive scheme for sustainable energy production, which is a financial mechanism for increasing the use of renewable energy in electricity, heat, and gas.[31]

The E.U. revised its Renewable Energy Directive in 2018 (2018/2001), establishing a new binding target for renewable energy of at least 33 percent for member countries with the possibility for upwards revision by 2023.[32] Part of the E.U.'s "Clean Energy for All Europeans" package, the new directive was created to help the EU meet its commitments under the Paris Agreement in reducing greenhouse gas emissions.[33] The directive required EU Member States (M.S.s) to create ten-year national energy and climate plans by 2019 to show how they plan to meet these new targets.[34] Every two years, M.S.s are to submit reports showing their progress on meeting these goals.[35] In its plan submitted for 2015–2016, the Netherlands set its binding target for the gross final energy consumption from renewable energy sources for 2020 to be at 14 percent and 16 percent by 2023.[36] Among the financial initiatives most applicable for encouraging renewables such as solar energy, the Dutch government highlighted its *Milieukwaliteit van de Elektriciteitsproductie* (M.E.P.), the Environmental Quality of Electricity Production, its Incentive Scheme for Sustainable Energy Production (S.D.E.+), its Energy Investment Relief Scheme (E.I.A.), its Green Projects Scheme, its Top Sector Energy program, and the Green Deals program. Nevertheless, as of 2017, the Netherlands was still struggling to meet its 2020 goals, with only 6.6 percent of energy sources derived from renewables.[37]

30 *The Netherlands National renewable energy action plan*, Directive 2009/28/EC, EUROPEAN COMMISSION, 11 (September 7, 2010).

31 *Ibid.*

32 Renewable Energy, *supra* note 27.

33 *Ibid.*

34 *Ibid.*

35 *Ibid.*

36 Netherlands Ministry of Economic Affairs and Climate, *Progress Report, Energy from renewable sources in the Netherlands 2015-2016,* Directive 2009/28/EC, EUROPEAN COMMISSION 4 (July 10, 2018).

37 "The Netherlands Still Trailing behind on EU Renewable Energy Targets." *DutchNews.nl*, 12 Feb. 2019, https://www.dutchnews.nl/news/2019/02/the-netherlands-still-trailing-behind-on-eu-renewable-energy-targets/.

8.2.2 The Environmental Quality of Electricity Production (M.E.P.) and the Incentive Scheme for Sustainable Energy Production (S.D.E.)

The M.E.P. initiative, operated between 2003 and 2006, provided subsidies to domestic producers of electricity from renewable resources in the form of a fixed feed-in premium for "feeding-in" to the grid.[38] The Dutch government ended this program because it was meeting its target of nine percent for 2010.[39] It was replaced by the Incentive Scheme for Sustainable Energy Production (S.D.E.+)[40] The S.D.E.+ was formed to achieve the Dutch government's new target of the reduction of CO2 emissions of 30 percent by 2020.[41]

The S.D.E.+ succeeded the M.E.P. from 2008 to 2010. This was a feed-in tariff scheme where a budget ceiling was published along with a base amount established for each technology. It was replaced by the S.D.E.+ in July 2011 to promote renewables in a more cost-effective way.[42] The S.D.E.+ is meant to offer long-term security by covering unprofitable higher costs of projects through a feed-in premium system where it pays the difference in cost between the cost of gray energy and sustainable energy over a period not exceeding fifteen years.[43] Under the S.D.E.+, the subsidy budget is "no longer [to] be allocated to individual technologies; instead the financial support is part of an overall subsidy platform, which means that all technologies compete against each other for the subsidy funding."[44] The S.D.E.+ has separate funding categories for wind, solar, etc., that are further subdivided. The subsidy increases with each phase.[45] This means that photovoltaic solar (P.V.) can compete with other technologies for subsidy funds and does not get "further differentiated support."[46]

38 Netherlands Ministry of Economic Affairs and Climate, *Progress Report, Energy from renewable sources in the Netherlands 2015-2016,* Directive 2009/28/EC, EUROPEAN COMMISSION 22 (July 10, 2018).

39 *State aid: Commission authorizes aid scheme to stimulate renewable energy (SDE) in The Netherlands,* EUROPEAN COMMISSION, (December 21, 2007) http://europa.eu/rapid/press-release_IP-07-1992_en.htm.

40 *Ibid.*

41 *Ibid.*

42 Netherlands Ministry of Economic Affairs and Climate, *Progress Report, Energy from renewable sources in the Netherlands 2015-2016,* Directive 2009/28/EC, EUROPEAN COMMISSION 21 (July 10, 2018).

43 *Ibid.*

44 Wilming, Wilhelm. "'SDE " Replaces the Former Subsidy of Renewable Energies in the Netherlands." *Sun & Wind Energy,* 6 Aug. 2013, https://www.sunwindenergy.com/news/sde-replaces-former-subsidy-renewable-energies-netherlands.

45 Bellini, Emiliano. "Another 3 GW on the Way for the Netherlands PV Market." *Pv Magazine International,* 29 Apr. 2019, https://www.pv-magazine.com/2019/04/29/another-3-gw-on-the-way-for-the-netherlands-pv-market/.

46 Wilming, *supra* note 44.

The S.D.E.+ scheme successfully adds solar capacity.[47] The S.D.E.+ program has already allocated 10.5 GW of solar P.V. capacity.[48] The subsidy and allocation scheme requires project realization within one to three years, after which allocations are withdrawn.[49] The Dutch electric grid is not designed to handle additional capacity. Regions where local grids are not designed for carrying large volumes of electricity could therefore experience disruptions in service.[50] Grid operators in the provinces of Groningen, Drenthe, and Overijssel claim that development and additional capacity is going too fast and that efforts to improve the grid cannot keep pace.[51] In response, the Dutch government issued new restrictions on the S.D.E.+ subsidies.[52] New projects must have preliminary approval from network operators and must formally request connection points and provide documentation that the capacity exists for their projects.[53] The solar industry, while acknowledging bottlenecks in the grid, is concerned that it is being specifically targeted in a way that may inadvertently disadvantage it.[54] There is also concern that solar P.V. will overtake agricultural land.[55] The Dutch Parliament continues to consider banning new solar P.V. ground-mounted projects, citing concerns to the landscape, nature, and farmland, and the country's small size,[56] as the industry soars—doubling between 2016 (2 GW) and 2018 (4.15 GW) because of the S.D.E.+[57]

The S.D.E.+ does not include projects of outputs less than fifteen kWp, but those systems are eligible to take advantage of net metering.[58] Net metering is responsible for the early development of solar in the Netherlands and it accounted for 38 percent of the solar market in 2018, approximately 550 MW.[59] Nevertheless, the Dutch government announced that it will be keeping its current net metering system until 2023, and then begin a gradual phase out of the scheme.[60] The government will also begin to phase out other fiscal incentives because it is expected that by that time "solar energy should be able to compete with fossil-fuel energy sources without tax incentives."[61] Projections are that

47 Bellini, *supra* note 45.
48 *Ibid.*
49 *Ibid.*
50 *Ibid.*
51 *Ibid.*
52 Martin, Jose Rojo. "Grid Bottlenecks Push Netherlands to Renewable Subsidy Restrictions." *PV Tech*, PV-Tech, 3 July 2019, https://www.pv-tech.org/news/grid-bottlenecks-push-netherlands-to-renewable-subsidy-restrictions.
53 *Ibid.*
54 *Ibid.*
55 Bellini, *supra* note 45.
56 Martin, *supra* note 52.
57 *Ibid.*
58 Wilming, *supra* note 44.
59 Bellini, *supra* note 45.
60 *Ibid.*
61 *Ibid.*

solar and wind will make up 60 percent of the Netherlands' capacity while coal is planned to be phased out by 2028.[62]

8.2.3 The Energy Investment Relief Scheme, Green Projects, Top Sector Energy (T.S.E.), Green Deals

While the S.D.E.+ and net metering are direct subsidies, the Dutch government has also developed alternative renewable support mechanisms that are meant to bring tax relief and collaboration among interested persons. The Energy Investment Relief Scheme offers entrepreneurs tax deductions for their investment in energy-saving technologies that use renewable energy.[63] The tax scheme is overseen by the *Rijksdienst voor Ondernemend Nederland*, R.V. (Netherlands Enterprise Agency), which assesses an item that is notified against the technical and administrative requirements of the Energy Investment Relief Scheme and issues a certificate if the requirements have been fulfilled.[64] The tax administration then determines what, if any, subsidy the entrepreneur receives.[65]

Under the "Green Projects" scheme, Dutch consumers have the option to purchase green securities or shares in a green investment fund, with the intention that the funds are raised in green projects that have been approved by the Dutch government.[66] Consumers receive tax relief and can obtain funds at a low interest rate for projects that qualify as a "benefit to the natural world and the environment."[67] To qualify, consumers must show that their project will be implemented in the Netherlands and that it cannot be completed without financial assistance; and they must obtain a *groenverklaring* (Green Declaration) from their bank.[68]

The Dutch government also formed the Top Sector Energy program (T.S.E.) to promote innovation as a guiding principle to drive the growth of certain sectors.[69] One of the nine top sectors chosen for this project is energy.[70] The idea is to promote growth in areas such as renewables on a global level so that companies

62 GlobalData Energy. "Future of Energy in the Netherlands: Managing the Move to Going Green." *Power Technology | Energy News and Market Analysis*, 18 Mar. 2019, https://www.power-technology.com/comment/future-of-energy-in-the-netherlands/.

63 Netherlands Ministry of Economic Affairs and Climate, *Progress Report, Energy from renewable sources in the Netherlands 2015-2016*, Directive 2009/28/EC, EUROPEAN COMMISSION 21 (July 10, 2018).

64 *Ibid.*

65 *Ibid.*

66 *Ibid.*

67 *Ibid.*

68 "Green Projects Scheme." *Business.gov.nl*, https://business.gov.nl/subsidy/green-projects-scheme/.

69 Netherlands Ministry of Economic Affairs and Climate, *Progress Report, Energy from renewable sources in the Netherlands 2015–2016*, Directive 2009/28/EC, EUROPEAN COMMISSION 22–23 (July 10, 2018).

70 *Ibid.*

no longer are motivated solely by rules and subsidies.[71] The Top Sector Energy scheme brings together various interested persons from industry, government, and academia to form a "top team" for each top sector in order to ease bottlenecks related to knowledge and innovation by forming pilot projects.[72] The program is funded both by the government as well as by private companies.[73] The ultimate focus of the T.S.E. is to meet the 2,050 goals stated in the National Energy Agreement "to achieve a fully sustainable energy supply and a CO2-reduction of 80–95% as opposed to the situation in 1990 by 2050."[74]

In 2011, the Dutch government developed the Green Deal plan to facilitate "initiatives by companies, civil-society organisations, decentralised authorities and citizens by eliminating obstacles (which may relate to existing legislation and regulations, or access to funding, for example)."[75] Energy is part of this initiative and by 2016, of the 208 deals undertaken by the scheme, 120 were energy-related.[76]

Given its limited space, geography, and weather, access to solar electrical energy can be elusive to the Dutch consumer. Forty-three percent of Dutch households cannot benefit from solar panels on their homes, because they live in apartments and do not own the building or have too much shade on their rooftop.[77] To address this and concerns of limited space on the ground, local communities and industry are coming up with ways to make solar P.V. more accessible. For example, Dutch engineers are in the process of constructing "what will be the world's largest archipelago of islands made up of sun-tracking solar panels"[78] on the Andijk reservoir in North Holland, containing 73,500 solar panels.[79] Another example is the company "WeShareSolar," the largest crowdfunding platform in the E.U. for community solar projects. WeShareSolar connects site owners who have suitable roofs without the funds to install solar P.V. with consumers who have the funds but do not have a suitable roof to install solar P.V.[80]

71 *Ibid.*

72 *Ibid.*

73 *Ibid.*

74 "Topsector in General." *Topsector Energie*, https://www.topsectorenergie.nl/en/topsector-general.

75 Netherlands Ministry of Economic Affairs and Climate, *Progress Report, Energy from renewable sources in the Netherlands 2015-2016*, Directive 2009/28/EC, EUROPEAN COMMISSION 24 (July 10, 2018).

76 *Ibid.*

77 "Crowdfunding for Community Solar Projects: The Netherlands." *UNFCCC*, https://unfccc.int/climate-action/momentum-for-change/financing-for-climate-friendly-investment/crowdfunding-for-community-solar-projects.

78 Boffey, Daniel. "Dutch Engineers Build World's Biggest Sun-Seeking Solar Farm." *The Guardian*, Guardian News and Media, 21 Apr. 2019, https://www.theguardian.com/world/2019/apr/21/dutch-engineers-build-worlds-biggest-sun-seeking-solar-farm.

79 *Ibid.*

80 "Crowdfunding for Community Solar Projects: The Netherlands." *UNFCCC*, https://unfccc.int/climate-action/momentum-for-change/financing-for-climate-friendly-investment/crowdfunding-for-community-solar-projects.

All of these economic incentives in the Netherlands are examples of planning to meet what one may call environmental needs. We have already seen in Chapter 5 that, despite these legal planning efforts, the Dutch public used litigation to demand further measures from its government to address the climate crisis.

8.3 Electricity transmission and distribution in the United States

Like the Netherlands, the United States has abundant renewable and nonrenewable energy resources. Renewable energy comes from sources that "are naturally replenishing but flow-limited," and "are virtually inexhaustible in duration but limited in the amount of energy that is available per unit of time."[81] The most common forms of renewable energy are biomass, hydropower, geothermal, wind, and solar.[82] Nonrenewable energy resources "do not form or replenish in a short period of time."[83] The most common forms of nonrenewable energy are crude oil, natural gas, coal, and uranium.[84] Although the amount of energy consumed has increased over time in the United States, the types of energy used have changed due to "advances in technology, energy resource discoveries, energy prices, societal pressures and other factors."[85] As of 2017, the electric power sector made up the primary energy consuming sector in the country, at 38.1 percent, followed by transportation (28.8 percent), industrial (22.4 percent), residential (6.2 percent), and commercial use (4.5 percent).[86] Fossil fuels made up 77.6 percent of U.S. energy production in 2017 whereas renewable energy was only 12.7 percent of that production.[87] Of that production, coal and crude oil decreased overall, and cleaner energy sources, natural gas, and renewables increased, wind and solar having reached record high production in 2017.[88] In April 2019, renewable energy surpassed coal for the first time in generating electricity in the United States,[89] largely because more renewable energy infrastructure is coming online as coal-fired power plants shut down at record rates.[90]

81 "U.S. Energy Information Administration - EIA - Independent Statistics and Analysis." *Renewable Energy Explained - U.S. Energy Information Administration (EIA)*, https://www.eia.gov/energyexplained/index.php?page=renewable_home.

82 *Ibid.*

83 *Ibid.*

84 *Ibid.*

85 King, Hobart. "History of Energy Use in the United States." *Geology*, https://geology.com/articles/history-of-energy-use/.

86 U.S. Energy, *supra* note 3.

87 *Ibid.*

88 *Ibid.*

89 Johnson, Stephen. "For First Time, U.S. Renewable Energies Surpass Coal." *Big Think*, Big Think, 4 July 2019, https://bigthink.com/politics-current-affairs/renewables-coal?rebelltitem=1#rebelltitem1.

90 *Ibid.*

Cleaner energy in the U.S.A. is a result of a mix of advances in technology, societal demands, and legal requirements. The United States' energy planning is divided among federal, state, and local governments. Resources are not equally available throughout the country, so one finds disparate attitudes, planning, and laws for renewables throughout the country. Coal-fired electricity generation has been in decline since its peak in 2008.[91] Local and regional trends vary, however. For example, coal in West Virginia makes up approximately 17 percent of the state's total G.D.P. and employs approximately 13,000 miners and is responsible for another 25,000 jobs throughout the state.[92] Support for renewable energy there is predictably lower.

Here, we see a concrete example of the interface of law and politics that was discussed in Chapter 6. Considered part of a "dying" industry, coal executives are trying to persuade law-makers that their demise is a national security issue.[93] As a reward for political campaign support from the coal industry, the Trump administration pledged $39 million for research to improve the efficiency of coal-powered electric plants.[94] Nevertheless, the industry was unable to convince the administration to utilize emergency powers under the Federal Power Act to impose a two-year moratorium on the closing of coal-fired power plants.[95]

By comparison, in California, sea level rise, coastal erosion, forest fires, water supply shortages, and poor air quality are a daily reality.[96] Even though California is the fourth largest producer of crude oil in the U.S.A., and second largest consumer of energy in the U.S.A., its per capita energy consumption ranks forty-eight out of fifty states because of its energy efficiency programs.[97] California ranks first in the U.S.A. for electricity generation from solar, geothermal, and biomass.[98]

The Clean Power Plan announced by former President Barack Obama in 2015 would have been the first regulation to set limits on carbon pollution from

91 U.S. Energy, *supra* note 3.

92 Hamilton, Chris. "Coal Industry Still Contributes Big to West Virginia." *Montgomery Herald*, 13 Aug. 2018, https://www.montgomery-herald.com/opinion/coal-industry-still-contributes-big-to-west-virginia/article_ed1804da-9eab-11e8-86a6-cb317335f5b4.html.

93 Roberts, David. "The US Coal Industry Is Going out, not with a Whimper, but with a Burst of Rent-Seeking." *Vox*, Vox, 26 Aug. 2017, https://www.vox.com/energy-and-environment/2017/8/25/16201218/us-coal-industry-handouts.

94 Picchi, Aimee. "Trump Says 'Coal Is Back," but Use of the Rock at a 41-Year Low." *CBS News*, CBS Interactive, 11 June 2019, https://www.cbsnews.com/news/trump-says-coal-is-back-coal-consumption-just-hit-a-41-year-low/.

95 "Plea for 2-Year Moratorium on U.S. Coal Plant Closings Lands with a Thud." *Institute for Energy Economics & Financial Analysis*, 22 Aug. 2017, http://ieefa.org/plea-2-year-moratorium-u-s-coal-plant-closings-lands-thud/.

96 "Climate Change Impacts in California." *State of California - Department of Justice - Office of the Attorney General*, 20 July 2017, https://oag.ca.gov/environment/impact

97 *Ibid.*

98 *Ibid.*

U.S. power plants.[99] Officially issued by the U.S.E.P.A., as a regulation under the Clean Air Act, the Clean Power Plan was to establish "national carbon dioxide emissions performance rates for existing coal- and gas-fired power plants," after which each state could "adopt its own plan—including enforceable emission limits—for its coal and gas plants"[100] to reduce emissions from electricity by 32 percent below 2005 by 2030.[101] If a state did not implement a plan as strict as (or stricter) than the federal plan, it would have been possible for the E.P.A. to regulate power plants directly in such states.[102] Planned to be introduced incrementally in 2022 with full effect in 2030, the Clean Power Plan regulation would have given credits to early adopters of renewable energy through a Clean Energy Incentive Program (C.E.I.P.)[103]

Private companies, the U.S. Chamber of Commerce (a private lobby group), and coal-producing states filed suit in federal court seeking an emergency stay and suspension of the permit requirements from the E.P.A. Clean Power Plan regulation.[104] After having its requests denied in lower courts, the fossil fuel industry interests were granted a temporary stay of the Clean Power Plan regulation requirements in 2016, but the entire matter was mooted by the change of executive administration later, after the national election in November 2016. In June 2019, the executive administration replaced the C.P.P. with the Affordable Clean Energy (A.C.E.) rule,[105] which has now also been legally challenged. These delays due to legal disputes at the interface of politics and law (see Chapter 6) suit polluting interests to continue *status quo ante* (see Chapter 4).

Within the frame of cooperative federalism, the failure of the federation to adopt a sufficient standard gave the states the incentive to act. That action has taken the form of thirty-eight states and the District of Columbia having enacted renewable portfolio standards (R.P.S.), which are also called "renewable energy standards" (R.E.S.)[106] Although varied from state to state, an R.P.S. is meant to establish incremental targets that legally require utility companies to source a certain amount of energy they generate or sell from renewable sources.[107]

99 "What Is the Clean Power Plan?" *NRDC*, 29 May 2019, https://www.nrdc.org/stories/how-clean-power-plan-works-and-why-it-matters?gclid=EAIaIQobChMI6veC_K3A4wIVjp-fCh0QWwbwEAAYASAAEgIrfPD_BwE.
100 *Ibid.*
101 *Ibid.*
102 *Ibid.*
103 *Ibid.*
104 "The Clean Power Plan." *Harvard Law Review*, vol. 129, 10 Feb. 2016, p. 1152.
105 Irfan, Umair. "Trump's EPA Just Replaced Obama's Signature Climate Policy with a Much Weaker Rule." *Vox*, Vox, 19 June 2019, https://www.vox.com/2019/6/19/18684054/climate-change-clean-power-plan-repeal-affordable-emissions.
106 "Renewable Energy Standards." *SEIA*, https://www.seia.org/initiatives/renewable-energy-standards.
107 *Ibid.*

California had established an R.P.S. program already in 2002, requiring 20 percent of all electricity retail sales be served by renewable resources in 2017, and further accelerated this requirement in 2015 to achieve 50 percent by 2030 with three-year compliance periods interim targets.[108] California's legislature further updated its R.P.S. requirements in 2018 by increasing the R.P.S. to 60 percent by 2030 and requiring all of the state's electricity to come from carbon-free resources by 2045.[109] By the end of 2017, retail sellers of electricity met or exceeded the updated interim 27 percent target and are expected to achieve their compliance requirements.[110] California's largest investor-owned utilities served 36 percent of their 2017 retail electricity customers with renewable power.[111] California legislation requires the California Energy Commission to prepare a biennial integrated energy report that assesses major energy trends and issues, and provides policy recommendations to conserve resources, diversify energy supplies, and protect the environment.[112] In West Virginia, by contrast, less than 5 percent of electricity generation comes from renewables.[113] Some of this is due to hydropower built in the 1930s which accounts for 2 percent of the state's net electricity generation, with biomass and solar photovoltaic generation each contributing less than 0.01 percent to the utility-scale electricity generation.[114] In 2009, West Virginia passed an R.P.S. standard that required investor-owned utilities and retail suppliers with more than 30,000 residential customers to acquire 25 percent of their retail electric sales from renewable energy resources by 2025. But in yet another episode of politics interfacing with law, in 2015 West Virginia became the first state to repeal its renewable energy portfolio.[115]

The value that states place on renewable energy resources such as solar are largely based upon their expected economic benefit. This is shown by the way in which state governments vary the use of renewables such as distributed solar, i.e., photovoltaic solar. In the United States, federal and state governments only set the parameters regarding economic incentives with renewables, because the U.S. economic theory is that the end user, i.e., the consumer, will ultimately dictate their direction.

108 "Renewables Portfolio Standard (RPS) Program." *RPS*, https://www.cpuc.ca.gov/rps/.

109 *Ibid.*

110 *Ibid.*

111 *Ibid.*

112 California Energy Commission. "Integrated Energy Policy Report - IEPR." *California Energy Commission*, California Energy Commission, https://www.energy.ca.gov/data-reports/reports/integrated-energy-policy-report.

113 "Renewable Energy." *Renewable Energy | WV Office of Energy*, https://energywv.org/wv-energy-profile/renewable-energy.

114 *Ibid.*

115 *Ibid.*

8.3.1 Solar photovoltaic (P.V.)

To facilitate the U.S. economic model of incentives for renewables like solar P.V., it is necessary to understand how an electric system operates. Electricity transmission in the U.S.A. is mostly generated remotely at large central stations, transmitted over distances with high-voltage lines, and then reduced in voltage for local distribution and delivery to customers.[116] The "core" of any electricity system is the grid, which is "a highly engineered network that coordinates the production and delivery of power to customers."[117] The U.S. power grid, created over a century ago, consists of more than 9,200 electric generating units with more than 1,000,000 MW of generating capacity connected to more than 300,000 miles of transmission lines.[118]

Transmission lines can be owned by investor-owned utilities, public power utilities, or cooperative-owned utilities and their "business and potential source of profits" depends upon the development and operation of transmission.[119] "Distribution" means delivering power from the transmission system to the end users of electricity by connecting the transmission system to secondary distribution lines that carry power to customers.[120] Distributed generation (D.G.) "refers to a variety of technologies that generate electricity at or near where it will be used, such as solar panel and combined heat and power."[121]

In a further demonstration of cooperative federalism, the Federal Energy Regulatory Commission (F.E.R.C.) regulates the transmission and wholesale sales of electricity in commerce between any combination of the fifty U.S. states.[122] State public utility commissions regulate the sale of retail electricity to consumers and approve physical construction of electric generation facilities.[123] Electricity markets are organized and managed by regional oversight entities that administer the wholesale market and encourage reliability in the form of Independent System Operators (I.S.O.s) and Regional Transmission

116 *E21 Initiative Phase II Report: On implementing a framework for a 21*st *century electric system in Minnesota,* GREAT PLAINS INSTITUTE, (December 2016), https://e21initiative.org/wp-content/uploads/2018/01/e21_Initiative_PhaseII_Report_2016.pdf, p. 69.

117 *Quadrennial Review Report, Chapter III: Modernizing the Electric Grid,* U.S. Department of Energy, (April 2015),https://energy.gov/sites/prod/files/2015/08/f25/QER%20Chapter%20III%20Electricity%20April%202015.pdf, pp. 3-4.

118 The Smart Grid: An Introduction, U.S. Department of Energy, https://energy.gov/sites/prod/files/oeprod/DocumentsandMedia/DOE_SG_Book_Single_Pages%281%29.pdf, p. 5.

119 Quadrennial Review, *supra* note 117.

120 *Ibid.*

121 "Distributed Generation of Electricity and Its Environmental Impacts." *EPA*, Environmental Protection Agency, 13 Mar. 2018, https://www.epa.gov/energy/distributed-generation-electricity-and-its-environmental-impacts.

122 "About FERC - What FERC Does" *FERC*, 14 Aug. 2018, https://www.ferc.gov/about/ferc-does.asp.

123 *Ibid.*

Organizations (R.T.O.s)[124] I.S.O.s and R.T.O.s usually cover multiple states and provinces in the U.S.A. and Canada and are meant to facilitate cooperation and nondiscrimination, promote reliability, manage interconnection and new loads, and increase transparency and protect against market abuses.[125] The North American Electric Reliability Corporation (N.E.R.C.) serves as the Electric Reliability Organization (E.R.O.) for the entire United States.[126] The N.E.R.C. is a non-profit international regulatory authority that sets standards for reliability and security for the United States, Canada, and parts of Mexico.[127]

Despite their many advantages, renewable technologies challenge the status quo, in particular the existing business model of utility companies with regard to transmission and distribution. Traditional players in electricity are facing a fundamentally different grid of the future, where newer technologies will eventually make for cleaner, more cost-effective operations.[128]

Not only does solar P.V. increase access to cleaner energy, but it also can help increase grid reliability. Although the U.S. electric system is 99.97 percent reliable, power outages cost the country a minimum of $150 billion per year.[129] While the U.S.A. has a reliable grid system overall, "when compared statistically to other nations, the U.S. grid does not necessarily meet … expectations." Improving the power delivery system through grid improvements "can only serve to improve the resiliency of the system, and help hasten recovery from weather-related outages." An uptick in weather-related outages combined with aging infrastructure and lack of leadership in the public and private sectors have hindered the U.S. electric grid from reaching its full potential. Because distributed generation can serve consumers without external lines, "it is far less vulnerable to the kinds of weather-related service interruptions that affect utility power systems." Recognizing the benefits of solar P.V., most states have enacted some type of compensation scheme or tariff to reward consumers for generating their own solar energy. In particular, the programs of net metering, feed-in tariffs, value-of-solar, and virtual metering are bringing about opportunities for consumers to take advantage of this technology.

124 Caspary, Jay. *Electric Transmission 101: Markets, ISO/RTOs and Grid Planning/Operations*, Environmental and Energy Study Institute, http://www.eesi.org/files/070913_Jay_Caspary.pdf.

125 *Ibid.*

126 *History of NERC*, North American Electric Reliability Corporation, (December 2012) http://www.nerc.com/AboutNERC/Documents/History_Dec12.pdf.

127 *Ibid.*

128 Quadrennial Review, *supra* note 117 at 3-4.

129 The Smart Grid: An Introduction, U.S. Department of Energy, https://energy.gov/sites/prod/files/oeprod/DocumentsandMedia/DOE_SG_Book_Single_Pages%281%29.pdf, p. 5.

8.3.2 Net metering

The rates charged by utilities are "typically set in a 'ratemaking' proceeding, at which a utility considers what the reasonable cost of the various components of delivering electricity is likely to be," where F.E.R.C. regulates prices of wholesale electricity in interstate commerce, and state utility boards set the state utility rates and other regulations of service.[130] Within the ratemaking procedure, net metering statutes are the most common economic incentive states provide for solar P.V. in the U.S.A. Net metering is a "billing mechanism that credits solar energy system owners for the electricity exported to the grid."[131] In net metering, solar energy producers, such as homeowners with rooftop solar that are connected to the grid, are able to "bank" as a credit each kilowatt-hour (kWh) of solar generation not consumed immediately, and allowed to use the banked credits to offset the consumption of energy provided by their electric utility during times when their rooftop systems do not meet their immediate energy needs.[132] For net metering to be effective, the electric distribution system must manage and control a two-way flow of power.[133]

Critics of net metering argue that net metering results in higher costs for all consumers, "by driving up electricity rates for everyone else."[134] This is because of the high rates they are paid and because they do not contribute directly to the costs associated with utilizing the grid. Electric utilities are required to buy this power, even though "it would cost them less to produce the electricity themselves or to buy the power on the wholesale market from other electricity providers."[135] Consumers with D.G. systems are still using the utility's grid infrastructure to buy and sell electricity.[136] Although rooftop solar consumers avoid paying for the utility's power they did not use, commercial profit-making electricity providers complain that rooftop solar consumers "also avoid paying for all the fixed costs of the grid that delivers power when they need it and/or takes the excess power they sell back to the utility."[137] These costs for the utility can include distribution, transmission, utility operating and maintenance

130 Hanna, Benjamin. "FERC Net Metering Decisions Keep States In The Dark." *B.C. Envtl. Aff. L. Rev.*, vol. 42, 2015, p. 138.

131 "Rooftop Solar." *SEIA*, http://www.seia.org/policy/distributed-solar.

132 White, Alexander. "Compromise in Colorado: Solar Net Metering and the Case for 'Renewable Avoided Cost.'" *U. Colo. L. Rev.*, vol. 86, 2015, p. 1096.

133 *Solar Energy and Net Metering*, Edison Electric Institute, (January 2016) http://www.eei.org/issuesandpolicy/generation/NetMetering/Documents/Straight%20Talk%20About%20Net%20Metering.pdf.

134 Warrick, Joby. "Utilities Wage Campaign against Rooftop Solar." *The Washington Post*, WP Company, 7 Mar. 2015, https://www.washingtonpost.com/national/health-science/utilities-sensing-threat-put-squeeze-on-booming-solar-roof-industry/2015/03/07/2d916f88-c1c9-11e4-ad5c-3b8ce89f1b89_story.html?utm_term=.acbff2e6b129.

135 Solar Energy, *supra* note 133.

136 *Ibid.*

137 *Ibid.*

expenses, utility administrative and general expenses, taxes, and other public benefit charges.[138]

Proponents of net metering argue that it makes solar more accessible, strengthens reliability, and reduces carbon emissions. Net metering has offset installation costs "by sharply lowering electric bills to offset the $10,000 to $30,000 cost of rooftop panels."[139] Furthermore, those in favor of solar P.V. argue that "solar users also save utilities money in assorted ways, including by reducing the amount of energy utilities must generate or buy from conventional power plants, lowering utilities' need for large capital investments and increasing the grid's resilience."[140]

It is difficult to determine whether net metering is beneficial to non-participants as the results of various studies are incomparable because they utilize different sets of variables.[141] Certainly it is benefitting everyone's climate, if not their bank account. Studies from state legislatures and regulatory commissions have generally found that most "net metering does not impose a significant cost to non-participants at current levels of penetration." California found that net metering imposes costs to all of its ratepayers while Mississippi found benefits for its population overall.[142] Although most U.S. states have net metering, the statutes and regulations vary as do the implementation of the statutes and regulations.[143]

8.3.3 Feed-in tariffs (F.I.T.s)

Less popular than net metering is the concept of feed-in tariffs (F.I.T.s). F.I.T.s require utilities or wholesale purchasers to purchase renewable energy power at rates set by the government.[144] F.I.T.s are more common in other parts of the world than they are in the U.S.A., where they are only mandated in six states.[145] That fact makes F.I.T.s a good marker from which to compare the economi-

138 Kenneth Rose and Karl Meeusen, *Reference Manual* (MARCH 22, 2006), U.S. Department OF Energy, https://energy.gov/sites/prod/files/Manual%20for%20Implementation%20of%20PURPA%20Standards%20in%20EPACT%202005%20(March%202006).pdf, p. 36.

139 Warrick, Joby. "Utilities Wage Campaign against Rooftop Solar." *The Washington Post*, WP Company, 7 Mar. 2015, https://www.washingtonpost.com/national/health-science/utilities-sensing-threat-put-squeeze-on-booming-solar-roof-industry/2015/03/07/2d916f88-c1c9-11e4-ad5c-3b8ce89f1b89_story.html?utm_term=.acbff2e6b129.

140 Leslie, Jacques. "Opinion | Nevada's Solar Bait-and-Switch." *The New York Times*, The New York Times, 1 Feb. 2016, http://www.nytimes.com/2016/02/01/opinion/nevadas-solar-bait-and-switch.html?_r=0.

141 Francisco Flores-Espino, *Compensation for Distributed Solar: A Survey of Options to Preserve Stakeholder Value*, Technical Report, National Renewable Energy Laboratory (2015) https://www.nrel.gov/docs/fy15osti/62371.pdf.

142 *Ibid.*

143 "Net Metering." *SEIA*, http://www.seia.org/policy/distributed-solar/net-metering.

144 Lieben, Ivan. "Making Renewable Energy FiT: A Feed-in-Tariff Certifying Body Could Accelerate Renewable Energy Deployment in the United States." *Nat. Resources & Env't.*, vol. 52, 2012, p. 161.

145 "U.S. Energy Information Administration - EIA - Independent Statistics and Analysis." *Feed-in Tariff: A Policy Tool Encouraging Deployment of Renewable Electricity Technologies - Today in Energy*

cally determined U.S. system to public interest government control in other countries, since F.I.T.s are performance-based rather than investment-based. The power generated by a utility customer's system is compensated at the rate set by the F.I.T. rather than the retail electricity rate.[146] By comparison, in net metering a utility customer is "effectively paid the retail rate for any generation that is fed back into the grid."[147]

Feed-in tariffs are not "taxes or tariffs as commonly understood in the United States" but are instead "a consumer funded subsidy" for renewable energy.[148] F.I.T.s work by having legally required long-term purchase contracts offered by the government to renewable energy generators "at wholesale rates that return to them the cost of installation plus a reasonable profit."[149] They succeed in increasing renewable energy generation because the government-set price "encourages production and use while the costs are passed on to the consumer."[150] Although feed-in tariffs can "raise costs for consumers who end up paying the bill," those in favor of the mechanism argue that in the long-term F.I.T.s help reduce dependency on fossil fuels, help utilities keep up with increased electricity demand, stabilize energy markets, and protect electricity grids from disruption by encouraging distributed generation.[151] Because they allow citizens and businesses to participate, feed-in tariffs "tend to reduce social opposition" to the development of renewable energy.[152]

In California, the Sacramento Municipal Utilities District (S.M.U.D.) F.I.T., which was created in 2010, has not been expanded because establishing F.I.T.s requires utilities to offer a set price to pay for power generated and the continuing drop in solar costs means that there is a risk that utilities may have to pay more than necessary for solar generated energy.[153]

8.3.4 Value-of-solar (V.O.S.) tariffs

Between West Virginia and California, one finds Minnesota. The value-of-solar (V.O.S.) tariff is an option in the U.S.A., but only in Minnesota and Texas.[154] Under

- *U.S. Energy Information Administration (EIA)*, https://www.eia.gov/todayinenergy/detail.php?id=11471.

146 *Ibid.*

147 *Ibid.*

148 Lieben, *supra* note 144 at 161.

149 *Ibid.* at 157.

150 *Ibid.* at 161. *Ibid.* at 161–164.

151 *Ibid.*

152 *Ibid.*

153 Trabish, Herman K. "RIP FITs: As US Feed-in Tariffs Fade, Adopting Elements Could Spur Solar Growth." *Utility Dive*, 18 July 2016, http://www.utilitydive.com/news/rip-fits-as-us-feed-in-tariffs-fade-adopting-elements-could-spur-solar-gr/422727/.

154 "Value-of-Solar Tariffs." *NREL.gov*, https://www.nrel.gov/state-local-tribal/basics-value-of-solar-tariffs.html.

a V.O.S., consumers continue to purchase all of their energy at the utility's retail rate but are separately compensated for their solar P.V. generation at a separate V.O.S. rate in dollars per kilowatt hour (\$/kWh).[155] The benefit of this approach is that it allows utilities to better understand customer load, timing, and volume.[156] It also helps utilities avoid purchasing energy from polluting sources, helps build capacity to meet peak energy needs, provides energy in the long-term at a fixed price, and reduces wear and tear on the electric grid and transmission system.[157] In addition, customers receive compensation based on the benefits and costs of their generation instead of a fixed rate that may span many regions.

Minnesota was the first U.S. state to adopt net metering in 1983 and it was the first state to approve V.O.S. in 2014.[158] The state officially adopted the V.O.S. in July of 2016.[159] Its net metering system applies to all investor-owned utilities, municipal utilities, and electric cooperatives.[160] Customers in the V.O.S. forfeit to the utility net excess generation, utilities choose whether to adopt V.O.S. or keep net metering, and the utility automatically obtains solar renewable energy credits (S.R.E.C.), with no S.R.E.C. credits going to the consumer.[161]

Minnesota's V.O.S. legislation requires that the energy valuation method be established by interested persons with experience and expertise in power systems, solar energy, and electric utility ratemaking.[162] Using the enabling statute, input from interested persons, and guidance from its own staff, the Minnesota Department of Commerce has designed a method for calculating the V.O.S. tariff for the Minnesota Public Utility Commission.[163] A unique component of the Minnesota method is known as "avoided environmental cost."[164] Determining the cost was based upon state and federal estimates, calculating the costs of carbon dioxide, particulate matter, and nitrogen dioxide[165]—a "significant feature" that "expressly includes the environmental value of solar within the customer compensation rate."[166]

The V.O.S. is rather new and utilities are not bound by it, as they can adopt V.O.S. on a voluntary basis.[167] The V.O.S. method "adds a layer of complexity,"

155 *Ibid.*

156 *Ibid.*

157 Farrell, John. Minnesota's Value of Solar (April 2014) https://ilsr.org/wp-content/uploads/2014/04/MN-Value-of-Solar-from-ILSR.pdf.

158 Miller, Addison. *A New "Sunshine State"? Evaluating Minnesota's Value of Solar Methodology*, Geo. Wash. J. Energy &Envtl. L. vol. 7, 2016, pp. 177, 180.

159 Jossi, Frank. "'Value of Solar' Gains Another Foothold in Minnesota." *Energy News Network*, 22 July 2016, http://midwestenergynews.com/2016/07/22/value-of-solar-gains-another-foothold-in-minnesota/.

160 U.S. Energy, *supra* note 3.

161 Farrell, *supra* note 157.

162 MINN. Stat. Section 216B.154 Subdiv. 10(e).

163 Farrell, *supra* note 157.

164 Miller, *supra* note 158. At 184.

165 *Ibid.*

166 *Ibid.*

167 Trabish, Herman K. "A Rising Tension: 'Value-of-Solar' Tariff Versus Net Metering." *A Rising Tension: 'Value-of-Solar' Tariff Versus Net Metering | Greentech Media*, Greentech Media, 10 Apr.

because its environmental value "becomes yet another obligation for regulators to oversee."[168] Being "the most controversial calculation," the avoided environmental costs are calculated carbon savings that "are different from actual savings to the utility, since no functioning price of carbon exists in Minnesota or the region."[169] Therefore, it is difficult to know whether the "carbon component is expensive or cheap relative to a transparent market price."[170] Moreover, in attempting to find common ground between the utility's avoided cost and retail rates, the V.O.S. rate, as it is currently is, "exceeds present retail rates for residential-end users."[171] As of March 2018, no utility had adopted V.O.S. in lieu of net metering.[172]

8.3.5 Community solar and virtual net metering

Twenty-five U.S. states have at least one community solar project online.[173] Community solar gardens are solar P.V. systems which "are larger, both by available generation capacity and acreage, and allow customers that are unable or choose not to have rooftop solar P.V. to participate in a solar program."[174] In community solar programs, instead of consumers having solar panels on their rooftops, utilities develop a solar farm or work with a third-party developer.[175] Consumers instead can buy a panel or a share of the solar community system.[176] Virtual net metering (V.N.M.) is a billing crediting system specifically for community solar, where consumers receive credits on their electric bill for excess energy produced by the consumer's share of the solar garden.[177] V.N.M. regulations vary by state and by utility.[178]

2014, https://www.greentechmedia.com/articles/read/A-Rising-Tension-Within-the-Solar-Industry-Value-of-Solar-Versus-NEM.

168 Miller, *supra* note 158 at 185.

169 Taylor, Mike. "The Minnesota Solar Experiment - Value of Solar: Part III." *Renewable Energy World*, 20 Dec. 2016, http://www.renewableenergyworld.com/articles/2016/12/the-minnesota-solar-experiment-value-of-solar-part-iii.html.

170 *Ibid.*

171 Miller, *supra* note 158 at 186.

172 *Ibid.*

173 "Community Solar." *SEIA*, http://www.seia.org/policy/distributed-solar/shared-renewablescommunity-solar.

174 Distributed Energy Resources (November 2016) http://pubs.naruc.org/pub/19FDF48B-AA57-5160-DBA1-BE2E9C2F7EA0, p. 46.

175 Ahlen, Jay. "The Consumer-Centric Utility." *Fortnightly*, https://www.fortnightly.com/fortnightly/2016/07/consumer-centric-utility.

176 *Ibid.*

177 Richardson, Luke. "Virtual Net Metering: What Is It? How Does It Work?: EnergySage." *Solar News*, EnergySage, 3 July 2019, http://news.energysage.com/virtual-net-metering-what-is-it-how-does-it-work/.

178 Belier, Larry. "What Is Virtual Net Metering?: Paradise Energy Solutions." *What Is Virtual Net Metering Comments*, https://www.paradisesolarenergy.com/blog/what-is-virtual-net-metering/.

8.3.6 Disruption and opportunity

Net metering, feed-in tariffs, value of solar, and virtual solar mean different things to different states. On one hand this is helpful, because each state is different in terms of geography, topography, climate, culture, and economy. All of these new and evolving schemes are making it difficult for utilities to continue with their traditional methods of doing business. Distributed generation "is turning the traditional model upside down by trending away from large, centralized generation connected to the interstate build transmission system, to building and integrating new resources at and connected to the distribution grid."[179] The rapid increase in distributed energy resources (D.E.R.s) means that "the current regulatory and utility models are a constraint to effectively address the growth of D.E.R. and its impacts on utility and regulatory frameworks."[180] Because distributed energy resources such as solar P.V. are becoming more prevalent, they reduce the need for utilities to expand operations.[181]

While it is possible for consumers to remove themselves from the grid, actually doing so at this point is complex, expensive, and impractical.[182] New technologies offer utilities a chance to evolve along with their customers. Creating a "consumer-centric" utility would make the most efficient use of distributed energy resources in conjunction with traditional generation, transmission, and distribution assets.[183] A consumer-centric utility also empowers consumers by facilitating new service offerings that "are tailored to local conditions and specific consumer preferences."[184] This helps utility providers to evolve while managing risk and allows consumers to make best use of distributed energy resources.[185]

8.4 Comparing the functions of economic incentives for environmental protection

To compare two cultures, in this case the United States and the Netherlands, does not mean choosing a winner or a loser. It does not mean that one country should follow the other. It is examining the issues and nuances that affect countries in a particular area and how that helps them function. What has been examined in this chapter is only one small slice of the issues surrounding the interjection of renewable, in this case solar P.V., into a country's electric grid. It

179 Distributed Energy Resources, 2016. http://pubs.naruc.org/pub/19FDF48B-AA57-5160-DBA1-BE2E9C2F7EA0, p. 42.

180 *Ibid.* at 59.

181 Leslie, *supra* note 140.

182 Aggarwal, Vikram. "Can I Go Off-Grid With Solar Batteries in 2019?: EnergySage." *Solar News*, EnergySage, 17 Dec. 2018, https://news.energysage.com/can-i-go-off-the-grid-with-solar-batteries/.

183 Ahlen, *supra* note 175.

184 *Ibid.*

185 *Ibid.*

is not meant to be exhaustive. Instead, it is meant to give the reader an overview of some of the technical and policy considerations that these countries are facing as they promote the use of new technologies through financial means.

The United States and the Netherlands share common cultural and democratic values. It should therefore be no surprise that in large part what motivates them to adopt renewable technologies is the overwhelming willingness or demand of their citizens for cleaner energy. Yet they face vastly different reasons from those who protest the adoption of renewables. In the United States, it is the coal producers and their constituents that are some of the main drivers of opposition to renewable energy on the grid. While the United States is full of vast spaces, in the Netherlands it is more of a concern of how renewables will affect the limited space available. It is also facing the inevitability of saying goodbye to its longstanding gas reserves, while the United States is booming on the recent discovery of its own. Both countries face a challenge of how to integrate renewables into a grid system that was not built to accept them.

While the Netherlands has an overarching obligation to the E.U. to reduce its emissions, the individual states in the United States do not have the same kind of pressure from the U.S. federal government. The Netherlands is motivated to reach certain targets along with several other Member States to meet the Paris Agreement goals, whereas the United States is slowly backing away from those agreed-upon goals. The Dutch government then is following through with these commitments by helping industry and local communities collaborate on methods to achieve this. In the United States, individual states and local communities are taking it on themselves to adopt more rigorous approaches to reducing carbon emissions by encouraging the use of renewables. In the Netherlands, it is more of a top-down approach, whereas the United States is taking more of a bottom-up approach.

In both countries, consumer engagement is a key in making any kind of renewable incentive plan work. As seen in the value of the solar program in Minnesota, utilities are not willing to take steps to embrace new programs if they are cost-prohibitive. Likewise, consumer awareness and access determine the success of these programs. In planning to integrate clean energy onto the grid, it is necessary for regulators to examine rate structures "so that they adequately value both capacity and energy."[186] Regulators in both countries should also recognize that, as the costs of renewables such as solar P.V. decrease, consumers "desire more control over their energy use" and "continue to demand safe, affordable electric service."[187] Therefore, it would help regulators to help

186 *Integrating the Grid,* Electric Power Research Institute, (February 10, 2014), http://www.epri.com/abstracts/Pages/ProductAbstract.aspx?ProductId=000000003002002733&Modedownload.

187 Ahlen, *supra* note 175.

create "consumer-centric" policies that give consumers greater access to a "pool of resources."[188]

In both the United States and the Netherlands, what ultimately drives the adoption of economic incentives is whether they serve a function. Despite their challenges, economic incentives are helping to reduce carbon emissions and are providing a path toward electric reliability and energy security. These economic incentives are helping both countries launch new industries and new jobs. Their ultimate success will come when these incentives are no longer needed and renewables such as solar P.V. become self-sustaining.

8.5 Conclusions

For most of the world's cultures today, life without electricity would be unthinkable. For decades the use of dirtier fuels such as coal was considered essential for electricity generation. Even if cleaner options like natural gas or renewables were available, their use was considered impractical and cost-prohibitive. Although this argument is still made in some circles, the reality is that cleaner energy options, including renewables, are not only more accessible, but also increasingly more cost-effective, especially if one includes long-term benefits to climate and human health. Technological advances and economic incentives, enabled by law, are making them more competitive. In the Netherlands, the Dutch citizens and the E.U. are driving forces in this effort. The United States is far more disparate in its motivations, in part due to its large size and the lack of clear direction on the federal level. Nevertheless, some U.S. states like Minnesota are taking a practical approach toward renewables, recognizing them as an integral part of the future for electricity.

Literature

Campbell, Richard J. "Weather-Related Power Outages and Electric System Resiliency." *Cong. Research Serv.* R42696, 2012 August 28. 29 July 2019. https://crsreports.congress.gov/product/pdf/R/R42696.

Caspary, Jay. "Electric Transmission 101: Markets, ISO/RTOs and Grid Planning/Operations." *Environ. Energy Study* Inst. 29 July 2019. http://www.eesi.org/files/070913_Jay_Caspary.pdf.

Coddington, Michael, O'Neil Rebecca, and Emma Stewart. "Distributed Energy Resources (DERs)." Presented at U.S. Department of Energy Grid Modernization Laboratory Consortium, January 1617, posted at Berkley Lab. 19 July 2019. https://emp.lbl.gov/sites/default/files/4._coddington_stewart_oneil_ders_updated.pdf.

Flores-Espino, Francisco. *Compensation for Distributed Solar: A Survey of Options to Preserve Stakeholder Value*. National Renewable Energy Laboratory, 2015, vii, 29 July 2019. https://www.nrel.gov/docs/fy15osti/62371.pdf.

188 *Ibid.*

Farrell, John. *Minnesota's Value of Solar: Can a Northern State's New Solar Policy Defuse Distributed Generation Battles?* Institute for Local Self-Reliance, April 2014. https://ilsr.org/wp-content/uploads/2014/04/MN-Value-of-Solar-from-ILSR.pdf

Hanna, Benjamin. "FERC Net Metering Decisions Keep States in the Dark." *BC Environ. Aff. Law Rev.* 42 (2015): 133, 138.

Hempling, Scott, et al. *Renewable Energy Prices in State-Level Feed-In Tariffs: Federal Law Constraints and Possible Solutions.* National Renewable Energy Laboratory, January 2010. 29 July 2019. http://www.nrel.gov/docs/fy10osti/47408.pdf.

Netherlands Ministry of Economic Affairs and Climate, Progress Report, Energy from renewable sources in the Netherlands 20152016, Directive 2009/28/EC, European Commission 4. 10 July 2018.

Rose, Kenneth, and Karl Meeusen. *Reference Manual and Procedures for Implementation of the 'PURPA Standards' In the Energy Policy Act of 2005.* U.S. Department of Energy, 22 March 2006. 29 July 2019. https://energy.gov/sites/prod/files/Manual%20for%20Implementation%20of%20PURPA%20Standards%20in%20EPACT%202005%20(March%202006).pdf.

Tomain, Joseph P. and Richard D. Cudahy. *Energy Law in a Nutshell.* West Academic, 2011.

Umberger, Allyson. "Distributed Generation: How Localized Energy Production Reduces Vulnerability to Outages and Environmental Damage in the Wake of Climate Change." *Golden Gate Univ. Environ. Law J.* 6 (20122013): 183, 193.

Van De Graaff, W.J. Evert, Lucia Van Geuns, and Tim, Boersma. *The Termination of Groningen Gas Production – Background and Next Steps.* Columbia SIPA Center for Global Energy Policy, July 2018. 29 July 2019. https://energypolicy.columbia.edu/sites/default/files/pictures/CGEP_Groningen-Commentary_072518_0.pdf.

9

ENVIRONMENTAL IMPACT ASSESSMENT SYSTEMS

Dennis Agelebe, Marek Prityi, and Jorge Ignacio García Nielsen

9.1 Introduction

The emergence of the concepts of Environmental Impact Assessments (E.I.A.s) and Environmental Impact Statements (E.I.S.s) can be seen as the result of efforts to devise tools to steer planning and decision-making processes with impact on the environment, health, and safety of the people.[1] The arrangement of the cultures compared in this chapter is chronological according to when each adopted some form of these tools. Thus, the U.S.A. is discussed prior to the regional E.U., and the E.U. prior to international law, because that was the chronological order of their development. From that, we can observe an important lesson for Type I comparative law in practice, and that is that individual states cannot only influence other individual states directly, but also international public law. In turn, one can then make Type II comparisons of how states align themselves with common commitments under international law in this area.

The underlying concept is that an E.I.S. or E.I.A. protects the environment by forcing a person to consider all environmental impacts *before* taking action, assuming that without having gone through the exercise of completing an E.I.S. or E.I.A. the person might not have recognized the environmental ramifications of the proposed project or process. The project or process could be from a wide array of possibilities, from construction of a new manufacturing plant to augmentation of one of the energy-producing operations described in Chapter 8, *supra*.

1 Caldwell, Lynton K. "Environmental Impact Analysis (E.I.A.): Origins, Evolution and Future Direction." *Impact Assessment*, vol. 6, no. 3-4, pp. 75, 76.

The E.I.A. was first adopted in the U.S.A. in the early seventies, and has now recorded nearly fifty years of experience.[2] The invention of the E.I.S. was preceded by the rapid urbanization and industrialization of Western countries, particularly after the Second World War. There was the need to rebuild, reinvest, and revive the economy by Western countries which resulted in the arbitrary exploitation of natural resources, thereby causing pollution, low quality of life, and environmental degradation. In response to calls for plans and actions that would ensure responsible and safe use of the environment, the U.S.A. enacted the E.I.S. legislation, thus becoming the first country to use it as a tool for protection of the environment.

Although the U.S.A. has not explicitly enshrined the precautionary principle in the law the way Germany has done, for example, the notion of the E.I.S. is seen to be the precautionary principle in practice. The great legal limitation of the E.I.S. in the U.S.A. is that the government has imposed the requirement only on itself and its contractors through the National Environmental Policy Act (N.E.P.A.) The original U.S. E.I.S. was not required for private persons before undertaking actions that might harm the environment.

In addition to the theory that the environment would be better protected when potential polluters are first required to demonstrate that their projects will not pollute or break the law, there are at least two legal advantages in requiring an E.I.A. before beginning or changing a structure or process. First, the great legal advantage of the precautionary principle and E.I.A. in practice is that it shifts the advocate's burden of proof from the state needing to prove that a person has violated the law to the person needing to prove that anticipated projects or processes will not violate the law. This is largely an evidentiary matter, but in an adversarial system such as that of the U.S. it is extremely important for the parties. Second, the E.I.A. adds a layer of standards to which a new operator or process needs to comply, albeit a self-imposed layer of standards. This assumes that it is always easier for discussion to occur among citizens, the operator, and the state when the object of discussion has been described in a standardized fashion.

Among all the forms of environmental protection tools available to states through the law, what is the function of the E.I.A.? Unlike controls that penalize wrongdoers after the fact, the E.I.A. was created to shift the burden from the state to the individual using the precautionary principle, and from proving that a harm had occurred after the fact to the proof that no harm would occur before the fact. The purpose of an E.I.A. is "to contribute to environmental awareness among proponents and competent authorities and to environmental protection by requiring the *ex ante* evaluation of the environmental impacts of a wide range

2 Glasson, John, et al. *Introduction to Environmental Impact Assessment: Principles and Procedures, Process, Practice, and Prospects*. Taylor and Francis, 2013, p. 3.

of public and private initiatives."[3] According to the International Association for Impact Assessment (I.A.I.A.), the E.I.A. has a dual nature: on the one hand, it is a tool to provide environmental information and, on the other hand, it is an "institutional procedure linked to the decision-making process of a planned intervention."[4] Following this understanding, the I.A.I.A. establishes four objectives for the E.I.A. First, it is to provide information for decision-making that analyzes the biophysical, social, economic, and institutional consequences of proposed actions. Second, the E.I.A. is to "[p]romote transparency and participation of the public in decision-making." Third, an E.I.A. identifies "procedures and methods for the follow-up (monitoring and mitigation of adverse consequences) in policy, planning and project cycles." And fourth, an E.I.A. contributes "to environmentally sound and sustainable development."[5]

It must be remembered, however, that the "E.I.A. is an aid to decision-making. It is not a substitute for decision-making, but it does help to clarify some of the trade-offs associated with a proposed development action, which should lead to more informed and structured decision-making."[6] There is broad consensus on the main objective of the E.I.A., which is to influence decision-making processes of regulatory and permitting bodies by providing environmental information on the consequences of development. The challenge has been to connect the E.I.A. process with its objective by effectively improving decision-making accordingly with the environmental information provided.

Many commentators have compared similarities in E.I.A. methods and core principles across the world.[7] There are cultural variations on the structure and implementation of the E.I.A., such as the differences between developed and developing countries, the form of government, as well as different traditions in the practice of law.[8] More specifically, these factors include the historical and geographical background of the countries, experience, and cooperation between state administrations, which is related also to historical and cultural acceptance of Western influence, participation in international and supranational organizations and international agreements, and, in the case of transitioning and developing countries, the extent of international assistance.[9]

3 Runhaar, H., et al. "Environmental Assessment in the Netherlands: Effectively Governing Environmental Protection? A Discourse Analysis." *Environmental Impact Assessment Review*, vol. 39, p. 13.

4 IAIA 2009. What Is Impact Assessment? : International Association for Impact Assessment. 1, https://www.iaia.org/uploads/pdf/What_is_IA_web.pdf (accessed on 20 August 2019).

5 *Ibid.*

6 Glasson, *supra* note 2 at 7.

7 Bremer, Nicolas. "Post-Environmental Impact Assessment Monitoring of Measures or Activities with Significant Transboundary Impact: An Assessment of Customary International Law." *Reciel*, vol. 26, no. 1, 2017, p. 80.

8 Abaza, Hussein, et al. *Environmental Impact Assessment and Strategic Environmental Assessment: towards an Integrated Approach*. UNEP, 2004, p. 17.

9 Cherp, Aleh. "Environmental Assessment in Countries in Transition: Evolution in a Changing Context." *Journal of Environmental Management*, vol. 62, 2001, pp. 357, 371.

Experience has exposed shortcomings in the E.I.A. procedures as well. They include the limitations in public involvement in decision-making procedures, in particular with respect to rural or otherwise isolated communities, insufficient understanding of E.I.A. procedures on the part of state administration and the public, inadequate legal and institutional framework coupled with unsatisfactory enforcement of laws, and a lack of coordination between various agencies at a national and local level.[10]

A third tool is the Strategic Environmental Assessment (S.E.A.) All of these tools—the E.I.S., E.I.A., and S.E.A.—can be described as "structured approaches for obtaining and evaluating environmental information prior to its use in decision-making."[11] The S.E.A. specifically has been defined as "formalized, systematic and comprehensive process of identifying and evaluating the environmental consequences of proposed policies, plans or programmes to ensure that they are fully included and appropriately addressed at the earliest possible stage of decision-making."[12] The purpose of an S.E.A. rests in identifying risks embedded in policies, plans, or programs in advance. It can be seen as an articulation of the need to extend the scope of the E.I.A. beyond mostly project-oriented E.I.A. applications.[13] The E.I.A. is more specifically "a process of evaluating the likely environmental impacts of a proposed project or development."[14] Apart from environmental impact, an E.I.A. should ideally consider also "inter-related socio-economic, cultural and human-health impacts, both beneficial and adverse."[15]

9.2 Environmental impact assessment in the Member States of the European Union

Within the European Union (E.U.), harmonization is achieved predominantly by the legal tools of regulations and directives. While regulations are directly applicable in all Member States and are binding in their entirety, directives are binding as to the result they aim to achieve, leaving the national authorities the discretion to decide on the forms and methods.[16] Even though the domestic

10 Abaza, *supra* note 8 at 16.

11 *Ibid.* at 6.

12 Convention of Biological Diversity, *Guidelines for Incorporating Biodiversity-Related Issues into Environmental Impact Assessment Legislation and/or Process and in Strategic Environmental Assessment, UNEP/CBD/COP/6/20*, pp. 93–94.

13 Morgan, Richard K. "Environmental Impact Assessment: The State of the Art." *Impact Assessment and Project Appraisal*, vol. 30, no. 1, 2012, pp. 5, 10.

14 Convention, *supra* note 12.

15 *Ibid.* At this stage it is also appropriate to mention the critique of such an integration approach related to the risk that when too many aspects need to be considered, some of them can be inevitably downplayed. See Morrison-Saunders, Angus, and Thomas B. Fischer. "What Is Wrong With EIA and SEA Anyway? A Sceptic's Perspective on Sustainability Assessment." *Tools, Techniques and Approaches for Sustainability*, 2009, pp. 221, 229.

16 Treaty on the Functioning of the European Union, Art.288; see also Craig, Paul P., and Búrca Gráinne De. *EU Law: Text, Cases, and Materials.* 2011.

laws of the individual Member States of the E.U. may differ, a Type II comparison can be made of the Member States' implementation of their common E.U. directives. E.I.A. procedures represent one of the most prominent examples of E.U. law shaping the national law of its Member States. In a further Type II comparison, one could also compare the performance of E.U. law itself to other signatory states of the Aarhus Convention on Access to Information, Public Participation in Decision-Making and Access to Justice in Environmental Matters[17] (Aarhus Convention), and the Convention on Transboundary E.I.A. (Espoo Convention).

The first E.I.A. Directive 85/337/EEC[18] was already in force in 1985. The implementation of the objectives of the E.I.A. Directive at the national level differed in various Member States. In some states, the provisions of the E.I.A. Directive were transposed into existing building and construction laws, while in others they were transposed into laws on protection of land and nature and eventually also separate E.I.A. laws were adopted.[19] The E.I.A. Directive stipulated the list of projects requiring an E.I.A. as mandatory, while leaving a certain discretion for Member States to determine whether screening for projects not named as mandatory is required. The first E.I.A. Directive was notably amended in 1997[20] to align it with the Espoo Convention adopted under the framework of the United Nations Economic Commission for Europe. Then the amendment from 2003[21] aligned the E.I.A. Directive with the provisions of the Aarhus Convention.[22] Gradually, the original E.I.A. Directive needed to be consolidated and replaced with a new E.I.A. Directive, reflecting the developments in the E.U. law and policy that emerged in the course of the previous thirty years. In 2014, E.I.A. Directive 2011/92/EU[23] was adopted.[24]

17 With respect to the Aarhus Convention it is appropriate to reiterate that parties to the Aarhus Convention are not only Member States of the E.U., but also the E.U. itself. The E.U. ratified the Aarhus Convention on 17 February 2005, while the selected E.U. member states ratified the Aarhus Convention on the following dates: Slovak Republic on 5 December 2005, Czech Republic on 6 July 2004, Germany on 15 January 2007, and Austria on 17 January 2005.

18 Directive 85/337/EEC of 27 June 1985 on the Assessment of the Effects of Certain Public and Private Projects on the Environment.

19 Katrlík, Radovan. *Zákon o posudzovaní vplyvov na životné prostredie: Komentár* (Environmental Impact Assessment Act: Commentary), 1st Ed Wolters Kluwer, 2016, p. 15.

20 Directive 97/11/EC of 3 March 1997 Amending Directive 85/337/EEC on the Assessment of the Effects of Certain Public and Private Projects on the Environment.

21 Directive 2003/35/EC of 26 May 2003 Providing for Public Participation in Respect of Drawing Up of Certain Plans and Programmes Relating to the Environment and Amending with regard to Public Participation and Access to Justice Council Directives 85/337/EEC and 96/61/EC.

22 European Commission, *Environmental Impact Assessment – EIA*, http://ec.europa.eu/environment/eia/eia-legalcontext.htm.

23 Directive 2014/92/EU of 16 April 2014 amending Directive 2011/92/EU on the Assessment of the Effects of Certain Public and Private Projects on the Environment.

24 Environmental Impact Assessment, *supra* note 22.

E.U. regulation of E.I.A. procedure is based on the procedure consisting initially of a screening and scoping stage, while creating a relatively ample space for public participation.[25] The purpose of the screening stage is to evaluate whether projects enumerated in Annex II of the E.I.A. Directive are to be examined in the form of E.I.A. To be examined, they need "to have significant effects on the environment."[26] The scoping stage provides developers with a possibility to ask the competent authorities about the information and its scope that should be submitted in E.I.A. report.[27] The purpose of the scoping stage is essentially "to identify which potential impacts are relevant to assess."[28]

Public participation represents a crucial part of the E.I.A. procedure. The value of public participation in E.I.A. procedures has been seen as the articulation of the ideas of "deliberative democracy, collaborative rationality and environmental justice."[29] Multiple arguments speak for the introduction of public participation as a constituent part of the E.I.A. procedure. It is for instance a fair expression of democratic government in decision-making and it enhances the probability that citizens' needs will be properly reflected in projects. Local knowledge and values may improve the final outcomes of the E.I.A. procedure.[30] Apart from gaining valuable inputs from local levels, the purpose of public participation in E.I.A. procedures rests also in public sharing and shifting power structures within the decision-making procedures.[31] According to the provisions of the E.I.A. Directive, the public should be given early and effective opportunities to participate in environmental decision-making, i.e. the public should have the opportunity to express its comments and opinions.[32] Essentially, the provisions on public consultation implement Article 6 of the Aarhus Convention. The E.I.A. Directive introduces the minimum threshold for the public participation, which should not be shorter than thirty days.[33] The E.I.A. Directive also implements the Aarhus Convention Article 9 (2) provisions on access to justice, which tie the access to justice to the criterion of a sufficient interest or impairment of right.[34] The access to justice rights complement the public participation rights and contribute to their enforcement. Their strengthening is crucial, as public

25 Directive 2014/92/EU of 16 April 2014 amending Directive 2011/92/EU on the Assessment of the Effects of Certain Public and Private Projects on the Environment, Art.4-8, Art.11, http://ec.europa.eu/environment/eia/pdf/EIA_Directive_informal.pdf (last updated Dec. 26, 2018).

26 *Ibid.* at 27.

27 *Ibid.* at 30.

28 Convention, *supra* note 12.

29 Morgan, *supra* note 13 at 5, 9.

30 Shepherd, Anne, and Christi Bowler. "Beyond Requirements: Improving Public Participation in EIA." *Journal of Environmental Planning and Management*, vol. 40, no. 6, 1997, p. 725.

31 Morgan, *supra* note 13 at 5, 10.

32 Directive 2014/92/EU of 16 April 2014 amending Directive 2011/92/EU on the Assessment of the Effects of Certain Public and Private Projects on the Environment, Art.6 (4).

33 Directive 2014/92/EU of 16 April 2014 amending Directive 2011/92/EU on the Assessment of the Effects of Certain Public and Private Projects on the Environment, Art.6 (7).

34 *Ibid.* at art.11.

participation is not only considered to be a part of E.I.A. procedure, but is also often defined by it. Still, one should not confuse the two. As the authors of the Aarhus Convention Implementation Guide mention, "while public participation is in fact a mandatory part of environmental assessment, an environmental assessment is not a mandatory part of a public participation procedure."[35] The rationale and spirit of the Aarhus Convention are in fact beneficial from the perspective of enforcement and implementation of the E.U. environmental law in general, which depends on the mobilization of the public. Public participation and access to justice rights are tools designed for this purpose.[36]

9.2.1 Central and Eastern Europe

The evolution of E.I.A. procedures in the countries of Central and Eastern Europe has undergone several stages. The first stage was comprised of the dissolution of Soviet rule, the dictatorial rule of the Communist Parties, and the reforms undertaken to establish market-based economies. Hence, the political conditions, level of socio-economic development, and related factors were a major influence in this regard. The rules and procedures were then transformed from the ones based predominantly on central planning to more transparent and independent procedures. The other important incentive was the preparation for the accession to the E.U., which is also inextricably connected with the implementation of the Aarhus Convention and the Espoo Convention.[37]

As noted by U.N.E.P. experts, the inclusion of public involvement in E.I.A. laws in Central and Eastern Europe was initially insufficiently detailed and therefore unenforceable.[38] The implementation of international standards has especially improved the public participation provisions and contributed indirectly also to the improvement of democratic standards.[39]

9.2.2 Slovak Republic and Czech Republic

When talking about the evolution of the E.I.A. procedure in the Slovak Republic, one needs to be aware of the developments during the common Czechoslovakian past. Shortly after the Velvet Revolution in 1989 the Federal Assembly adopted the *Zákon č.*17/1992 *Zb. o životnom prostredí* (Act on the Environment), which in

35 United Nations Economic Commission for Europe, *The Aarhus Convention: An Implementation Guide*, https://www.unece.org/fileadmin/DAM/env/pp/Publications/Aarhus_Implementation_Guide_interactive_eng.pdf pp. 122, 127.

36 Marty, Michael. "Die Erweiterung des Rechtsschutzes in Umweltangelegenheiten – Anmerkungen Zum Umwelt-Rechtsbehelfsgesetz." *Zeitschrift Für Umweltrecht*, vol. 3, 2009, pp. 115, 117.

37 Abaza, *supra* note 8.

38 *Ibid.* at 28, 32–33.

39 Hayward, Tim. *Constitutional Environmental Rights*. Oxford Univ. Press, 2012, p.144.

its para. 20–26 included the provisions on assessing the impacts on the environment. The Act on the Environment anticipated the adoption of more detailed laws, but the Czechoslovak Federal Republic was in a short period of time dissolved and two successor States were established, the Slovak Republic and the Czech Republic.[40]

In the Slovak Republic, the underlying preconditions for E.I.A. procedure are laid down in the *Ústava Slovenskej republiky* (Constitution). The right to favorable environment enshrined in Article 44 of the Constitution should be read in conjunction with Article 20 (3) of the Constitution, which states that the exercise of property rights should not infringe health, nature, cultural heritage, and the environment beyond the scope stipulated by the law.[41] In 1994, at the time when the Slovak Republic was not yet a Member State of the E.U., the first law on E.I.A. was adopted. Given the need to adjust the laws to meet the E.U. standards, the *Zákon č.24/2006 o posudzovaní vplyvov na životné prostredie*, a new law on E.I.A., was adopted in 2005. Subsequently, the E.I.A. law was repeatedly amended, reflecting the amendments of the E.U. E.I.A. Directive and responding to the comments posed by the European Commission.[42] As was noted in an interview with officials working for the Slovak Ministry of the Environment, the compliance of national E.I.A. law with E.U. legislation was gradually ensured also by the means of infringement proceedings initiated by the European Commission.[43] As one may assume, the E.I.A. law of the Slovak Republic (and essentially also the E.I.A. law of the Czech Republic) follows the general pattern consisting of screening and scoping stage, public participation stage, expert stage, and final opinion.[44] Except for certain provisions, E.I.A. procedure in the Slovak Republic and the Czech Republic is governed by the general laws on administrative procedure.[45]

The progress and improvements related to the accession to the E.U. were noticeable mainly in the field of the rights of the public, namely that the public should have early and effective possibilities to participate in environmental decision-making and corresponding access to justice rights.[46] Currently, the E.I.A. law in the Slovak Republic operates with the terms the *verejnosť* (public) and *dotknutá verejnosť* (the concerned public). In this way, the E.I.A. law reflects the provisions of the E.I.A. Directive and the Aarhus Convention. The public under the E.I.A. law involves essentially everyone.[47]

40 Katrlik, *supra* note 19 at 16.

41 *Ibid.*

42 *Ibid* at 16-17.

43 Interview with an official working for the Slovak Ministry of the Environment undertaken in Bratislava, Slovakia, on 3 April 2017.

44 See Slovak E.I.A. law No. 24/2006 (*Zákon č.24/2006 o posudzovaní vplyvov na životné prostredie*).

45 § 64 of the Slovak E.I.A. law (*Zákon* č.24/2006 *o posudzovaní vplyvov na životné prostredie*); see also Katrlik, *supra* note 19 at 297-99.

46 *Ibid* at 16-17.

47 *Ibid*. at 33-34.

This is in line with the so-called "any person principle" applied by the Aarhus Convention.[48]

The concerned public consists of persons affected by any activities involving the environment and expressing their interest to participate in the procedure. These people include for instance the people neighboring with the lands affected by environmental impacts or people and organizations with genuine interest in environmental decision-making. For example, environmental N.G.O.s are automatically considered to be organizations with a genuine interest in the E.I.A. procedure.[49] The rationale of the Aarhus Convention states that the objectives listed in the articles of incorporation of the organization referring to the interests of environmental protection determine whether an organization should be considered a member of the concerned public.[50] To be recognized as a member of the concerned public, an interest for this purpose encompasses not only a legal interest, but also a factual one. Moreover, the category of potentially affected interests covers a broad spectrum of interests ranging from violations of social rights or the right to healthy environment to unspecified interests in decision-making procedure. The category of the concerned public is closely connected to the public participation rights under Article 6[51] and access to justice[52] rights under Article 9 (2) of the Aarhus Convention.[53] In fact, as was mentioned in an interview with officials working at the department of the environment at the local level, E.I.A. procedure has become more complex and bureaucratic and the officials observed a tendency to make use of access to justice rights and to push the matters before a court. Considering the complexity and bureaucratic character of the E.I.A. procedure, the officials complained about being

48 United Nations Economic Commission for Europe, *The Aarhus Convention: An Implementation Guide*, 55.

49 Katrlik, *supra* note 19 at 33–34. In this regard it might be also mentioned that, in the opinion of the Aarhus Convention Compliance Committee, environmental N.G.O.s are considered to have the ability to exercise the rights under the Aarhus Convention in a more effective manner than individual members of the public. See United Nations Economic Commission for Europe, *The Aarhus Convention: An Implementation Guide*, 58.

50 Marty, *supra* note 36 at 115, 120.

51 As noted by the authors of the Aarhus Convention Implementation Guide, "[e]ven if the Aarhus Convention does not establish an E.I.A. regime per se, its article 6 does establish a kind of review of the environmental impacts of particular activities, where decision-making in relation to them takes place." See United Nations Economic Commission for Europe, *The Aarhus Convention: An Implementation Guide*, 128.

52 The access to justice rights and a consequent and clearer implementation of Art. 9 of the Aarhus Convention in the Slovak Republic has been recently improved by adopting the Administrative Court Procedure Act (*Správny súdny poriadok*) in 2015. Its § 42 titled "Interested public" (*zainteresovaná verejnosť*) provides access to justice rights to members of the concerned public who have been in the position of the parties in administrative procedures. See Katrlik, *supra* note 19 at 172.

53 United Nations Economic Commission for Europe, *The Aarhus Convention: An Implementation Guide*, 57.

overwhelmed with formal matters resulting in a lack of time to focus on substantive aspects.[54]

The *Zákon č.100/2001 o posuzování vlivů na životní prostředí* (E.I.A. law in the Czech Republic) also makes the distinction between the *veřejnost* (public) and the *dotčená veřejnost* (concerned public), even though the definitions are slightly different from the Slovak law. While the public is understood as one or more persons (under the "any person" principle), the concerned public is specified in a manner including a person affected by a decision adopted in a subsequent procedure and legal person of private law other than a profit-making company created for the purpose of protecting the environment or public health functioning for the period of at least three years or that is supported by at least 200 signatures.[55] The differentiation between the public and the concerned public becomes more relevant in the *navazující řízení* (subsequent procedures),[56] in which only the concerned public has a full-fledged set of rights.[57]

The rights of the public in environmental decision-making in the Slovak Republic are quite significant. However, their extent differs depending on the category of the public. For instance, in the S.E.A. procedure, the public has a right to submit written opinions when a document is subjected to S.E.A. assessment.[58] The rights of the concerned public from the early stages of the S.E.A. procedure are regulated in a separate § 6a of the E.I.A. law. The "concerned public," is defined in this provision as a natural person order than eighteen years of age, a legal person, or a civic inititative. A "civic initiative" is a legal status for "natural persons older than eighteen years of age who submitted a signed common standpoint on a proposal of a strategic document. The term "civic initiative" is a literal translation of the relevant provision of the EIA law. It is neither an NGO nor a variation of the German "*Verein*", but is rather an ad hoc initiative legally recognized only in the EIA or SEA context. The

54 Interview with officials working for the department of the environment undertaken in Nové Zámky, 4 April 2017.

55 § 3 (h)-(i) of the Czech law on E.I.A., Act 100/2001 (*Zákon* 100/2001 *Sb. ze dne 20. února 2001 o posuzování vlivů na životní prostředí*). Even though the definition of the concerned public in the Czech context is different than in the Slovak one, it meets the criteria stipulated by the Court of Justice of the European Union in the case C-263/08 Djurgården, in which the court stipulated that procedural rights, including mainly access to justice rights in the pertinent case, should be open also local associations with the number of members lower than 2,000, as it was stipulated in the Swedish legislation at that time. See C.J.E.U., *Djurgården-Lilla Värtans Miljöskyddsförening v Stockholms kommun genom dess marknämnd*, C-263/08 of 15 October 2009.

56 Czech law on E.I.A. procedure includes an explicit list of subsequent procedures. E.I.A. procedure serves as the basis for the subsequent procedures, including for instance urban planning procedure. See Frank Bold Society, *Změny v zákoně o EIA od 1. ledna 2018* (Changes in E.I.A. Law since 1 January 2018).

57 Frank Bold Society, *Účast spolku v řízeních* (Participation of Associations in Procedures), https://frankbold.org/poradna/kategorie/obcanska-sdruzeni-zalozeni-fungovani-cinnost/rada/ucast-v-rizenich-podle-zvlastnic.

58 Katrlik, *supra* note 19 at 50–54.

rights of the concerned public in this respect range from the possibility of submitting written opinions to participation in consultations and public hearings within the S.E.A. procedure.[59] The rationale of stipulating the rights of the concerned public in a separate section has been justified by the need to have a clear, transparent, and coherent legal framework compatible with the spirit of the Aarhus Convention.[60]

The same justification can be applied also in the case of § 24 of the E.I.A. law, which represents the core provision from the perspective of the public and the concerned public in the current legal framework regulating the E.I.A. procedure in the Slovak Republic. In line with the assumption that the active disclosure of information is one of the crucial preconditions for the active involvement of the public in the E.I.A. procedure, the provision imposes information disclosure obligations on the bodies of state administration. Hence, the organs of state administration are obliged to provide information of sufficient quality in a timely and transparent manner.[61] The rights of the concerned public are tied to the submission of substantiated opinions, comments, or suggestions as well as to the lodging of an appeal by a member of the public against one of the decisions adopted in the E.I.A. procedure. This right to file an appeal is not made conditional upon prior participation in the stages preceding the adoption of the decision.[62] The provision is compatible with the jurisprudence of the Court of Justice of the European Union (C.J.E.U.), which in its decision C-570/13 concluded that also members of the concerned public, such as neighbors of the land where the projects are being planned, need to have the possibility to lodge an appeal and challenge the decisions adopted in E.I.A. procedure.[63] In the *Djurgården* case, the C.J.E.U. made clear that the participation in decision-making and review procedure are two distinct stages and the participation in the first stage should not have an impact on the conditions for access to the review procedure.[64]

By becoming a party to the E.I.A. procedure, the concerned public attains also the right to participate as a party in the subsequent permitting procedures, such as the urban planning procedure or the building permission procedure.[65] Considering the need to take into account the impacts of projects on the environment, the results of the E.I.A. procedure also become relevant for adopting

59 *Ibid.* at 54–58.

60 *Ibid.* at 55.

61 § 24 (1) of the Slovak E.I.A. law (*Zákon č.24/2006 o posudzovaní vplyvov na životné prostredie*).

62 § 24 (2)-(4) of the Slovak E.I.A. law (*Zákon č.24/2006 o posudzovaní vplyvov na životné prostredie*).

63 C.J.E.U., Karoline Gruber vs. Unabhängiger Verwaltungssenat für Kärnten et al., C-570/13 of 16 April 2015, para.50–52.

64 CJEU, Djurgården-Lilla Värtans Miljöskyddsförening v Stockholms kommun genom dess marknämnd, C-263/08 of 15 October 2009, para.38.

65 § 24 (2)-(3) of the Slovak E.I.A. law (*Zákon č.24/2006 o posudzovaní vplyvov na životné prostredie*).

decisions and permits in the permitting procedures.[66] The E.I.A. stage is only one of several stages leading to the realization, implementation, and subsequent enjoyment of projects.[67] In the Slovak Republic, the participation in specified parts of the E.I.A. procedure leads to the automatic participation in subsequent permitting procedures.[68] The legal framework in the Czech Republic aims to create balance and explicit connection between the E.I.A. procedure and related subsequent procedures, such as the urban planning procedure or the building permission procedure.[69] The objective is to create one coherent procedure and to avoid duplicity of procedures.

Even though the legal framework regulating E.I.A. in the Czech Republic does not include a separate provision on the rights of the public to participate in environmental decision-making as the Slovak E.I.A. law does, the provisions on public participation and related access to justice rights are included throughout the law, marking various stages of the E.I.A. procedure.[70] This appears to be essentially in line with the approach taken by the Aarhus Convention, and reflects the discussion about access to justice within (and after) the E.I.A. procedure as described in the jurisprudence of the C.J.E.U.

9.2.3 Germany

The *Umweltverträglichkeitsprüfungsgesetz* (German law on environmental impact assessments) has undergone evolution similar to that of its Slovak and Czech counterparts, owing to a large extent to the need to make it consistent with the corresponding E.U. legislation. This is unsurprising in a Type II comparison. The German approach to E.I.A. follows the rationale of treating the E.I.A. procedure as a precursor and a part of subsequent administrative law procedures, the result of which rests in granting a permit.[71] As such, the E.I.A. procedure is not an independent administrative law procedure, but a pre-step that needs to be taken before a permit for the project is issued.[72] This coherent approach creating

66 See for instance § 39, §66, §140c of the Slovak Act on Urban Planning and Building and Construction Procedure Act (*Zákon 50/1976 o územnom plánovaní a stavebnom poriadku*).

67 Supreme Court of the Slovak Republic, judgment from 14th of Mai 2013, 1Sžp 1/2010 (*Rozsudok Najvyššieho súdu Slovenskej republiky zo 14. mája 2013, sp. Zn. 1 Sžp* 1/2010).

68 Katrlik, *supra* note 19 at 154.

69 See for instance § 3 (g) of the Czech law on E.I.A., Act 100/2001 (*Zákon 100/2001 Sb. ze dne 20. února* 2001 *o posuzování vlivů na životní prostředí*). In this regard it is also relevant to mention that the countries attempting to consider environmental impact in a single procedure with one public participation stage should pay attention to the effectiveness of public participation and its conduct in a timely manner.

70 See for instance § 6 (8), § 7 (6)-(9), § 9c of the Czech law on E.I.A., Act 100/2001 (*Zákon 100/2001 Sb. ze dne* 20. *února 2001 o posuzování vlivů na životní prostředí*).

71 Paragraph 4 of the German E.I.A. law (Umweltverträglichkeitsprüfungsgesetz) states explicitly that *"[d]ie Umweltverträglichkeitsprüfung ist unselbständiger Teil verwaltungsbehördlicher Verfahren, die Zulassungsentscheidungen dienen."*

72 Pu'nder Hermann, and Dirk Ehlers. *Allgemeines Verwaltungsrecht.* De Gruyter., 2016, p. 538.

an explicit connection between the E.I.A. procedure and subsequent procedures resembles the rationale taken by the Czech E.I.A. legislation.

To undergo E.I.A., projects need to meet the criteria stipulated in Annex 1 of the E.I.A. law. The objective of the screening stage is to evaluate whether a certain project should be subjected to E.I.A.[73] The implications of the projects for the environment are decisive in this regard.[74] The elements of continuous public participation are present in the law when describing respective procedural steps.[75] Reflecting the spirit of the Aarhus Convention, the prominent role is given to the concerned public, which is represented by environmental *Vereine* (N.G.O.s). The criteria for the recognition of environmental N.G.O.s as a part of the concerned public are enshrined in § 3 of the *Umwelt-Rechtsbehelfsgesetz* (Environmental Appeals Act). To be recognized as a member of the concerned public, an environmental N.G.O.'s primary goal shall rest in promoting and supporting environmental protection. Its activities shall be genuinely recognized as pursuing non-profit objectives as stated in the Fiscal Code of Germany and it shall be active for a period of at least three years, which resembles Czech legislation.[76] German E.I.A. law rests also in the stipulation of criteria on "internal democratic structure" of environmental N.G.O.s. The democratic structure of the N.G.O.s is evaluated based on whether the members have voting rights during the general meeting of the organization.[77] To be considered a member of the concerned public, the demonstration of the ability of N.G.O.s and their organizational structure to fulfill their statutory tasks in a competent manner is required. For this reason, the record of previous activities, performances, and

73 See § 5-7 of the *Umweltverträglichkeitsprüfungsgesetz* (German E.I.A. law).

74 Philipp-Gerlach, Ursula, and Andreas Lukas. "Die UVP-Vorprüfung in Der Rechtsprechung Und Praxis." *Zeitschrift Für Umweltrecht*, 2014, pp. 548, 549.

75 As project develops and eventual changes are made to the project in the course of the EIA procedure, public participation shall take place on these occasions as well. See § 22 or § 30 of the *Umweltverträglichkeitsprüfungsgesetz* (German EIA law).

76 See § 3 of the *Umwelt-Rechtsbehelfsgesetz* (Environmental Appeals Act).

77 In 2016, a procedure before the Aarhus Convention Compliance Committee was initiated by the organization WWF Germany. WWF Germany argues that the criteria required in Germany for N.G.O.s to be recognized as members of the concerned public are too strict and in particular the requirement referring to "internal democratic structure" excludes from the scope of this provision organizations active in the environmental field other than non-profit organizations, such as foundations or limited companies, which do not offer their members full membership with voting rights. WWF is in Germany registered only as a foundation, i.e. its members pay fees, but the requirement of internal democratic structure is missing. For this reason, it does not meet the criteria for having a full-fledged set of rights. According to WWF, this might violate Art. 3 (4) of the Aarhus Convention, which stipulates that "associations, organizations and groups" should be able to obtain the status of recognition. Art. 2 (4) of the Aarhus Convention argues for the widest range of N.G.O.s to get public participation and standing rights. As of August 2018, this communication is still pending. See § 3 (1) (5) of the Environmental Appeals Act; see also WWF, *Communication to the Aarhus Convention Compliance Committee*, https://www.unece.org/fileadmin/DAM/env/pp/compliance/C2015-137_Germany/Communication_Germany_WWF_resubmitted_10.02.2016.pdf.

memberships need to be shown.[78] With respect to public participation, German law on E.I.A. recognizes that the access to information is the crucial precondition for an early, effective, and meaningful public participation. For this reason, it introduces provisions on information disclosure followed by the provisions enabling the concerned public to submit comments on, suggestions for, and objections to the project.[79]

The compliance with public participation rights is strengthened through access to justice rights. Access to justice rights in environmental matters in Germany are enshrined in the Environmental Appeals Act, the objective of which has been to transpose Article 9 of the Aarhus Convention and corresponding E.U. legislation at a national level. The Environmental Appeals Act also enables one to challenge the decisions adopted within the E.I.A. procedure.[80] German law in general has struggled to fully accommodate and implement the review procedures and access to justice mechanisms as required by the Aarhus Convention and corresponding E.U. legislation.[81] Not only the procedures before the Aarhus Convention Compliance Committee, but also, and mainly, the C.J.E.U.'s jurisprudence, have brought the German legislation closer to the spirit of the Aarhus Convention and E.U. standards.

As was observed already in the 1980s, the effectiveness of E.I.A. law depends on the political will committed to the objective of environmental protection.[82] Experience shows that the effectiveness and relevance attributed to E.I.A. procedures also depends on the approach of state authorities, including the judiciary. Depending on their approach, they might undermine or strengthen the effectiveness of the E.I.A. procedures.[83] The corrections made by the C.J.E.U. in this regard might be seen as a demonstration of the positive contribution of the E.U. and the dialogue between national courts and the C.J.E.U. to the improvement and better enforcement of national environmental law.

With respect to E.I.A. procedure, two decisions of the C.J.E.U. are of main importance. In the *Trianel* case, the C.J.E.U. clarified the scope of access to justice rights for environmental N.G.O.s in the E.I.A. procedure and strengthened them by stating that E.U. legislation governing the E.I.A. procedure precludes national

78 § 3 (1) – (3) of the *Umwelt-Rechtsbehelfsgesetz* (German Environmental Appeals Act).

79 § 19–§ 22 of the *Umweltverträglichkeitsprüfungsgesetz* (German E.I.A. law).

80 § 1 (1) of the *Umwelt-Rechtsbehelfsgesetz* (German Environmental Appeals Act).

81 The access to justice in German administrative law has been to a large extent shaped by the so-called impairment of rights doctrine (*Schutznormtheorie*), which is based mainly on § 42 (2) and § 113 (1) of the German *Verwaltungsgerichtsordnung* (VwGO) (Administrative Court Procedure Act). The impairment of rights doctrine connects the legal standing in administrative law cases with violation of their individual rights by the administrative act. See Marty, *supra* note 36 at 115, 116.

82 Caldwell, *supra* note 1.

83 Eckard Rehbinder, Ziele, Grundsätze, Strategien und Instrumente, in Gründsätze des Umweltrechts, 4th Ed. Klaus Hausmann & Dieter Sellner Eds., Erich Schmidt Verlag, 2012, pp. 135, 192.

legislation that does not permit environmental N.G.O.s to challenge decisions authorizing projects with a potential to have significant impact on the environment.[84] The *Trianel* case clarified the channels for environmental N.G.O.s to gain access to justice in E.I.A. procedure.

Another clarification of access to justice rights by the C.J.E.U. concerned § 4 of the Environmental Appeals Act. The *Altrip* case, which shed light on the scope of review of decisions adopted within the E.I.A. procedure, addressed the issue of whether it is possible to challenge E.I.A. procedure in cases where it took place but was inadequate. The C.J.E.U. held that decisions should be challenged not only in cases when no E.I.A. assessment was carried out, but also in cases when it was carried out, but it was inadequate or irregular.[85]

Moreover, the C.J.E.U. also put certain limitations on the possibility of the court to refuse to recognize impairment of a right in a situation when the contested decision would not have been different without that procedural defect. When evaluating the nature and seriousness of the procedural defect in question, the C.J.E.U. held that one of the major aspects to consider is the impact on public participation rights.[86] The failure to pay adequate attention to public participation rights and the failure to provide the access to information and public participation possibilities fall in the category of absolute procedural mistakes. Their explicit inclusion in the text of § 4 of the Environmental Appeals Act in the form of the *Altrip* amendment might be attributed to the decision of the C.J.E.U. of the same name.[87] According to the interview undertaken with officials working for the German Ministry of the Environment, these matters had been still reviewed in practice by courts even before the relevant case and amendment. However, thanks to the C.J.E.U.'s judgment, the legislation was made more precise, which has enhanced legal certainty and predictability. Also, the respective bodies of state administration and courts can better work with the legislation.[88] Hence, not only a failure to conduct E.I.A., but also procedural deficiencies in the E.I.A. procedure, can justify the right of the concerned public to challenge the decisions adopted within that procedure. As noted above, the concerned public, including the affected people and environmental N.G.O.s, might argue that they have been affected in their own rights by such actions, as required by § 42 (2) of the *Verwaltungsgerichtsordnung* (VwGO),

84 C.J.E.U., Decision C-115/09 from 12 May 2001, *Bund für Umwelt und Naturschutz Deutschland*, Landesverband Nordhein Westfalen e.V. v. Bezirksregierung Arnsberg, para. 50, 59.

85 C.J.E.U., Decision C-72/12 from 7 November 2013, Gemeinde Altrip v. Land Rheinhald-Pfalz, para.38.

86 C.J.E.U., Decision C-72/12 from 7 November 2013, Gemeinde Altrip v. Land Rheinhald-Pfalz, para.57.

87 Fellenberg, Frank. "Weiter Frischer Wind aus Luxemburg – Zu Den Klagemöglichkeiten Im Umweltrecht." *Neue Zeitschrift Für Verwaltungsrecht*, 2015, pp. 1721, 1722.

88 Interview with officials working for the German Ministry of the Environment in Berlin, Germany, on 9 July 2018.

the Administrative Court Procedure Act.[89] In cases of uncertainty about the relevance and gravity of procedural mistakes attributable to the outcome of the E.I.A. procedure, it is presumed that procedural mistakes did influence the outcome of the E.I.A. procedure.[90]

9.2.4 International approach

In a clear exercise of Type II comparative law in practice, many cultures of the world that had not yet instituted a domestic environmental law regime had the concept of the E.I.A. introduced to them through their common membership in international conventions, treaties, declarations, and agreements that had hitherto not instituted an environmental law regime. Under international environmental law, recognized principles provide the needed processes to be applied by law to activities related to the environment. Over the years of their evolution and application, most of these principles have become principles of customary international law, having fulfilled conditions stated in Article 38 of Statute the International Court of Justice. One such principle is the Environmental Impact Assessment principle.[91] The E.I.A. can be defined in broad terms as a systematic study or evaluation for identifying and predicting potential effects of a proposed project, plan, or program on the environment.[92] Seen from this point of view, the E.I.A. has evolved into one of the few mechanisms the results of which are publicly recognized.[93] As such, "E.I.A. is not in itself a permitting or authorization process, [it should be perceived rather as] a tool for decision-making."[94] The universal reputation of the E.I.A. concept rests in its value as an environmental management instrument.[95]

An international consensus on the core attributes of the E.I.A. procedure has gradually developed in the course of time. In 1987, the United Nations Development Programme (U.N.E.P.) adopted a (non-binding) note entitled "Goals and Principles of Environmental Impact Assessment" that sets forth

89 Philipp-Gerlach, Ursula, and Andreas Lukas. "Die UVP-Vorprüfung in Der Rechtsprechung Und Praxis." *Zeitschrift Für Umweltrecht*, 2014, pp. 548, 551.

90 Fellenberg, *supra* note 87 at 1722.

91 The International Court's 2010 judgment in Pulp Mills on the River Uruguay is the most significant authority on E.I.A. in general international law. It is the first occasion on which an international court has held that prior assessment of transboundary impacts is not merely a treaty-based obligation but a requirement of general international law.

92 G. Piero & S. Clare, "European Strategy on Invasive Alien Species: Convention on the Conservation," available at https://books.google.com.ng/books?isbn-9287154880 (last accessed 4 June 2018).

93 Abaza, *supra* note 8 at 6.

94 United Nations Economic Commission for Europe, *The Aarhus Convention: An Implementation Guide*, 127.

95 Morgan, *supra* note 13 at 5, 6.

certain general principles and goals that should be detailed at a national, regional, or even international level.

The U.N.E.P. principles require that states establish criteria to determine whether activities are likely to have an impact on the environment and whether they should be subjected to E.I.A. According to the U.N.E.P. principles, the E.I.A. should meet certain minimum standards, including the evaluation of environmental impacts of the project, consideration of possible alternatives, as well as the identification and description of mitigation and adaptation measures.[96]

The need to undertake E.I.A. for projects with potentially significant impact on the environment was reiterated also in principle 17 of the (non-binding) Rio Declaration on Environment and Development (Rio Declaration) in 1992.[97] Subsequently, the United Nations Conference on Environment and Development (U.N.C.E.D.) in Rio de Janeiro in 1992 formally adopted the E.I.A. Principle 17 of U.N.C.E.D.'s Agenda 21 states: "Environmental Impact Assessment as a national instrument shall be undertaken for proposed activities that are likely to have a significant adverse impact on the environment and subject to a decision of a competent authority." Other international conferences have helped create awareness on the devastating effect of wanton exploitation of the environment without understanding potential dangers to humans, wildlife, and our ecosystem.

Moreover, other principles enshrined in the Rio Declaration are reflected in the E.I.A. procedures as well. For example, principle 10 stipulates the need to introduce public participation and consultation possibilities in decision-making procedures, and principle 15 sets forth the precautionary principle, which is important from the perspective of risk assessment within E.I.A. procedure.[98] E.I.A. procedures should involve risk assessment methods to determine the effects of the project, including the evaluation of potential ecological deterioration or loss of ecological resources, followed by monitoring of impacts and other similar steps.[99]

Apart from the precautionary principle, the polluter pays principle is present in the E.I.A. procedures as well, especially with respect to mitigation and adaptation measures and steps, placing the responsibility for environmental costs of development projects on project proponents.[100] The precautionary principle together with the polluter pays principle are the core principles forming the basis of E.U. environmental law and policy.[101]

The E.I.A. is also implied in Millennium Development Goal 7 (A), which mentions the need to "integrate the principles of sustainable development

96 U.N.E.P., *Goals and Principles of Environmental Impact Assessment, Preliminary Note Issued on January 16, 1987*, 1-3.
97 The Rio Declaration on Environment and Development (1992).
98 Abaza, *supra* note 8 at 31.
99 Abaza, *supra* note 8 at 53.
100 *Ibid.* at 56.
101 Treaty on the Functioning of the European Union, Art.191.

into country policies and programmes and reverse the loss of environmental resources."[102]

Apart from soft law documents, the need to conduct E.I.A. has been emphasized also in the realm of binding international law. A Type II comparison could be made to ascertain how various cultures implement their commitments to international conventions. The obligation to monitor activities with a potential risk to the marine environment, coupled with the obligation to publish reports with the results of such monitoring and the obligation to assess the effects of such activities on the marine environment, is enshrined in the United Nations Convention on the Law of the Sea.[103]

The references to E.I.A. procedures to evaluate the impact on biological diversity are made in Article 14 of the Convention on Biological Diversity.[104] The United Nations Framework Convention on Climate Change sees the E.I.A. procedure as one of the necessary steps to ensure the minimization of negative impacts of projects on climate change and the effectiveness of adaptation measures related thereto.[105]

It has been suggested that certain E.I.A. practices, associated mainly with transboundary E.I.A. procedures,[106] are part of customary international law. This was even explicitly reiterated in the case from 2010 concerning a pulp mill on the Uruguay River between Argentina and Uruguay, which was tried before the International Court of Justice.[107] The obligation to exchange information was recognized as a part of customary international law.[108] Further international law references to E.I.A. have been made for instance in the Convention on Transboundary Environmental Impact Assessment, the Convention on Wetlands of International Importance, and the Protocol on Environmental Protection to the Antarctic Treaty.[109]

102 Abaza, *supra* note 8 at 10.

103 United Nations Convention on the Law of the Sea, Art. 204–206.

104 Convention on Biological Diversity, Art.14.

105 United Nations Framework Convention on Climate Change, Art.4 (f).

106 The most significant international treaty is the Espoo Convention on Transboundary E.I.A. However, first transboundary E.I.A. requirements were elaborated in the treaties on the protection of the marine environment, such as the Kuwait Convention adopted already in 1978. Similar requirements might be found also in the Madrid Protocol to the Antarctic Treaty of 1991 or the North-American Agreement on Environmental Cooperation of 1993. See Bremer, *supra* note 7 at 81–82.

107 The International Court of Justice held that a practice: which in recent years has gained so much acceptance among states that it may now be considered a requirement under general international law to undertake an environmental impact assessment where there is a risk that the proposed industrial activity may have a significant adverse impact in a transboundary context, in particular, on a shared resource. I.C.J. 20 April 2010, Pulp Mills on the River Uruguay (Argentina v. Uruguay), [2010] I.C.J. Rep. 14 ('Pulp Mills'), para.204; see also Bremer, *supra* note 7.

108 See for instance the Lake Lanoux Arbitration case (France vs. Spain) from 1957; see also Bremer, *supra* note 7 at 80, 83.

109 Morgan, *supra* note 13 at 5, 6.

9.3 The Nigerian environmental regulatory system

After independence in 1960, Nigeria did not see the need for national legislation regulating activities relating to the environment, even regarding the oil exploration that began before independence. However, the 1987 incident of illegal dumping of toxic wastes in Koko (in today's Delta State) made the then Nigerian Military Government promulgate the Harmful Wastes Decree which lays out the legal framework for the control of disposing toxic and hazardous waste into any environment within the territory of Nigeria. With legislation in place, the need for an institutional structure became apparent, which led to the establishment of a regulatory body, the Federal Environmental Protection Agency (F.E.P.A.) in 1988. Other sectoral regulations were made. For example, the Department of Petroleum Resources (D.P.R.), within the Ministry of Petroleum Resources and understanding the short- and long-term effects of continuous exploitation, exploration, and production of petroleum resources on the environment, introduced the Environmental Guidelines and Standards for the Petroleum industry in Nigeria (E.G.A.S.) of 1991.

Though the E.G.A.S. made E.I.A. a mandatory enforcement tool in the monitoring of activities in the petroleum industry, in 1992 the E.I.A. Act made it mandatory that a prior assessment of likely environmental impacts be conducted before the approval and commencement of any proposed projects in both public and private sectors.[110]

With the return of democratic rule in 1999, the responsibilities of F.E.P.A. were taken over by the Federal Ministry of Environment, which started making new environmental policies and widening the scope of environmental regulation. In 2007, the F.E.P.A. Act was repealed by the National Assembly and replaced with the National Environmental Standards and Regulations Enforcement Agency (Establishment) Act. The National Environmental Standards and Regulations Enforcement Agency (N.E.S.R.E.A.) was created and replaced the Federal Environmental Protection Agency.[111] The scope of the N.E.S.R.E.A. covers almost every area of the environment and has twenty-four regulations under its administrative authority that adequately empower it to enforce extant standards in the protection of the environment.

9.3.1 Nigeria's environmental impact assessment framework

The N.E.S.R.E.A. is responsible for enforcing the E.I.A. Act in both the private and public sectors in Nigeria. The Act seeks to achieve three basic goals, which are:[112]

110 Section 14, E.I.A. Act 1992, LFN 2004.

111 Muhammed Tawfiq Ladan, Review of NESREA Act 2007 and Regulations 2009–2011: A New Dawn in Environmental Compliance and Enforcement in Nigeria, 8 Law Env't & Dev. J., 2012, pp. 116, 119, 121.

112 Section 2, E.I.A. Act, LFN 2004.

1. To consider prior environmental effects. Before any person or authority takes a decision to undertake or authorize the undertaking of any activity that may likely or significantly affect the environment, prior consideration of its environmental effects should first be taken.
2. To promote the implementation of appropriate procedures to realize the above goal.
3. To seek the encouragement of the development of reciprocal procedures for notification, information exchange, and consultation in activities likely to have significant trans-state (boundary) environmental effects.

Three regulatory components constitute the procedural legal framework for E.I.A. in Nigeria: The E.I.A. Act, the N.E.S.R.E.A. Act, and the E.I.A. guidelines made by the N.E.S.R.E.A. Under the E.I.A. Act, a proponent is required to get the approval of the N.E.S.R.E.A. before embarking on a project. A "project" as defined by section 63(1) of the Act is "a physical work that a proponent proposes to construct, operate, modify, decommission, abandon, or otherwise carry out or a physical activity that a proponent proposes to undertake or otherwise carry out." Projects are classified by using these six criteria: magnitude of project; extent or scope; duration and frequency; associated risks; significance of impacts; and availability of mitigation measures for associated and potential impacts identified.

The three categories of activities under which an assessment is classified by the N.E.S.R.E.A. are (see Figure 9.1):

Category 1: Activities requiring full and mandatory E.I.A. It is either an activity that is on the list or an activity that is outside the list but Initial Environmental Evaluation (I.E.E.) shows E.I.A. is required.[113]

Category 2: Unless within the Environmentally Sensitive Area (E.S.A.), full E.I.A. is not mandatory.

Category 3: Activities that are likely to have beneficial impacts on the environment.

In the E.I.A. Act, "environment" means the "components of the Earth"[114] and includes:

a) land, water, and air, including all layers of the atmosphere;
b) all organic and inorganic matter and living organisms; and
c) the interacting natural systems that include components referred to in paragraphs (a) and (b).

113 Section 12, E.I.A. Act, LFN 2004.

114 The E.I.A. Act does not have an "interpretation" section, but from Section 12 this meaning can be inferred.

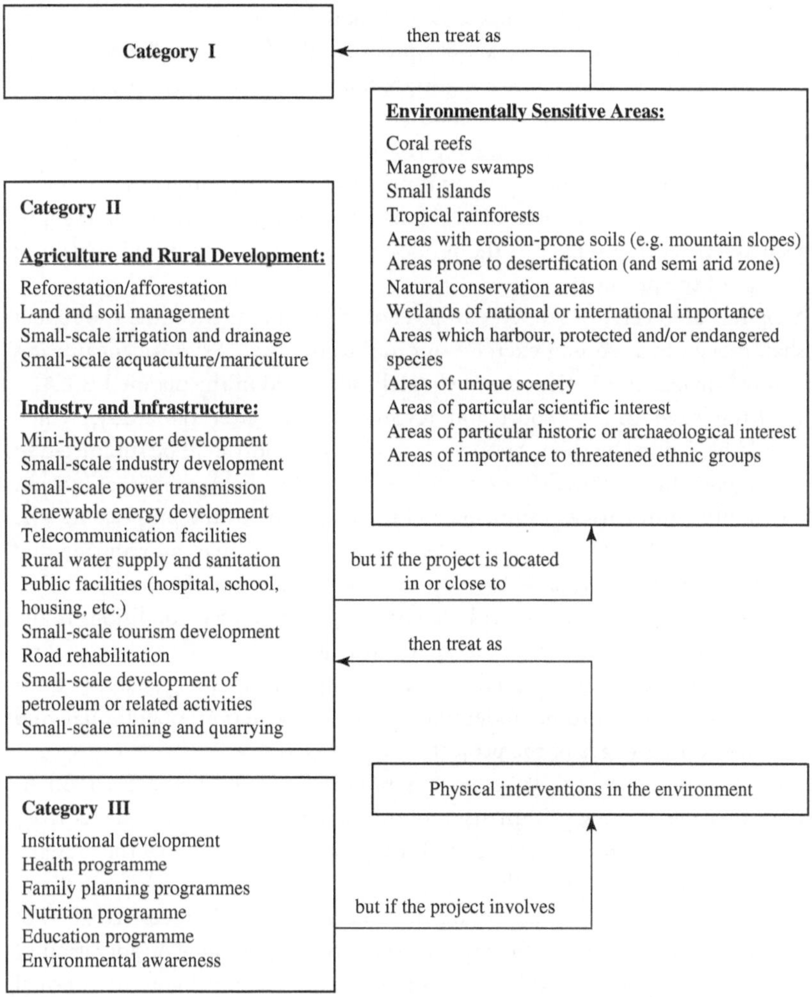

FIGURE 9.1 N.E.S.R.E.A. Guidelines. Source: Office of the Director General of the N.E.S.R.E.A.

The N.E.S.R.E.A. has produced E.I.A. guidelines for projects of various sectors (see Figure 9.1), plus a general E.I.A. procedure for all sectors. The procedure involves nine stages for an E.I.A. to be adjudged conclusive (see Figure 9.2):[115]

1. Project proposal: a proponent embarking on any major project shall notify the N.E.S.R.E.A. formally by submitting a project proposal. The proposal shall

115 Section 15, E.I.A. Act, LFN 2004.

include all relevant information needed for the next stage. Submission of a project proposal to the N.E.S.R.E.A. indicates that the E.I.A. process has formally commenced. Upon receipt of a proposal, the Agency shall enter into its register the proposal and issue a registration number and supply the proponent with all necessary documents needed to facilitate a smooth E.I.A. process.

2. Screening: after receiving and registering the proposal, the Agency will conduct the I.E.E. in order to appropriately assign the project into one of the three categories of activities. After studying the proposal, the Agency provides a written report to the proponent within twenty working days of receipt of the proposal.
3. Scoping: on receiving the screening report from the Agency, the proponent shall carry out a scoping exercise in order to be sure all significant impacts and reasonable alternatives are adequately addressed in the planned E.I.A.
4. Drafting final E.I.A. report and review process: after the study, which should be funded by the proponent, a draft final report shall be submitted to the Agency for review. After reviewing the report, the Agency shall within six months summarize comments and forward same to the proponent. The proponent shall effect recommended amendments as may be required and include them in the final E.I.A. report. The Agency has the prerogative to approve or disapprove a project based on various stipulated conditions in the E.I.A. Act. In a situation where approval is granted, a follow-up program (e.g. a mitigation and compliance monitoring plan) may be established for specific assignments to be undertaken in the construction, operational, and commissioning stages of the project.
5. Final E.I.A. report: the proponent is required to submit the final E.I.A. report to the Agency within six months of the receipt of comments from the Agency. Failure to meet up with the deadline may lead to the Agency requesting a new E.I.A. study.
6. Decision-making: a technical committee headed by the Director General of the Agency is responsible for approving E.I.A.s. Upon due consideration of the final report submitted by the proponent, the technical committee shall consider and approve the issuance of an E.I.A. certificate.
7. Project implementation: after certification, the proponent is expected to embark on the project in accordance with all the specifications contained in the final E.I.A. report. However, if the project is not commissioned within the period during which the certification is valid, the Agency may require a new E.I.A. study.
8. Mitigation compliance monitoring (pre-commissioning): in the course of implementing the project, the Agency shall have unfettered access to monitor the progress of the project to ensure compliance with all proposed mitigation and project specifications.
9. Environmental auditing (post-commissioning): the Agency shall conduct periodic assessment of the positive and negative impacts of the project with the aim of improving the E.I.A. process.

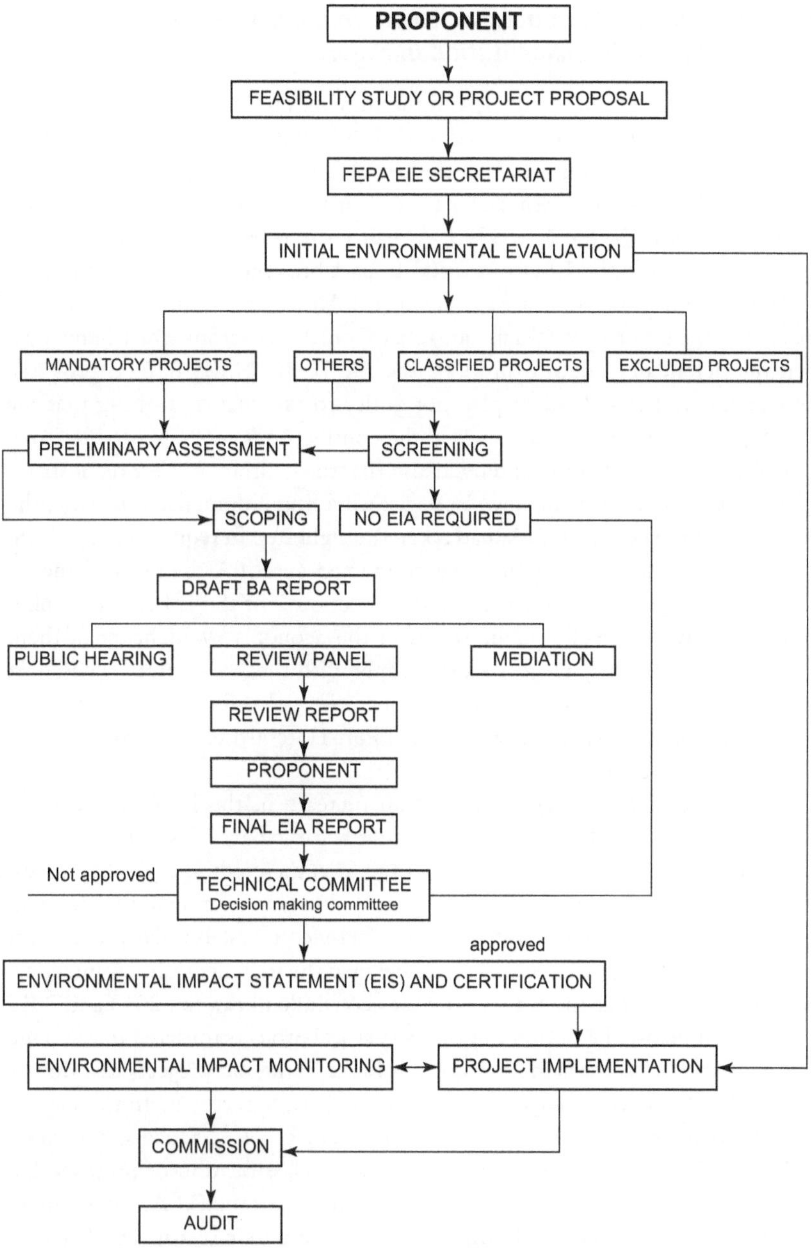

FIGURE 9.2 Environmental Impact Assessment flow chart. Source: Federal Ministry of Environment.

9.3.2 Environmental impact assessment implementation and gaps in implementation in Nigeria

Many of the lapses in enforcement and monitoring of E.I.A. in Nigeria can be traced to the disproportionate attention that environmental activists and regulators give to the upstream sector of the petroleum industry. There is very low awareness in other sectors, especially in the construction sector. In the case of *FRN v. O.K Isokariari*, the N.E.S.R.E.A. had written letters three times to the defendant requesting the Environmental Impact Statement (E.I.S.) of the project. An Environmental Impact Assessment (E.I.A.) is a precondition to the commencement of any project which can impact on the environment in any way. When the defendant failed to produce the E.I.S., the N.E.S.R.E.A. took the matter to court. The defendant pled not guilty to the charge, arguing that the Rivers State Government who awarded the contract to it was supposed to carry out the E.I.A. The company contended also that the contract was for reconstruction and not the building of a new market. Delivering judgment in the case, the court held that the N.E.S.R.E.A. had taken the right step in requesting the E.I.S. and that the defendant ought to have conducted an E.I.A. before commencing the construction. The court noted that "section 2 of the E.I.A. Act applies to both the private and the public sector of the economy. Whichever of them intends to undertake any project must comply with the E.I.A. Act." The court convicted O.K Isokariari & Sons Ltd. of the criminal offense of commencing the construction of a new town market, old Port Harcourt town, Port Harcourt without conducting an E.I.A.

The practice of E.I.A. has witnessed an increase in the level of participation by both private and public sectors since it became part of Nigeria's body of law. However, there are gaps existing that threaten the credibility of the process and reduce its efficiency as a very important tool for environmental protection. Typically for a developing country, strict adherence to established processes is rare. In the case of Nigeria, having acknowledged the increase in E.I.A. practice, studies have shown that the E.I.A. processes concluded between 2010 and 2015 are very much below the number of infrastructural projects initiated in the same period. The public sector is more notorious for carrying out projects that do not fulfill the E.I.A. approval requirement than the private sector is. In the private sector, the majority are not aware of the mandatory E.I.A. process, and those who know of E.I.A. are not conversant with the classification of projects that fall under the E.I.A. Act. Nevertheless, compliance with the E.I.A. requirement in the private sector is generally higher than in the public sector. Further gaps exist in two major aspects: administration and public participation. Regarding administration, there are problems in implementation, politicization of E.I.A., and the enforceability of the Act.

Implementation of environmental law in Nigeria is generally loose, uncoordinated, and fraught with conflict of roles, standards, and responsibilities among the different organs and agencies of government. The conflicts are in forms of inconsistences, overlaps, and duplications in laws and regulations mandating different statutory and supervisory organs, ministries, departments, and agencies. In the case of the E.I.A., there has been a dichotomy of administrative powers between the N.E.S.R.E.A. and the Department of Petroleum Resources (D.P.R.) While the N.E.S.R.E.A. is empowered by law to have the powers to implement E.I.A.s in the private and public sector, the D.P.R. has arrogated for itself the same authority to approve E.I.A.s for the petroleum sector. The D.P.R. has its own guidelines and standards under the E.G.A.S. which differ from the guidelines of the N.E.S.R.E.A. Such inconsistency and overlapping in the E.I.A. process is seen to be national and another one that appears to be sectoral has resulted in the lack of transparency and ineffective coordination of E.I.A. across all sectors.

Administration of the E.I.A. also provides an example of the negative interface of politics with law as discussed in Chapter 6. Political interference in the general administration of the N.E.S.R.E.A. has weakened the agency, thereby making it slack in its statutory responsibilities. It has also precluded the agency from showing the required courage in confronting proponents of projects in the private and public sector who have the political influence to circumvent procedures to insist on respecting the rules and guidelines for E.I.A. For example, the nominee for the Administrator of the E.P.A. in the United States must be confirmed by the U.S. Senate after a public hearing and before his or her appointment takes effect. But in Nigeria, appointment of the Director General of the N.E.S.R.E.A. is through the exercise of executive powers only, which gives the federal government a great latitude of political influence over the agency.

Enforcing the Act depends more on the discretion of the Agency than on a mandatory process. For instance, the Act did not specify who can bring a suit before a court for the enforcing of a E.I.A. In the case of *Douglas v. Shell Petroleum Development Company Ltd and Others*,[116] the plaintiff sought a declaration against the commissioning of a gas project by the defendants without complying with the E.I.A. Act. The court held that the plaintiff lacked *locus standi* to institute the action. The lack of provisions in the Act that would empower individuals to call for an E.I.A. to be initiated or to challenge non-compliance in court leaves an individual with only the option of enforcing a fundamental right under chapter iv of the Nigerian Constitution. But as it was emphasized in the *Douglas* case, environmental rights are not listed in the Constitution.

The final gap to be considered is the low level of public participation. The E.I.A. Act provides for public participation.[117] It clearly recognizes the importance of public inputs in the E.I.A. review process, spelling out the procedure for notifying the public on the intended action and modalities for filing feedback or reactions by way of comments. Furthermore, details of the stages where the public can be involved such as public display, mediation, and review panel are

116 Unreported Suit No. FHC/CS/573/931.
117 Sections 25, 39, 41 E.I.A. Act, LFN 2004.

provided in the Act. However, after the E.I.A. certificate has been issued, the public is cut off from the monitoring and follow-up stage, thereby diminishing the credibility of the entire process. Also, project proponents are not mandated to meet with the local communities before embarking on the project and, though they are mandated to interact with the communities during the E.I.A. process, there is no provision for any penalty for failure to consult the public. In addition, lack of effective communication between the public and the government makes it difficult for infringements to be reported to the government. The root of the problem of lack of effective communication can be traced to poor education. The lack of knowledge in environmental matters makes it difficult to know what to report and how to report it.

9.3.3 Social impact of environmental impact assessments in Nigeria

The introduction and practice of E.I.A. has contributed immensely to environmental protection in Nigeria. Even if the level of enforcement is presently below what it should be, several projects have been either screened or assessed since the inception of E.I.A. in Nigeria. Some of such projects include: Mobil Usari Production Facility in Akwa Ibom State; Western Metal Products Company (W.E.M.P.C.O.) in Ikom, Cross River State; Mobil Natural Liquefied Gas Project in New Finima, Rivers State; and Mobil Capital Dredging of Approach Channel Off-shore Bonny in Rivers State.[118]

Apart from the interest of the environment, which is the primary component of E.I.A., the process also contributes to developmental planning as the post-approval process is linked to the process of approval of projects in conformity with the Urban and Regional Planning Act. Where the project is industrial in nature (after E.I.A. approval by the N.E.S.R.E.A.) the Federal Ministry of Industries may have to give its own approval in addition to the approval by the department for development control.

The pace of development since the return to democratic rule has increased consciousness of social issues connected to projects being initiated by the private and public sectors. This necessarily attracts civil society organizations' interests in how projects will affect the environment and population. Presently, some mega-projects have been assessed not solely to know their environmental impacts but also to understand their social impacts. For example, the E.I.A. Report for the Second River Niger Bridge Project was entitled "Environmental and Social Impact Assessment for the Second River Niger Bridge Project." The belief that environmental impact cannot be severed from social impact is further reinforced by the present practice of international money lenders supporting projects in developing

118 Nwoko, Chris O. "Evaluation of Environmental Impact Assessment System in Nigeria." *Greener Journal of Environmental Management and Public Safety*, vol. 2, no. 1, 2013, pp. 22–31.

countries. They insist on an environmental and social impact assessment for every project to which they will commit finance. This has also contributed to the pursuit of sustainable development goals (S.D.G.) as developing countries that draw funds from international lenders are constrained to satisfy terms agreed on by the lenders.[119] In the petroleum industry, operators have increased their level of compliance with the E.I.A. guidelines provided in the E.G.A.S. due to strict monitoring by the Department of Petroleum Resources. Most of the operators have been publishing highlights of their environmental performance.

The E.I.A. process in Nigeria is still evolving and not integrated as a systematic law, responsive to the introduction of all projects over which the enabling Act empowers it. Enforcement and monitoring are at a low level because regulators do not have the required capacity to cover the rate of development in an expansive country like Nigeria. Within the areas where it is expected that enforcement should be effective, it has been discovered that projects with very high potential of negative environmental impact have been completed in operation without an E.I.A. For example, the partial deregulation of the downstream sector of the petroleum industry has attracted investors to building tank farms for storage of petroleum products. Recently, due to reports of pollution caused by leakage, residents of communities hosting tank farms are being endangered and the government is investigating if the owners carried out the mandatory E.I.A. required by law, to determine the effects of such projects on their communities. An increase in the public's orientation for stronger participation will strengthen enforcement and assist in monitoring E.I.A.s in all sectors. People can only participate in those matters of which they are aware. The present practice of restricting publication in print media of a proposed project that will undergo E.I.A. is an exercise that merely fulfills the letter of the law but does not fulfill the intent of enabling public participation.

9.4 The Chilean framework for the Environmental Impact Assessment System

Chilean environmental law is established under the Constitution, but is interpreted together with economic rights, as was discussed in Chapter 5.[120] The

119 It has been the policy of a number of international organizations that E.I.A. be part of conditions for international assistance. The Organisation for Economic Cooperation and Development (O.E.C.D.) recommended that member governments adopt E.I.A. procedures and methods in the process of granting aid to developing countries (O.E.C.D., 1992). In 1989 the World Bank ruled that E.I.A. for major projects should normally be undertaken by the borrower country under the Bank's supervision. (The World Bank (1999) has recently updated its guidance on E.I.A.) The United Nations Environment Program (U.N.E.P.) also made recommendations to Member States regarding the establishment of E.I.A. procedures and established goals and principles for E.I.A. It subsequently issued guidance on E.I.A. in developing countries (U.N.E.P., 1988).

120 It has to be noted that the E.I.A.S. was implemented in Chile in 1994. The law and the regulatory body described onwards are the same instruments of that time, but their current date

Chilean legal system established an Environmental Impact Assessment System (E.I.A.S.) in Law No. 19.300 and was regulated by the Supreme Decree (S.D.) N° 95 until 2013, and since 2014 by S.D. N° 40 of the Environment Ministry. For the purposes of functional comparative legal analysis, the following section "lays out the essentials" for Chile's E.I.A.S.

9.4.1 General context

The E.I.A.S. focuses on the determination and evaluation of all the environmental impacts of a project or activity, both those that impact upon ecosystems and those that impact upon human groups or sociocultural values. When such impacts are significant,[121] the law requires the owner of the project or activity to submit an Environment Impact Study (E.I.S.), and to establish appropriate mitigation, reparation, or compensation measures to address the impacts.

Law 19.300 provides a definition for E.I.A.S. as follows: "a process under the authority of the Environmental Assessment Service that is based on an Environmental Impact Study or an Environmental Impact Declaration, and that determines whether an activity, the environmental impact of an activity, or a project complies with the existing norms."[122] This law requires a license for projects and activities matching the descriptions contained in the law, when they exceed certain magnitudes.[123] It must be noted that, despite its relevance, the environmental license, as regulated in Chilean law, is only one of the many licenses that a project or an activity needs to acquire before beginning. Finally, environmental law regulates what it calls "Instruments of Environmental Management," one of which is the E.I.A.S.[124]

9.4.2 The Chilean Environmental Impact Assessment System

In practice, when someone is operating under the Chilean E.I.A.S., one must check whether the project or activity is listed among the ones that require an evaluation, based upon Article 10 of Law 19.300 and Article 3 of the Regulatory Body (S.D. 40/2013 EM). In this manner of listing, the Chilean system follows general international practice, with some differences on the type of projects included, and no case-by-case consideration at all.

obey to the slight modifications they have experienced throughout the years, which have not affected the overall structure of the E.I.A.S. and which are not relevant for this research.

121 When the impacts are not significant, the owner of the project must submit to the E.I.A.S. only an Environmental Impact Declaration, which not consider any mitigation, reparation, or compensation measures, but must include a legal compliance plan.

122 Chile, R. D. Aprueba Ley Sobre Bases Generales del Medio Ambiente. *In:* PRESIDENCIA, M. S. G. D. L. (ed.) *Ley 19.300.* Available at http://www.leychile.cl/Navegar?idNorma=30667. Article 2 (J).

123 *Ibid.* Article 10.

124 *Ibid.* 2° Title.

If the project is listed, the executor of a project or an activity described in the environmental law must submit an E.I.S. or an Environmental Impact Declaration (E.I.D.), depending upon whether its execution will generate significant adverse impacts on the environment, according to the rules provided in the law itself and its regulatory body.[125]

One of the first and main difference between an E.I.S. and an E.I.D. is that the E.I.S. implies a mandatory and automatic public participation process, but the E.I.D. has only the possibility to include public participation, depending on whether the community asks for it, and whether the project generates negative externalities or social benefits. These norms also establish the content for the E.I.S. and E.I.D., with a detailed description of every element to be considered by the executor, also providing that failure to perform elements in the content will result in the rejection of the E.I.S. or E.I.D.[126]

In any case, the law and regulatory body provisions assure comprehensive contents for the E.I.S. and E.I.D. The amount of environmental information that an E.I.S. provides turns it into an effective tool to discover the environmental assets of the areas surrounding projects' sites. When the E.I.S. or E.I.D. is presented to the proper authority—regional or national according to the scope of the project—it is communicated to all the technical organs of the state with environmental competences related to the project under assessment.[127] These technical organs review and discuss the E.I.A. of a project, and assess the quality of the contents of the E.I.S. or E.I.D., having the power to request more information or to reject the E.I.S. or E.I.D. due to lack of substantive material.[128]

Nonetheless, the state organs participating in the E.I.A. of a project must provide their technical opinions within the environmental technical competences they hold, and also under the regulations regarding the E.I.A.S. provided by the law and its regulatory body.[129] The technical process of assessment concludes with a comprehensive report containing the relevant details of the E.I.S. or E.I.D. and the technical opinions given by the state's organs during the assessment.[130] This report also includes a recommendation to grant or to reject the environmental license for the project under assessment.[131] Until this stage, the E.I.A.S. is a detailed regulation and a technical system that acts as a guide for the process, while the technical organs of the state apply their environmental competences to assess the conformity of project's E.I.A. with environmental regulations.

125 *Ibid.* at Article 4.
126 *Ibid.* Articles 12-19.
127 *Ibid.* Articles 9 and 35.
128 *Ibid.* Article 36.
129 *Ibid.* Article 35.
130 *Ibid.* Article 44.
131 *Ibid.* Article 56 (l). The report considers this recommendation only for projects assessed from year 2010 onwards.

The final stage of the decision of whether to grant or reject an environmental license is, however, performed by the Environmental Commission.[132] Representatives of State Ministries, appointed under political criteria, form this Commission.[133] The committee in charge of ruling on the claims against the environmental license being granted or rejected is also formed by law-appointed State Ministers.[134] Neither the Environmental Commission nor the Ministers Committee is bound to accept the recommendation contained in the technical report described above, except by those aspects that are expressly "regulated aspects" according to Article 9 of Law 19.300. One may conclude that the E.I.A.S. as described leaves a gap between the assessment process, which is carried out by the technical organs, and the decision-making stage, performed by political bodies, reflecting an internal and structural interface of law and politics, as discussed in Chapter 6. This tension is usually reflected in the assessment of large-scale projects where the technical requirements conflict with the political contingencies, decoupling the decision-making process from the environmental information provided by the E.I.A.

Finally, it is difficult to find sufficient reasons to reject an E.I.S. or E.I.D., since the Chilean E.I.A.S. was created as a process to approve projects rather than reject them. The reasons are that, on one hand, the E.I.A.S. is a self-regulatory instrument under the control of evaluators. They define their projects in the way they want and decide the amount of information that they are willing to make public. And on the other hand, the E.I.A.S. is an "incremental" procedure, meaning that during the evaluation process the proposed project will be improved based on the observations given by the different authorities at each step.

The following section provides three issues with the Chilean E.I.A.S.'s ability to assure substantive and procedural effectiveness, each of which is connected with cultural and social issues: biodiversity offsets, improper segmentation of projects, and public participation.

9.4.3 Biodiversity offsets

Among the significant environmental impacts recognized by the law is the generation of "significant adverse effects on the quantity and quality of renewable natural resources."[135] To them we must also add the circumstance of "[l]ocation in or near … resources and protected areas, priority sites for conservation, protected wetlands and glaciers, susceptible to be affected, as well as the environmental value of the territory in which it is intended to locate."[136]

132 *Ibid.* at Article 59.
133 *Ibid.* at Article 86.
134 *Ibid.* at Articles 77–81.
135 *Ibid.* at Article 11 (b).
136 *Ibid.* at Article 11 (d).

Mitigation measures are defined as those for which the purpose is to avoid or reduce the adverse effects of the project or activity (Article 98 of the new E.I.A.S. Regulation, S.D. N° 40). Examples include measures that prevent or completely avoid the significant adverse effect, by not executing a work or action, or any of its parts; measures that minimize or diminish the significant adverse effect, by means of an adequate limitation or reduction of the extension, magnitude, or duration of the work or action, or of any of its parts; and measures that minimize or diminish the significant adverse effect, through technological and/or management measures considered in the design.

Reparation or restoration measures are defined as those whose purpose it is to replenish or replace one or more of the components or elements of the environment to a quality similar to which they had prior to the damage caused or, if this is not possible, to reestablish their basic properties.[137] However, in many cases, once measures to avoid, minimize, or repair such impacts are established, the remaining measures are not only inevitable for the development of a project or activity, but are also irreversible and permanent. In such cases, mitigation or reparation measures will not be established. It is only possible to establish compensation measures, that is, those measures intended to produce or generate an alternative positive and equivalent effect to an identified adverse effect.[138]

As an example, the complexity of a concept like "biodiversity" hinders the evaluation of a proposed project's impacts. The evaluation of impacts usually focuses only on the "components" of biodiversity—what is present and with what abundance. The process should contemplate two other aspects: "structure," that is, how the biological units are organized in time and space; and "functionality" or "key processes," that is, which physical, biological, or anthropic processes play a relevant role for the creation or maintenance of biodiversity. The 2013/2014 Policy Reform ("the Chilean Policy Reform") tries to include all these impact assessment parameters.[139]

Finally, biodiversity offsetting or compensation (the Chilean E.I.A.S. does not distinguish between them) consists of the effective conservation of biodiversity as a result of actions designed to replace the significant adverse impacts caused by a project, once the appropriate measures have been taken to avoid such impacts or, if this is not possible, to mitigate or repair them. This is referred to as the "mitigation hierarchy principle," which means first avoid, then mitigate, then restore, and only then offset. The objective of compensation is to achieve no real loss of biodiversity, which is called the "no net loss principle." Thus, any

137 *Ibid.* at Article 99 of the new Environmental Impact Assessment System Regulation, S.D. N° 40.

138 *Ibid.* at Article 100 of the new Environmental Impact Assessment System Regulation, S.D. N° 40.

139 We are using this two year reference, since the regulation was published on December 2013, but started to be mandatory on 2014, as we will have the chance to explain later on this work.

compensation action must demonstrate a conservation result that is measurable and additional.[140]

With regard to the concept of "equivalence" or, in the present context, "ecological equivalence," it is necessary to take into account the following key aspects for its determination:

a) The definition of detailed biodiversity objectives (for example: population of plants and animals, types of communities, properties of an ecosystem, etc.);
b) The development or selection of appropriate indicators or metrics;
c) The identification of an appropriate baseline to calculate losses and gains, as well as what needs to be addressed;
d) Temporality; and
e) Uncertainties, both in the evaluation and in the results of the compensation.[141]

However, there are more than 100 different methods around the world to quantify and compare the loss and gain of species, habitats, and ecosystem services, and there is little legislation or guidance that provides clarity on what the principle of "no net loss" (or, even better, a net gain) actually means in practice.[142] In this sense, in the matter of biodiversity offsetting measures, the "like for like" principle is also applied, that is, the requirement that a direct relationship should exist between the environmental objects affected by a certain project, and the objects on which the compensation will be implemented (that is, the same type of biodiversity).

The Chilean Policy Reform tries to take into account all of these different aspects of biodiversity offsetting as described above. Article 100 defines "compensation measures" as those intended to produce or generate an alternative positive effect and equivalent to an identified adverse effect, which cannot be mitigated or repaired. The measures shall include, among others, the replacement of natural resources or elements of the environment affected by others of similar characteristics, class, nature, quality, and function.

Thereafter, Chilean Policy Reform Article 101 provides some rules regarding the places where the measures must be adopted, by saying that the effects of the mitigation and repair measures must occur in the areas or places where the significant impacts on the elements of the environment occur or are generated. The compensation measures will be carried out in the areas or places where the

140 "B.B.O.P. Overview – 2009." *Forest Trends*, https://www.forest-trends.org/publications/bbop-overview-2009/.

141 Quetier. "Assessing ecological equivalence in biodiversity offset schemes: Key issues and solutions," *Bio. Conservation*, vol. 114(12), 2011.

142 "B.B.O.P. Overview – 2009." *Forest Trends*, https://www.forest-trends.org/publications/bbop-overview-2009/.

significant impacts are presented or generated or, if this is not possible, in other areas or places where they are effective.

Since the content of these new regulations is provided in only two articles, without further details, the Environmental Assessment Service and the Ministry of Environment published an official "Guideline for Biodiversity Offsetting within the E.I.A.S." in 2014, explaining with a little more detail the concepts and ideas introduced above. The non-binding Guideline provides some definitions to terms that are not scientifically used in the Chilean Policy Reform. For example, "class" can be understood as a level of biodiversity (genes, species, ecosystems). A compensation measure must be specified at the same class of biodiversity. In addition, the nature, quality, and characteristics can be interpreted as a reference to any of the three attributes of composition, structure, and function of biodiversity.

A review of the projects processed in the E.I.S. shows that, although there is a general association between what is understood as mitigation, reparation, or compensation, the criteria to identify a measure as one class or another is not univocal and is often qualified in different ways. Thus, the most frequent measures related to the impacts of loss of vegetation or flora correspond to the transplantation (also called "rescue and relocation") of flora and reforestation with new plants in other areas than those in which the impact of loss by the cutting or extraction of vegetation or flora takes place. Although the rescue and relocation measure is always classified as compensation, transplantation is sometimes classified as mitigation[143] and other times not classified at all.[144]

Something similar happens with the impacts on fauna. A recurrent measure to deal with the loss of habitat, or the impact on significant populations of species or on species classified in some category of conservation, is the "rescue and relocation" of individuals of the affected species. This measure is generally understood as mitigation, but there have also been cases in which it is characterized as compensation.[145]

The confusion is significant. If the harm is to individuals of the species that are being saved, the objective of the measure will be to reinsert them into another habitat, thus mitigating the impact of the proposed project. This qualification as a mitigation measure focuses on the individual. It does not consider the relevance of the displacement nor the effects where the displaced individuals will be reintroduced. On the other hand, it could be understood that habitats are being irreversibly altered, and then compensation measures are more appropriate.

Neither before the Policy Reform nor afterwards has any sort of definition of the concept of "equivalence" been provided in the regulation, despite the fact

143 *Proyectos "Depósito de relaves en Pasta Sector 5," y "Manejo y Disposición de RISES de combustión del Complejo Termoeléctrico Ventanas."*

144 *Proyecto "Planta Desalinizadora Minera Candelaria."*

145 *Proyectos: "Planta Desalinizadora Minera Candelaria"; y "Depósito de relaves en Pasta Sector 5.*

that it is the core concept around which biodiversity offsetting is built. There are also no guidelines regarding the indicators or metrics of biodiversity offsetting; not even the components to be used in an equation. This omission allows private consulting companies that prepare E.I.S.s and E.I.D.s to determine the features of these metrics as they see fit for their own needs.

9.4.4 Improper segmentation of projects

There are two relevant regulations in Chile regarding the improper segmentation of projects from the environmental perspective. The first, Article 11 of Law 19.300, provides that:

> Proponents may not knowingly segment their projects or activities in order to vary the evaluation instrument or to avoid entry into the Environmental Impact Assessment System. It will be the responsibility of the Superintendency of the Environment to determine the infraction of this obligation and require the proponent, after a report from the Environmental Assessment Service, to adequately enter the system. The provisions of the preceding paragraph shall not be applied when the proponent confirms that the project or activity corresponds to one whose execution will be carried out in stages.

The second, Article 3 (k) Law 20.417, states that the Superintendency of the Environment may:

> Oblige, the proponents, after a report from the Environmental Assessment Service, to properly enter the Environmental Impact Assessment System when they have segmented their projects or activities with the purpose of avoiding the requirements of paragraph 2 of Article 11 of Law No. 19.300.

Article 11 of Law 19.300 was established within the environmental assessment of projects procedure, but the competency to sanction remains with the Environmental Superintendency, an institution that has the competency to monetize all environmental assessment instruments, one of which is the environmental permit that concludes the E.I.A. process.

Citizens can force projects into the E.I.A. system through formally complaining to the Environmental Superintendency. One of the acts that causes a project to be forced into the E.I.A. system is the improper segmentation of projects. Companies can, however, usually offer a "compliance program" in order to start to comply with the norms that were violated. The acceptance of these programs cannot be challenged by communities or citizens.

This regulation was introduced due to environmental abuse in real estate projects that were permanently avoiding entering the E.I.A. System by staying below the threshold of homes or hectares. From that experience, a series of essential elements have been determined, including "proponent," "segmentation," "knowing," and "elude or vary."

Article 11 uses the word "proponent." By carefully comparing different applicable norms, we would conclude that proponents consist of those persons who are in the E.I.A.S. at some stage of project evaluation but who have not yet obtained a permit. Further, a proponent is the person who proposes an E.I.S. or E.I.D. However, the Superintendence of Environment applies the concept in a broader way by including the holders of projects or permits, and even persons before they propose any E.I.S. or E.I.D. The law does not give a definition of the segmentation or division concepts. One of the basic principles of any environmental assessment system is the integration principle, which requires evaluating projects based on the overall nature of the proposed activity, not by segments of any kind, including aspects, parts, or territories.

The law uses the concept "knowingly" to imply "the author has knowledge of the facts constituting the infraction, as well as its unlawful significance." The foregoing use of the term "knowingly" has been confirmed by jurisprudence. The Court of Appeals of Santiago, in Case No. 35.693–2014, confirmed by the Supreme Court in Case No. 3.014–2013, pointed out that the concept "knowingly" in the division of projects should be understood as meaning that "[] has acted fraudulently, in bad faith, by breaking up his project, which is not proven in this procedure or in any other." The Court of Appeals of Santiago, in the same case, ruled the same.[146]

The term "elude," according to the R.A.E., means "to avoid with cunning a difficulty or obligation" or "not to take into account something, inadvertently or intentionally." It seems that, according to this, the norm prohibits a cunning aimed at abusing a form of action that allows avoiding the entrance to the S.E.I.A. In this sense:

> what would be sanctioned would be the use of atypical juridical acts or businesses devoid of a specific cause that, organized in such a way, simulate an action with the purpose of preventing a project or activity or a part of this, be evaluated in the S.E.I.A. and thus obtain a specific profit.[147]

Following this line of argument, the circumvention of the S.E.I.A. would be a form of fraud. The term "vary" means to use several E.I.D.s simultaneously. Observing from the administrative practices of the Chilean Superintendence of Environment, we can derive five main criteria that orient the application of improper segmentation of projects: piercing the veil, functional unit, special–temporal analysis, execution in stages, and public participation. So far, there is no jurisprudence in Chilean Environmental Courts regarding these criteria.

146 No. 40,665-2013, cumulative to the Rol: No. 42,513-2013, confirmed by the Supreme Court in Case No. 11,982-2013.

147 *Ibid.*

In simple terms, the theory of "piercing the veil" (of legal personhood) makes it lawful for the courts, at times, to dispense with the external form of legal persons and, subsequently, to penetrate their interior in order to reveal the underlying subjective interests behind them and reach the people and goods that are protected under the veil. The Superintendence of Environment has applied the theory to some degree, when it investigated the corporate structures and commercial relationships between the involved parties in order to determine, mainly, the intentionality in the configuration of the facts that give rise to the infringement.

The interdependence of one or more parts of a project, known as its "functional unit," generates the need to evaluate it in a joint and unitary manner especially as regards its economic purpose. Spatial characteristics of the works or parts of the project, and temporary issues of activities related to it, such as certain procedures performed prior to the start of works, can be taken into consideration when determining a possible segmentation. A fourth criterion is separated in Chile, although it would not be separated at the international level: execution of the project by stages. At the international level, the fact that a project can or should be executed in stages is not a differentiating criterion to establish whether or not the segmentation of a project occurs. The Chilean case would be different, according to Article 11, § 2 of Law 19.300, if it is proven that the project will be executed in stages.

Although the E.I.S. and E.I.D. regard the public participation procedures differently, they are similar in that the result of the public participation process is not binding for the authorities. The document that contains all the public observations must be considered only by the E.A.S. and properly answered, but not followed by it. This has resulted in a significant amount of litigation.

9.5 The elements of an effective Environmental Impact Assessment

The E.I.A. is a tool widely spread across national legal frameworks,[148] but little is known about its real contribution to sustainability. As regards the oil and gas sector, for example:

> E.I.A. has become an established component of regulatory provisions for off-shore oil and gas development. Yet, whilst such provisions are a welcome reflection of a desire to promote environmental accountability within the industry, relatively little is known about the contribution that E.I.A. has made to date.[149]

148 Sands, Philippe, et al. *Principles of International Environmental Law*. Cambridge University Press, 2018. p. 601.

149 Barker, A., and C. Jones. "A Critique of the Performance of EIA within the Offshore Oil and Gas Sector." *Environmental Impact Assessment Review*, vol. 43, pp. 31–39.

Some elements of the E.I.A. structure operate properly, such as data collection and presentation, but others like problem-solving and best alternative proposals do not.[150] These findings question E.I.A.'s efficiency when it comes to sustainable decision-making.[151]

After a comprehensive review of the literature, researchers Chanchitpricha and Bond have found that "effectiveness can be divided into four categories: procedural, substantive, transactive, and normative."[152] Following the purpose of this research our focus has been on the procedural category which "considers the principles and practice of the impact assessment process," and addresses whether E.I.A. processes conform to established provisions and principles. Some of the factors are political context and active public participation.

We have also considered the substantive category that is related "to the achievement of the agreed objectives of the impact assessment tool,"[153] such as, for example, the support of well-informed decision-making that results in environmental protection. Regarding the decision-making process, they stated that effectiveness is achieved when it considers "all mandatory documents produced as part of the [Strategic Environmental Assessment/Impact Assessment] process coupled with continuous dialogue between the parties involved in the process of informed decisions on the final version of the programme."[154] The authors state that:

> In summary, the factors influencing substantive effectiveness are regulatory frameworks on implementing impact assessment in the decision making process, mechanisms in decision-making context, availability of stakeholder and public participation, and the communication credentials of the impact assessment report based on its quality and accuracy.[155]

9.6 Comparing conclusions across cultures

As this chapter shows through comparisons across cultures in Africa, the Americas, and Europe, the E.I.A. system has been one of the most widely used environmental management instruments across the world. However, its contributions to sustainable development in each country remain often unrecorded or unreported and, consequently, unknown. Nonetheless, both the procedural and substantive factors of E.I.A. systems are evolving in each culture to determine their effectiveness. One of the most important topics that is under discussion and continuous improvement is the role and scope of the public participation process.

150 *Ibid.* at 33
151 *Ibid.* at 33.
152 *Ibid.* at 66.
153 *Ibid.*
154 *Ibid.* at 70.
155 *Ibid.* at 69.

Each culture is addressing known gaps and specific challenges that undermine E.I.A. effectiveness. The E.U. and Nigerian E.I.A. systems have screening mechanisms and require alternative analysis, but the Chilean E.I.A. system does not. In addition, there seems to be a trend in which politicization of E.I.A. systems results in increasing demand for more technical resolutions of environmental assessments and approval of projects.

The main objective of the Chilean E.I.A. is to influence decision-making processes, by providing environmental information on the consequences of a project's execution. The achievement of this purpose can be assessed considering whether all mandatory documents produced as part of the Strategic Environmental Assessment and Impact Assessment process coupled with continuous dialogue between the parties involved in the process inform decisions on the final version of the program.[156] Considering these elements, it is important to highlight that the Chilean framework for E.I.A. mandates no connection between the environmental information provided in the E.I.A. and final decision-making.

In addition, the E.I.A.S. is structured in a manner that allows the decision-making stage not to consider the environmental information provided by the E.I.A. process. The E.I.A.S. has a gap between the assessment process, which is carried out by the technical organs, and the decision-making stage, which is performed by political bodies. The structure of the Chilean E.I.A. focuses on the production of environmental information to make the particular decision to grant or reject an environmental license. In Chile, the E.I.A. remains isolated from state environmental planning, except insofar as the Chilean E.I.A.S. requires a license for certain activities. Finally, a specific procedure to legally avoid E.I.A. system, called "Letter of Project Pertinence," allows companies to ask the Environmental Assessment Service whether a project must enter the E.I.A. system or not. This process is carried out without any public participation process, and is done despite the competencies of the Environmental Superintendency to force projects to enter the E.I.A. system.

The regulation in Chile for partitioning projects within the E.I.A. system is quite unique, however, and remains an underutilized mechanism, since it is very difficult to prove the intention of the proponent. Nonetheless, at the same time, the incremental relevance about this specific institution is a sign of the legal culture of this country, where confidence among authorities, communities, and other interested persons is nearly non-existent, and focus falls on dispute resolution as an alternative to environmental planning and evaluation. This characteristic is not observed, by comparison, at the same level in E.U. countries, where the E.I.A. system is comparatively well-regulated, and the public participation process is very well-established.

156 Chanchitpricha, C., and A. Bond. "Conceptualising the Effectiveness of Impact Assessment Processe." *Environmental Impact Assessment Review*, 2013, pp. 43, 65–72, 70.

Finally, one can say that environmental law in Chile is "*seizado*," but only in part. Even when it is attempted to solve environmental problems through the E.I.A. process, it almost always results in litigation before the Environmental Courts. The analysis of the selected E.U. Member States indicates that, even though E.I.A. as such is not a self-standing permit procedure,[157] there is a growing tendency to perceive the E.I.A. procedure and subsequent permit procedures together and thus more coherently. Such coherence reduces conflicts in the laws applicable to various stages of E.I.A. procedures.

When looking at public participation and access to justice rights associated with the E.I.A. procedure, one cannot understate the influence of the E.U. legislation and the Aarhus Convention. This is visible in particular when looking at the rights of the concerned public. Even though there are certain distinctions among the approaches of the cultures discussed regarding the concepts and rights of the public and the concerned public, the legislation of the selected countries in this regard also overlaps, as the respective E.I.A. laws share a common basis. Still, it is necessary to perceive the E.I.A. procedures and their implementation in practice in the light of social, political, and cultural determinants and influences.[158] The Aarhus Convention's impairment of rights doctrine, which influenced E.U. legislation, is a challenge to a country like Germany with a historically conservative approach to access to justice rights. The harmonization in this regard has not been an easy process, requiring clarification from the C.J.E.U. Considering that the rights of the public and the concerned public are nowadays firmly established in all of the cultures that we have discussed, there are reasons to conclude that E.I.A. procedure as applied in the E.U. countries represents essentially a viable tool for better enforcement and implementation of environmental law.

Even though Nigerian E.I.A. laws at the first glance seem like a perfect example for the viability of a legal transplantation from one culture to another,[159] the use of imported E.I.A. legislation can also prove to be rather improper and dysfunctional. The formalized procedures adopted in Western countries ensuring neutrality, impartiality, and to a certain extent predictability sometimes turn out to be rather inflexible, overly bureaucratic, and not properly responding to the needs of legal cultures in developing countries.[160] A study conducted under the auspices of the U.N.E.P. reported that the process of evolving a domestic legal and institutional framework has, overall, a better track record

157 United Nations Economic Commission for Europe, *The Aarhus Convention: An Implementation Guide*, p. 127.

158 Morgan, *supra* note 13 at 5, 10.

159 Yang, Tseming, and Robert V. Percival. "The Emergence of Global Environmental Law." *Ecology Law Quarterly*, vol. 36, 2009, pp. 625, 626–30.

160 Abaza, *supra* note 8 at 19.

in developing countries than a process imposed from outside.[161] This implies and fosters the introduction of E.I.A. systems and procedures from the bottom, based on non-statutory procedures and pilot studies.[162] Such an approach was for instance applied in Nepal, where the elaboration of a national E.I.A. system was identified as a priority by the National Planning Commission. Even though E.I.A. guidelines were prepared in collaboration with the International Union for Conservation of Nature (I.U.C.N.), their development and implementation were based on interdisciplinary and inter-sectoral cooperation and coordination conducted under the auspices of the Environmental Core Group, consisting of government representatives, N.G.O.s, and private sector representatives. In this way, it was possible to develop indigenous E.I.A. expertise without involving outside experts to steer the development of E.I.A. legislation.[163]

The quality of democracy, cultural traditions, education, literacy, and gender inequality significantly influences certain aspects of E.I.A. procedural stages involving public participation.[164] In general, these legal and procedural guarantees might be less common in developing countries than in developed ones.[165] The requirements imposed by multilateral financial institutions, such as the World Bank or the Inter-American Development Bank, have also served as an incentive to improve public participation provisions in E.I.A. laws in the borrowing countries of the developing world.[166]

Literature

Glasson, John, et al. *Introduction to Environmental Impact Assessment: Principles and Procedures, Process, Practice, and Prospects.* Taylor and Francis, 2013.

Kravchenko, Svitlana. "The Myth of Public Participation in a World of Poverty." *Tulane Environ. Law J.* 23 (2009): pp. 33–55.

Ladan, Muhammed Tawfiq. "Review of NESREA Act 2007 and regulations 20092011: a new dawn in environmental compliance and enforcement in Nigeria." *Law Env't & Dev. J.*8 (2012): 116, 119, 121.

Nwoko, C.O. 2013. "Evaluation of Environmental Impact Assessment System in Nigeria." *Greener J. Environm. Manage. Public Saf.* 2.1: 022–031. ISSN: 2354–2276.

Philipp-Gerlach, Ursula, and Andreas Lukas. "Die UVP-Vorprüfung in Der Rechtsprechung Und Praxis." *Zeitschrift für Umweltrecht* (2014): vol. 10, pp. 548, 551.

161 On the other hand, the U.N.E.P. study mentions also that the Nigerian E.I.A. system was originally based closely on the Canadian E.I.A. system, so the success of various approaches might depend on case-by-case basis. See Abaza, *supra* note 8 at 25.

162 *Ibid.* at 21.

163 *Ibid.* at 22.

164 Public participation has been a constituent part of E.I.A. procedures since the 1970s. However, its full potential is often not achieved. See Abaza, *supra* note 8 at 66.

165 *Ibid.* at 17.

166 Still, these requirements are often formulated in vague language. See Kravchenko, Svitlana. "The Myth of Public Participation in a World of Poverty." *Tulane Environmental Law Journal*, vol. 23, 2009, pp. 33, 36.

Piero, G., and S. Clare, *European Strategy on Invasive Alien Species: Convention on the Conservation*. 4 June 2018. https://books.google.com.ng/books?isbn-9287154880.

Sands, Philippe, et al. *Principles of International Environmental Law*. Cambridge University Press, 2018.

Shepherd, Anne, and Christi Bowler. "Beyond the Requirements: Improving Public Participation in EIA." *J. Env. Planning Manage.* (1997): vol. 40, no. 6, pp. 725–738.

10

ENVIRONMENTAL CRIME AND ENFORCEMENT

Mrinalini Shinde[1]

10.1 Introduction

In this concluding chapter of the book, we face a function that is not difficult to describe, but is contentious to apply. The functions of criminal law are well-known and even considered to be the most basic functions of law in many cultures. Criminal law is a system of sanctions against legal persons whose conduct hurts the public and usually at the same time an individual to such an extreme that the state must step in. When it does, it uses state tools to deter the individual who committed the misconduct, to deter society generally from committing such a misconduct, to punish the wrongdoer so as to placate those who adhere to the law, to rehabilitate the wrongdoer or to remove him, her, or it from society because he, she, or it is undeterrable and not capable of rehabilitation. When do these theories of sanction apply to environmental wrongs? Unlike the sanction of more direct harms against the person such as battery or murder, one might well argue that all environmental harm is a harm to society as a whole and therefore requires the state, on behalf of the public, to exercise its tools to sanction all who harm the environment. Opponents of that concept might argue that, because many environmental harms are indirect, they cannot meet the requirement of harmful intent, and therefore none can be considered criminal. As time goes on, that latter argument is losing ground.

Aside from the technical legal requirements necessary to characterize behavior as "criminal," there is a large measure of discretion exercised by prosecutors in deciding whether to take an action. That discretion is influenced by a number of factors, some of which are discussed in Chapter 6 on the interface of politics and law. In addition, public cultural expectations can influence a choice to prosecute, as we saw in Chapter 7 on public participation in environmental law. How all these factors—and more—result in criminal prosecution is the object of study of criminologists. Given that the prosecution of environmental crimes is relatively

1 The views and opinions presented in this chapter belong entirely to the author, and they do not, in any manner or form, represent the views and opinions of the UNU-EHS, the UNFCCC secretariat, or any of their affiliated bodies, organizations, or agents.

new, it is necessary to consider green criminology first in order to understand environmental crimes.

This chapter consists of two parts: first, it is a literature review in green criminology and environmental crime, specifically including environmental crimes committed by corporations; and second, it is a comparison of environmental law as enforced in three legal cultures: the United States, India, and Germany. The chapter begins with a discussion of that which constitutes environmental harm, and how the character of environmental harm is treated differently in criminal law. This is then followed by reference to the criminal behavior of corporations, their management, and employees so as to understand the factors that result in offender behavior, along with a review of literature relating to the sentencing and punishment of corporations convicted of environmental crimes. Due to green criminology being a relatively new field of scholarly inquiry, and due to most literature on the subject having been written by scholars in the U.K., U.S.A., and Australia, there is a disproportionate representation of views and examples based on these jurisdictions. The second part of the chapter seeks to compare the nexus between criminal enforcement against air pollution and the respective air quality in the three jurisdictions of the U.S.A., India, and Germany.[2] Germany and the U.S.A. differ from India in terms of per capita wealth, and Germany is differentiated from India and the U.S. on the basis of its legal system.

10.2 Understanding the offenders

The profiles of environmental offenders could range from sole operators to multinational entities (M.N.E.s), but the maximum harm to the environment is perpetrated by corporations and state actors.[3] All criminal acts are not seen as equally deviant in the eyes of the public, and deviancy is determined by characters such as seriousness of the crime, the irreparability of harm, the social status of the offender, the status of the victims, and access to information about that

2 "The Economist Intelligence Unit." *Country Analysis, Industry Analysis – Market Risk Assessment*, https://www.eiu.com/topic/democracy-index. The assumption while studying all three jurisdictions of India, the U.S.A., and Germany countries is that, as democracies, the data that they make available in the public domain is *prima facie* true, due to it emerging from political systems based on public accountability. India, the world's largest democracy, and the U.S.A., the oldest modern democracy, are both common law nations with very different political structures and economic status, making for an important comparison. However, according to the latest Democracy Index for the year 2017 by the Economist Intelligence Unit, both India and the U.S.A. are classified as "flawed democracies," while Germany, with a civil law system and Europe's largest economy, is classified as a "full democracy" by the same index.

3 Cao, Anh. "The Conceptual Compatibility Between Green Criminology and Human Security: A Proposed Interdisciplinary Framework for Examinations into Green Victimisation." *Critical Criminology*, vol. 24, 2016, pp. 413, 419.

crime.[4] This deviancy associated with criminal behavior does not usually apply to environmental crime because society tolerates corporation depredations of the environment as long as they contribute to a legitimate business enterprise. For environmental crime to be seen as deviant, it must involve the personalization of that offense, both by dispersing information about the specific perpetrator of the crime, and those who are specific victims.[5] The nature of M.N.E.s, and the multiple decision-making personnel involved in the perpetration of environmental crimes, makes it difficult to identify a single person who is seen as the face of that criminal operation.

Through a questionnaire, Martin et al. asked 487 participants to rate seven types of "environmental transgressions" using seventeen rating scales in order to understand the social perceptions relating to environmental crimes.[6] They found that "[p]eople consider that environmental crime is wrong in general terms and that it reflects the badness of whoever behaves in that way" and is an indictment of the offender's character, but "that certain circumstances may lead individuals who are not essentially bad (including participants or even ourselves) to behave illegally in environmental terms."[7] This partly explains the disparity between environmental crimes and other criminal behavior, in terms of how the public perceives it as being criminal behavior.

Michael Lynch et al. conducted an empirical study using data from the U.S. Environmental Protection Agency's (U.S.E.P.A.) summary of criminal prosecutions for a period of thirty-one years from 1983, which revealed that the rate of criminal prosecutions for environmental laws was a low average of eighteen cases per year,[8] and if "these figures were considered alongside the number of potential violators and opportunities for violation, an estimated likelihood of criminal prosecution of eight in 1,000,000 was obtained" leading to an estimate of less than 0.006 percent.[9] They assert that this estimation, coupled with the average of 8.7 months of prison served by offenders, implies, among other things, that a rational offender would not be deterred from committing environmental crimes in pursuit of financial gain,[10] which is central to corporate behavior.

Studying corporate environmental crime involves not just the study of the firms, but of the people who take the decisions involved in the commission

4 McKie, Ruth E. et al., "Carbon Crime in the Voluntary Market: An Exploration of Modernization Themes Among a Sample of Criminal and Non-Criminal Organizations." *Critical Criminology*, vol. 23, 2015, p. 473.

5 *Ibid.*

6 Martin, Ana M. et al., "The Relationship Between Moral Judgments and Causal Explanations of Everyday Environmental Crimes." *Social Justice Research*, vol. 26, 2013, pp. 320, 335.

7 *Ibid.* at 337.

8 Lynch, Michael. "The Weak Probability of Punishment for Environmental Offenses and Deterrence of Environmental Offenders: A Discussion Based on U.S.E.P.A. Criminal Cases, 1983–2013." *Deviant Behavior*, vol. 37, 2016, p. 29.

9 *Ibid.* at 30.

10 *Ibid.*

of the environmental offenses at stake, because almost all modern justice systems involve attribution of liability to individuals, either under the responsible officer doctrine, or strict liability of management, or satisfying the requirement of knowledge during commission of the crime.

10.3 Prosecution

When environmental crimes are to be decided by juries, it is often difficult to make a case for the prosecution owing to the reality of white-collar environmental offenders who tend to be businesspeople with education and social respectability and who usually do not have criminal records.[11] Their employment status also makes it difficult for juries to ascertain the financial motives behind the commission of those crimes. The fact that they are not perceived as having the "typical" characteristics of a criminal leads to advantages at the sentencing stage.[12] Jurors identify with those at the managerial level much better than they would if the accused were at the CEO level. Additionally, the absence of identifiable victims makes it likelier that juries will balance the equities on the otherwise good citizen defendant side. It is imperative to focus on and prove the motivation and intent of white-collar environmental offenders to successfully prosecute them.[13] As for the prosecution of a corporation as a whole, the fact that corporations have access to large amounts of finances at their disposal makes their prosecution proportionally expensive for the state through its regulatory agencies, the police, and investigative agencies, resulting in prosecutorial discretion choosing only the most serious or "winnable" cases.[14] Also, when criminal law uses the standard of proving guilt beyond reasonable doubt, the diffused nature of environmental pollution, and its multi-layered harms and scope of victimization, makes it difficult to attribute the guilt entirely to one corporation in areas where multiple offenders could be operating, and the ability to gather evidence can be challenging for investigating agencies.[15]

10.4 Penalties

A major motivation behind criminal law enforcement is the "just deserts model" under which criminals "deserve" the punishment they are assigned for choosing to commit an offense (since natural persons are rational actors in the absence of mitigating factors). The just deserts model that is often applied to street crimes

11 Cassidy, Kevin M. "The Role of Motive in White Collar Environmental Crimes." *National Resources & Environment*, vol. 23, 2009, pp. 37, 38.

12 *Ibid.*

13 *Ibid.*

14 White, Rob. "Prosecution and Sentencing in Relation to Environmental Crime: Recent Socio-Legal Developments." *Crime L. & Soc. Change*, vol. 53, 2010, p. 366.

15 *Ibid.* at 372.

does not seem to extend to corporate crimes, despite the fact that "corporations that survive tend to be efficient and economically rational,"[16] as they make decisions in the interest of their profits, bond ratings, shareholders, etc., satisfying the criterion for rationality under criminal law.

10.4.1 Fines

Monetary fines provide deterrence against environmental crimes, as can be seen in cases of water pollution compliance in which the imposition of additional fines was accompanied by a reduction in toxic discharges, by both compliant and non-compliant units, indicating effective general and specific deterrence. Studies relating to federal criminal fines under the Clean Water Act were seen to result in both general and specific deterrence as opposed to administrative fines, which resulted in only specific deterrence.[17] On the other hand, studies which found that, in cases of the heaviest monetary penalties imposed by the U.S.E.P.A. over seven years, there was a negligible relationship between the number of penalties and discharge of pollutants.[18] Additional factors such as the racial composition of the locations where violations occur significantly affect the amount of the penalty with black census tracts receiving lower penalties than white census tracts, and areas with a higher number of lower-income Hispanic residents receiving lower penalties than areas with higher-income non-Hispanic residents.[19] So the deterrence value of monetary fines is also a function of where the fine was imposed, given the influence that location has on the amount of the penalty. If one were to extrapolate this conclusion to transnational corporations, there is an obviously lower deterrence value when committing environmental offenses in developing states where the amount of fines would be lower. Barett's study of deterrence found, in the cases of thirty-seven facilities that were fined for environmental violation in the state of Michigan in the U.S.A., that there was "a modest effect on the degree of noncompliance in the quarter immediately following a monetary penalty."[20]

While the majority of the discussion above has been on the deterrent value of criminal fines, they can also produce restorative outcomes, which in the environmental context has usually been accomplished through tort claims, or with the application of the polluter pays principle. During their study of the New South Wales Land and Environment Court, Walters and Westerhuis observed that, when the needs of a community or nature were negotiated as part of the

16 Russell, Steve, and Michael J. Gilbert. "Truman's Revenge: Social Control and Corporate Crime." *Crime, Law and Social Change*, vol. 32, 1999, pp. 59, 60.

17 Barrett, Kimberly L., et al. "Monetary Penalties and Noncompliance with Environmental Laws: a Mediation Analysis." *SpringerLink*, Springer U.S., 17 Dec. 2017, p. 3.

18 *Ibid.* at 4.

19 *Ibid.*

20 *Ibid.* at 15.

calculation of fines, then paying the fines toward the community or for research and protection of a particular affected species served a restorative justice process as well.[21]

10.4.2 *Incarceration*

Incarceration is seen as particularly ineffective as punishment for environmental offenders as it does not incapacitate them, since environmental crimes take place not as personal offenses by the CEO or director, but through multiple levels of employees, who can continue to receive directions from the imprisoned officer, and does not necessarily address whether the offense will continue, and the purpose of imprisonment based on incapacitation falters for environmental criminals.[22] However, if imprisonment is used for the purpose of retribution, imprisonment does create a stigma associated with criminal behavior.

The other purpose of imprisonment—deterrence—is of two types, collective and individual deterrence. Collective deterrence is usually not achieved when imprisoning environmental offenders, since only a few members of society are in a position to commit those crimes in the first place; however, corporate criminals when conducting the cost-benefit analysis of such behavior were "more likely to be deterred from acting by the prison terms."[23] As for the purpose of individual deterrence, U.S. and Brazilian recidivism rates indicate that individual deterrence is not achieved through imprisonment.[24] A study of criminal sentencing in environmental offenses found that, in Belgium, there is a trend to suspend execution of prison sentences, even if awarded. If there is no credible threat of harsh punishment after non-compliance after the initial warning or fine, deterrence cannot be achieved in cases of future offenders.[25]

10.5 Interaction with green criminology

Environmental law as a discipline of law is relatively young, its full-fledged life only about half a century old, from the 1970s onward. As was discussed in Chapter 3, in the U.S.A., for example, environmental harm was generally argued in tort claims or property law cases, under the common law.[26] This changed

21 Walters, Reece. "Green Crime and the Role of Environmental Courts." *Crime L. & Soc. Change*, vol. 59, 2013, p. 286.

22 Wolff, Rafael. "Environmental Crimes and Imprisonment: Does Prison Work to Prevent and Punish Environmental Criminals?" *Pace*, 2018, p. 112.

23 *Ibid.* at 113.

24 *Ibid.* at 120.

25 Billiet, Carole. "How Real Is the Threat of Imprisonment for Environmental Crime?." *Eur. J. L. Econ.*, vol. 37, p. 196.

26 Lazarus, Richard. "The Greening of America and the Graying of United States Environmental Law: Reelections on Environmental Law's First Tree Decades in the United States." *Va. Envtl. L.J.*, vol. 20, 2001, p. 77.

dramatically with the introduction of several key pieces of state and federal legislation in the 1970s, which virtually created the discipline of environmental law and entrusted the state with responsibility of ensuring the protection of the environment.[27] This was then followed by the strengthening of institutions of environmental protection such as the U.S.E.P.A., and, in the 1980s, legislative amendments resulted in environmental felony violations introducing criminal law into the practice of environmental law.[28] India, on the other hand, having been a British colony, was subject to British laws that assumed dominion over its natural resources, especially forests and minerals, and the "environmental laws" which applied to those resources, such as the Forest Act of 1927, were not intended with the primary purpose of environmental protection as a public resource, but the maintenance and preservation of forests as reserves to serve the interest of the British Empire. Similar to the United States, modern environmental law with both civil and criminal provisions was introduced in India only in the 1970s with subsequent amendments.

An area of legal practice which till very recently in history was primarily a part of private civil law, based in property and tort, is currently a branch of criminal law requiring that environmental crime be studied in a manner akin to the study of other forms of crime in criminology. Whereas criminology in general is a discipline that studies crime, justice, and punishment, based in empirical research of societies, individuals, and their relationship with criminal justice institutions and the law:[29]

> green criminology can be defined as a framework of intellectual, empirical and political orientations toward primary and secondary harms, offenses and crimes that impact in a damaging way on the natural environment, diverse species (human and non-human) and the planet.[30]

Green criminology involves the study of environmental justice from the local to the global level, and the examination of the "ecological consequences that arise from the crimes of the powerful."[31] An evolved version of green criminology, as suggested the green criminologist Rob White, is "eco-global criminology:" a framework to connect the ecological, transnational, and criminal justice analysis of environmental crime, and based on "collaboration involving individuals,

27 *Ibid.* at 77.
28 *Ibid.* at 89.
29 See Bosworth, Mary. *What Is Criminology?* Oxford University Press, 2012, pp. 5–6.
30 Ruggiero, Vincenzo. "Green Criminology and Crimes of the Economy: Theory, Research and Praxis." *Int'l J. Crime, Just & Soc. Democracy*, vol. 59, 2015, pp. 8, 9.
31 Arrigo, Bruce. "The Human Consequences of Ecological Violence and Corporate Victimization: Public Sector Psychology and Green Criminology." *Int'l J. Offender Therapy & Comp. Criminology*, vol. 59, 2015, p. 227.

groups and countries from different geographical regions" for research and investigation.[32]

Conflict theory in criminology suggests that, within society, those possessing social power decide what is viewed as criminal in the eyes of society and the law, resulting in the criminalization of behaviors generally observed in those lacking that social power, such as immigrants, people of color, members of labor unions, and political protesters.[33] The alignment of state and corporate interests normalizes wrongdoings that affect certain people or species, whose victimization is not seen as a loss to the economy. Since an increasingly small number of people control the majority of the economy, the power of the state, including and especially the criminal justice system, expresses itself in accordance with economic control.[34] This nexus between the state and corporations creates situations wherein the poor are criminalized for accessing nature against the interests of capital owners, as seen in instances of the criminalization of indigenous communities when accessing natural resources for their subsistence.[35] Brisman points out that the threat posed by the nexus between the state and the form of the economy functioning under that state is often manifested in the criminalization of dissent of public opposition to economic projects, and, along with causing specific environmental damage, it fosters a culture where the public does not wish to participate in making decisions regarding the environment.[36] In certain cultures, prosecution of environmental crimes is only permitted after administrative sanctions have taken place, if it can be proven that bringing the charge is in public interest,[37] an assumption that implies that some environmental harms are not meant to be of public interest. One might also argue that indirectly an economic model based upon unlimited consumption will criminalize acts which do not require constant consumption such as do-it-yourself recycling of greywater or scavenging for food and discarded building materials, while artificially creating social sanction and stigma against those who engage in such acts, due to the fundamental dissonance between unlimited growth economics and conservation.[38]

32 White, Rob. "The Four Ways of Eco-Global Criminology." *Int'l J. Crime, Just & Soc. Democracy*, vol. 6, 2016, pp. 8, 9.

33 Vito, Gennaro F., and Jeffrey R. Maahs. *Criminology: Theory, Research, and Policy*. Jones & Bartlett Learning, 2007.

34 White, Rob. "Critical Criminology and the Struggle Against Climate Change Ecocide." *Critical Criminology*, vol. 23, 2015, p. 385.

35 Lynch, *supra* note 8 at 148.

36 Brisman, Avi. "The Violence of Silence: Some Reflections on Access to Information, Public Participation in Decision-Making, and Access to Justice in Matters Concerning the Environment." *Crime L. & Soc. Change*, vol. 59, 2013, p. 300.

37 Sahramaki, Jina. "Prevention of Environmental Crime through Enforcement—Finland and Sweden Compared." *Scandinavian Stud.* , vol. 16, 2015, pp. 41, 50.

38 McClanahan, Bill. "Green and Grey: Water Justice, Criminalization, and Resistance." *Critical Criminology*, vol. 22, 2014, p. 410.

10.6 Environmental harm and its victims

Environmental harm, whether inside or outside the ambit of harm that is criminalized by the state, has a distinct body of victims, such as local citizens and wildlife, including, in certain cases, unborn and future generations. Harm consists of primarily that which is defined by law as causing "social injury" and only secondarily outside of these defined parameters.[39] When we expand the characterization of environmental harm, we breach the traditional notions of what is an acceptable degree of exploitation based upon unequal systems of power.[40] Similar to other criminology scholars, Reece Walters argues for a shift from the anthropocentric notion of what is acceptable environmental harm in the interest of human progress (as put forth by liberal ecologists who aim for the minimization of said harm), to using the "deep ecology" approach which accounts for harm against all non-human species as well. She also criticizes the "conservation ethics approach" for focusing on conserving nature primarily for the benefit of human well-being, use, and interest.[41]

Environmental crimes as defined by law are usually acts or omissions in contravention of existing laws, and subject to criminal sanction, with victims being persons, the environment, or both.[42] Since crimes are prosecuted by the state, owing to a crime being a social injury, green criminology applies extra-legal definitions of crime because the state is often a facilitator or perpetrator of environmental harm, and that harm then often falls outside the purview of state prosecution.[43] Additionally, while environmental harm is criminalized under various offenses based on human harms such as fraud, bribery, and violation of permits, the victimization from environmental harm extends to flora, fauna, and stability of ecological systems; this victimization is not the basis of social injury when defining environmental crime.[44]

Risks and harm are closely interconnected when studying green criminology, since potential risks of harm and the decisions taken regarding those risks determine whether the harms will occur. The European Union Agency for Law Enforcement Cooperation (E.U.R.O.P.O.L.), in its Serious and Organized Crime Threat Assessment list, classifies environmental crime and energy fraud as "emerging threats."[45]

Critical criminologists have argued that the reduction of social harm should be a major driving force of law as opposed to reduction of crime, essentially

39 Brisman, *supra* note 36.

40 Walters, Reece. "*Toxic Atmospheres Air Pollution, Trade and the Politics of Regulation*," *Critical Criminology*, vol. 18, 2010, p. 308.

41 *Ibid.* at 315–316.

42 White "Prosecution," *supra* note 14 at 366.

43 *Ibid.*

44 Huisman, Wim. "Offender Decision Making in Corporate and White Collar Crime." *The Oxford Handbook of Offender Decision Making*, 2017, p. 20.

45 *Ibid.*

making a case for criminal justice to include harm which has not been fully criminalized.[46] They argue that several of those harms are incommensurate with existing quantitative standards.[47] The study of environmental harm requires examining the victimization, and, in the case of environmental harm, victims include both the biotic and abiotic components of the environment, as individuals or collectively; victimization may manifest itself over time like in cases of genetic disorders caused by pollution.[48] There is a general perception in society and in criminal law enforcement that environmental crimes are "victimless" and therefore somehow less serious offenses, due to the absence of the traditional perpetrator–victim model to which criminal law enforcement is accustomed.[49]

Environmental crimes by corporations (also called "green-collar crimes") include a vast range of crimes including improper disposal of hazardous waste and intentional pollution of water sources.[50] The victims of these crimes can be geographically and temporally far removed from the perpetrators of that crime, such as mercury pollution in the Great Lakes before 1970, which forty years later continues to result in residents suffering from mercury poisoning, especially the members of the indigenous peoples living in the Aamjiwnaang First Nation on the banks of the Great Lakes. In an important study across Europe on the NOx emissions in excess of legal limits released by cars with cheating devices installed, it was also found that 3.5 percent of all premature deaths due to exposure to PM2.5 in Europe could be attributed to the excess NOx emissions from the vehicles in question, expanding on the ideas of causation, reasonable foreseeability of harm, and the determining of knowledge and intention when committing corporate environmental crimes. This finding further illustrates the complex and lengthy causal chain in determining corporate liability for types of death that in other criminal scenarios would be considered "violent."

Actions that are considered to be harmful are derived from social consensus regarding that harm, and this consensus relies upon state machinery to penalize that harm. The exercise of political power in its interface with criminal prosecution often results in prosecutions that benefit those with political influence.[51] For example, since the production and disposal of goods is based on individual purchase, there is a lack of communal ownership over the harms associated with production and disposal of those goods, and an externalization of harms. An example of this externalization is the shifting of manufacturing processes and

46 *Ibid.*

47 *Ibid.*

48 Jarrell, Melissa. "Few and Far between: Understanding the Role of the Victim in Federal Environmental Crime Prosecutions in the United States." *Crime L. & Soc. Change*, vol. 61, 2014, p. 565.

49 Wright, Glen. "Conceptualising and Combating Transnational Environmental Crime." *Crime L. & Soc. Change*, vol. 61, 2014, p. 565.

50 *Ibid.*

51 *Ibid.*

waste disposal to developing countries with lower regulatory restrictions for goods that are consumed by developed countries, but the developing nations are held responsible for escalation of global pollution associated with the manufacturing process, resulting in an externalization of harm by transnational manufacturers and distributors in various other cultures.[52] This is supported by public perception surveys carried out in the U.S.A. which have consistently shown that white-collar crimes are viewed less seriously than violent street crimes. In a more recent study relying on a 2005 national survey it was found that the American public continues to hold violent street offenders as more deserving of harsher punishments, though two-thirds of those surveyed felt that equal or more resources should be devoted to white-collar crimes.[53]

A very small fraction of environmental harm is characterized as "criminal" and harms such as fishing and deforestation are legal, albeit controlled. While the majority of harms are suffered by the poor, those responsible for the harm almost always tend to be for-profit corporations, who escape from criminal sanctions due to the priorities of the criminal justice system.[54] Criminology concerning white-collar crime applies corporate crime theory to conservation criminology in order to effectively study harm perpetrated by environmental corporate crime and the criminal behavior of corporations.[55] The next section of this chapter will delve into the literature on white-collar crime which is applicable to green criminology.

10.7 Corporate criminality and the environment

> [C]orporate criminality remains a curious concept. As artificial creatures of the law, corporations per se have no emotions or culpable mental states. Nor are they subject to incarceration … To use the hoary phrase, there is "no soul to damn, no body to kick."[56]

Holding corporations criminally liable is in no way a radical concept; the Nuremberg trials for war crimes resulted in the court imposing criminal liability on corporations such as Krupp and IG Farben[57] for the decisions of their directors. A review of the literature on green criminology and white-collar crime will aid in ascertaining the scope of corporate criminality in environmental offenses.

52 *Ibid.*
53 Holtfreter, Kristy. "Public Perceptions of White-Collar Crime and Punishment." *J. Crim. Just*, vol. 36, 2008, p. 50.
54 *Ibid.*
55 *Ibid.*
56 Luna, Erik. "The Curious Case of Corporate Criminality." *Am. Crim. L. Rev.*, vol. 46, 2009, p. 1507.
57 IG Farben or '*Interessen-Gemeinschaft Farbenindustrie AG*' was the then conglomerate consisting of companies which today operate as BASF Aktiengesellschaft, Bayer AG, Hoechst Aktiengesellschaft and the Agfa-Gevaert Group. Encyclopaedia Britannica.

In the U.S.A., criminalization of environmental violations developed from the 1980s onward, in order to address the reality that civil penalties were viewed by violators as a part of production costs and would be passed on to the consumer; civil penalties also lacked the negative publicity and stigma associated with criminal sanction.[58] Civil penalties are expressed as monetary sums which must be paid for violations of a statute or regulations, usually imposed and administered by an agency or body authorized by law, but these violations are not prosecuted as criminal offenses and are not intended to achieve the goals of penalization, but usually intended to correct any undue benefits that a person might have accrued out of non-compliance.[59] Moreover, in the U.S.A. there are tax ramifications, disclosures to shareholders and government agencies, for example, for corporations and other persons when paying criminal fines that are not the same as when paying civil penalties. Some have concluded that in fact the otherwise oxymoronic combination of "civil" with "penalty" would not need to exist.

A 1996 study by Alexander and Cohen found that, in the ten-year period between 1984 and 1994, in the U.S.A., 22 percent of all federal corporate crimes were environmental crimes[60] (twenty-six convictions).[61] The U.S.A. combines corporate criminal liability with vicarious liability through the "responsible corporate officer" doctrine when prosecuting the corporation. Negligence liability is often exercised through the responsible corporate officer doctrine.[62] The "responsible corporate officer" doctrine has been critiqued as it can result in situations wherein the lowest ranking officer who is directly involved in the last step of the offense is convicted, whereas higher ranking officers and directors can be acquitted for lacking knowledge of the offense, which does little to address the corporate culture which enables such offenses in the first place.

Imposing corporate criminal liability is an effective way of controlling pollution-related offenses, as for example in the oil transportation industry, which has led to an increase in prevention measures, owing to the deterrent value of potential penalties for oil spills. However, criminal liability imposed through fines and economic consequences is ineffective due to the costs being shifted to the consumer, similar to civil penalties.[63] Limiting punitive responses to fines is also ineffective in cases where the harm perpetrated by a company far exceeds the collectible assets of the responsible corporate officer, or even the company as a whole, as seen in the case of the gas disaster in Bhopal, India, or the long-term effects of asbestos-containing products manufactured by the Johns Manville Corporation.[64]

58 Edwards, Sally. *Environmental Crime and Criminality: Theoretical and Practical Issues.* Taylor & Francis, 2016.

59 Belden, Roy S. *Clean Air Act.* American Bar Association, Section of Environment, Energy, and Resources, 2011, p. 136.

60 Alexander, Cindy & Mark A. Cohen. "New Evidence on the Origins of Corporate Crime," 17 *Managerial & Decision Econ.* 1996, pp. 421, 422.

61 *Ibid.* at 426.

62 Arlen, Jennifer. "The Potentially Perverse Effects of Corporate Criminal Liability," *J. L. Stud.* 1994, pp. 833, 866.

63 *Ibid.*

64 *Ibid.*

By applying traditional criminological concepts to green criminology, the creation of offender databases of criminals and using technology for "offender profiling" would make more sense when applied to corporate crimes, given the ease with which their personnel and day-to-day work patterns can be identified and investigated, but we see that most profiling technology is used for non-corporate crimes.[65]

10.8 Comparing the U.S.A., India, and Germany

The questions that this chapter seeks to answer while comparing these jurisdictions and their respective legal cultures are:

1) How do India, the U.S.A., and Germany differ in their treatment of environmental crimes?
2) Is there a nexus between the criminal justice processes in these respective jurisdictions and their level of environmental harm, and, if so, what is its nature?

I will employ the following matrix to compare the nexus between criminal prosecution and levels of pollution in the U.S.A., India, and Germany. I will first examine the goals with which each jurisdiction introduced its criminal provisions, describing the legislation introduced toward those goals and the outcomes from that legislation. The outcomes will be assessed using available data from the respective government agencies' reports on crimes prosecuted or registered for air pollution offenses, and the coinciding performance of each country with respect to air quality using data published by the World Bank.

Since the comparison pertains to states, I have relied on the use of data published by the federal and national agencies of the governments of those three states. The most significant limitation of this data is that it is derived or released by government agencies according to their relative resources and priorities, resulting in there being a vast difference in the lengths of time for which the data is available, depending on each individual jurisdiction.

The U.S.E.P.A. maintains a Summary of Criminal Prosecutions that can be used to find the summary of each individual criminal case prosecuted in every fiscal year starting from 1983, referenced according to the federal statutes under which it was prosecuted. Using the search fields for each year from 1983 up to 2017, and under the Clean Air Act (C.A.A.),[66] I computed the number of criminal prosecutions under each statute for that time period.[67]

65 *Ibid.*

66 The Clean Air Act (42 U.S.C. § 7401).

67 United States Environmental Protection Agency, "Summary of Criminal Prosecutions." *U.S.E.P.A.*, Environmental Protection Agency, https://cfpub.U.S.E.P.A..gov/compliance/criminal_prosecution/index.cfm?bct=19&pt=1. In using data on federal prosecution, I have

India's National Crime Records Bureau (N.C.R.B.) is the agency of the Central Government that is responsible for reporting the data on crimes committed throughout the country. However, I found that data on environmental crimes was available only for the years 2014–2016, a relatively small time period from which it is difficult to adequately infer any trend or behaviors, and the data released does not specify the number of punishments or nature of the defendants. Therefore, an application under the Right to Information Act[68] was made to the Public Information Officer of the N.C.R.B.,[69] which resulted in the N.C.R.B.'s response dated August 31, 2018, explaining that they did not collect or possess data regarding environmental crimes prior to 2014, and that the latest available data was for 2016.[70]

The *Bundeskriminalamt* (B.K.A.), the Federal Criminal Police Office of Germany, has been publishing data via annual reports and datasets of police crime statistics detailing the nature and quantum of crimes committed across the country, including the specific data on crimes relating to air and water pollution. However, these reports are currently available only for the periods 2001–2005 and 2009–2017 on the B.K.A. website, and only the data for those periods will be used for comparison in this chapter.[71]

10.8.1 United States

A justified criticism of criminal sanctions of environmental crimes in the U.S.A. is that overcriminalization would lead to a trivialization of criminal law; as Cohen puts it: "[i]f every action that harms society is a crime, the criminal law loses its one distinguishing characteristic—the moral stigma associated with being labelled a criminal."[72] However, deterrence is not affected by the enactment of criminal statutes alone, but the threat of prosecution, and in the U.S.A. specifically prosecutors play a major role in the adjudication of crime, often through

intentionally omitted referring to data for the U.S.E.P.A. rate for U.S. states and their prosecution of pollution crimes, with the awareness that doing so might result in a far deeper ambit of study which would reveal relevant comparison between states but would be outside the scope of the comparison between nations in this chapter.

68 Right to Information Act (Act No. 22/2005).

69 The text of the application under the Right to Information Act, 2005, addressed to the N.C.R.B., is attached as Appendix 1, and the confirmation of receipt of the application dated August 7, 2018 is attached as Appendix 2.

70 I considered potentially using an additional approach to inferring data on criminal prosecutions from judgments or decisions of the courts, but the "Legal Search" function in the Indian search engine "Manupatra" was not able to provide judgments from the subordinate criminal courts where criminal prosecution are initiated.

71 Bundeskriminalamt, *Police Crime Statistics*, available at https://www.bka.de/EN/CurrentInformation/PoliceCrimeStatistics/policecrimestatistics_node.html.

72 Cohen, Mark. "Environmental Crime and Punishment: Legal/Economic Theory and Empirical Evidence on Enforcement of Federal Environmental Statutes." *Crim. L. & Criminology*, vol. 82, 1992, p. 1104.

the pursuit of plea bargaining.[73] Prosecutorial discretion determines which cases are chosen for criminal prosecution, and only the cases with the best evidence and which are most likely to result in a conviction are those that make it to the trial stage. According to a study of prosecutions for pollution between 2005 and 2010 in the U.S.A., 96 percent of defendants prosecuted had committed a minimum of one aggravating factor, such as being repeat offenders, deception, or operating outside the regulatory scheme. Not only do these findings reveal that prosecutors in the U.S.A. prosecute only environmental crimes with aggravating factors, almost as a rule it also reveals that prosecutorial discretion leads to not prosecuting crimes without aggravating factors, possibly due to the wish to avoid overcriminalization.[74]

In the 1970s, environmental crimes were prosecuted rarely, and usually as misdemeanors.[75] In 1976, the U.S.E.P.A. published its first Agency guidelines to pursue criminal sanctions with more focus for the enforcement of the Clean Air Act which continued to be supported by the Carter administration.[76] A rise in public concern regarding the environment in the 1980s and 1990s led to an increased number of criminal prosecutions in the U.S.A., along with increased penalties for convictions.[77] This period was also marked by an increase in staffing of agents in the U.S.E.P.A., whose staff began to include special agents equipped with firearms and responsible for investigations and arrests, field lawyers, and litigation attorneys.[78] In 1982, the U.S.E.P.A. and the U.S. Department of Justice (D.o.J.) established dedicated units for the investigation and prosecution of environmental crimes, creating the necessary infrastructure for greater criminal enforcement of the Clean Water Act (C.W.A.)[79] and the C.A.A. In 1987, felonies were included under the C.W.A.[80] The introduction of environmental felonies resulted in a shift from the willful or negligent mental state requirement to a standard of knowledge for felonies and negligence for misdemeanors.[81] As for the knowledge requirement, specific and direct knowledge of a crime is not necessary to prove guilt under the C.A.A. The C.A.A. allows for knowledge to

73 Haugh, Todd. "Sox on Fish: A New Harm of Overcriminalization," *Nw. U. L. Rev.* 2015, pp. 835, 842.

74 *Ibid.*

75 Jarrell, *supra* note 48 at 568.

76 *Ibid.*

77 Harig, Lisa. "Ignorance Is Not Bliss: Responsible Corporate Officers Convicted of Environmental Crimes and the Federal Sentencing Guidelines." *Duke L. J.*, vol. 42, 1992, p. 165 (describing the increase in fines and incarceration for environmental crimes in the 1990s).

78 *Ibid.*

79 Clean Water Act, 33 U.S.C. § 1251 et seq.

80 United States Department of Justice, *Historical Development of Environmental Criminal Law*, available at https://www.justice.gov/enrd/about-division/historical-development-environmental-criminal-law.

81 Uhlmann, David. "Environmental Crime Comes of Age: The Evolution of Criminal Enforcement in the Environmental Regulatory Scheme." *Utah L. R.*, vol. 4, 2009, p. 1235.

be established using circumstantial evidence, as well.[82] Along with the C.A.A, provisions under Title 18 of the United States Code (U.S.C.)[83] are used in the prosecution of environmental offenses[84] such as the penal provisions for "aiding and abetting,"[85] conspiracy,[86] false statements,[87] mail fraud,[88] wire fraud,[89] and obstruction of justice.[90] Under § 7401(c) of the C.A.A.,[91] the primary goal of the statute "is to encourage or otherwise promote reasonable Federal, State, and local governmental actions, consistent with the provisions of this chapter, for pollution prevention." Toward preventing this pollution of air, the U.S.E.P.A. established the U.S. National Ambient Air Quality Standards (N.A.A.Q.S.), which detail the permissible amounts of different types of pollutants in outdoor air quality. The annual permissible mean $PM_{2.5}$ is 12 $\mu g/m^3$ under the N.A.A.Q.S.[92]

Section 7413 (c) of the C.A.A.[93] provides for the federal enforcement of the C.A.A. with the imposition of criminal penalties on violators, which penalizes violations with varying amounts of fines under Title 18 of the United States Code (U.S.C.), or imprisonment, or both, with repeat convictions for the same offense resulting in the doubling of fine amounts and prison terms.[94] Specifically, the release of hazardous pollutants in violation of existing permits into the ambient air with the knowledge that such release may place "another person in imminent danger of death or serious bodily injury shall, upon conviction, be punished by a fine under title 18 or by imprisonment of not more than 15 years, or both."[95] If such a violation is committed by an organization, upon conviction each violation shall attract a fine of not more than one million U.S. dollars. The act provides for the prosecution of organizations, defining organization to mean "a legal entity, other than a government, established or organized for any purpose, and such term includes a corporation, company, association, firm, partnership, joint stock company, foundation, institution, trust, society, union, or any other association of persons,"[96] and the term "person" to include "responsible corporate officer."[97]

82 Mullikin, Nancy. "Holding the Responsible Corporate Officer Responsible: Addressing the Need for Expansion of Criminal Liability for Corporate Environmental Violators." *Golden Gate U. Envtl. L. J.*, vol. 3, 2010, p. 423.
83 Crimes and Criminal Procedure, Title 18, U.S.C.
84 *Ibid.* at 9–10.
85 18 U.S.C. § 2.
86 18 U.S.C. § 371.
87 18 U.S.C. § 1001.
88 18 U.S.C. § 1341.
89 18 U.S.C. § 1343.
90 18 U.S.C. § 1512.
91 42 U.S.C. § 7401.
92 United States Environmental Protection Agency, NAAQS Table, available at https://www.U.S.E.P.A..gov/criteria-air-pollutants/naaqs-table.
93 42 U.S.C. § 7413 (c).
94 42 U.S.C. § 7413 (c)(1).
95 42 U.S.C. § 7413 (c)(5)(A).
96 42 U.S.C. § 7413 (c)(5)(E).
97 42 U.S.C. § 7413 (c)(6).

TABLE 10.1 Crimes prosecuted under the C.A.A. in the U.S.A.: 1983–2017.

Year	*No.*	*Year*	*No.*	*Year*	*No.*	*Year*	*No.*
		1990	2	2000	10	2010	13
		1991	2	2001	7	2011	24
		1992	1	2002	5	2012	18
1983	0	1993	5	2003	21	2013	28
1984	0	1994	5	2004	18	2014	25
1985	0	1995	6	2005	12	2015	15
1986	2	1996	8	2006	19	2016	15
1987	4	1997	11	2007	21	2017	18
1988	2	1998	11	2008	8		
1989	3	1999	13	2009	7		

Source: U.S.E.P.A. database

As per the U.S. Sentencing Commission, in 2015, the federal crimes most commonly committed by organizational offenders such as corporations were environmental crimes, comprising 33.2 percent of total crimes. Of these, 70 percent were related to water, and 5 percent were related to air.[98] As suggested by the green criminology theorists above, environmental crimes are not seen as criminal in the manner violent crime or street offenses are, due to nature of the offense and the offenders. This is confirmed by a study of sentencing in environmental criminal cases in the U.S., where researchers found a "green gap in punishment." The green gap refers to the observed difference in proportion of sentencing that environmental crimes vis-à-vis non-environmental crimes, such as sentences with more probation (72 percent as opposed to 57 percent), and fewer prison terms (2 percent as opposed to 9 percent).[99]

Using the U.S.E.P.A. database containing the summary of criminal prosecutions, the following is a compilation of federal criminal prosecution under the C.A.A. in the U.S. from 1983 to 2017.[100]

As can be observed from Table 10.1, there has been a marked increase in the prosecution of crimes under the C.A.A. in the U.S.A., up from zero in 1983 to eighteen in 2017, with the highest number of prosecutions being in 2013 (twenty-eight). While the increase in prosecution has not been consistent, with diverse data points over the years as observed in Figure 1, the overall trend line over time is positive, indicating an increased willingness and ability of the federal government to criminally prosecute violations under the C.A.A.

98 United States Courts, *Environmental Crime Tops List by Organizations* available at http://www.uscourts.gov/news/2016/12/22/environmental-crime-tops-list-organizations.

99 Cochran, Joshua. "Court Sentencing Patterns for Environmental Crimes: Is There a "Green" Gap in Punishment?" *Quantitative Crim.*, vol. 34, 2018, p. 51.

100 US U.S.E.P.A., *Summary of Criminal Prosecutions.*

10.8.2 India

The Air (Prevention and Control of Pollution) Act of 1981[101] was enacted in 1981, using the power of the Central Government under Article 253 of the Constitution of India, to legislatively implement India's commitment under international treaties or agreements, which in this case was implementing the decisions at the U.N. Conference on the Human Environment in Stockholm, 1972. The Air Act established the Central Pollution Control Board (C.P.C.B.) and State Pollution Control Boards (S.P.C.B.s), listing their functions and responsibilities, and the procedures for their functioning.

In Indian law, the goals of a statute, for the purpose of statutory interpretation, can be gleaned from the Preamble. The Preamble of the Air Act invokes the Stockholm Conference decision "to take appropriate steps for the preservation of the natural resources of the earth which, among other things, include the preservation of the quality of air and control of air pollution," and describes it as an "Act to provide for the prevention, control and abatement of air pollution."[102] In order to quantify these goals for this chapter, the air quality standard which the Air Act seeks to achieve and maintain, according to the current National Ambient Air Quality Standards for $PM_{2.5}$ (μg/m³), annually for all areas (industrial, residential, rural, and ecologically sensitive), is 40 μg/m³.[103] This ambient air quality standard can be deemed to be the intended goal for $PM_{2.5}$ under the Air Act.

The Air Act contains a "command-and-control"[104] type of criminal enforcement which imposes prohibitions on the release of pollutants in excess of the S.P.C.B. standards, giving the S.P.C.B. authority to approach a criminal court in order to seek the restraining of a person likely to violate those standards.[105] Under § 37 of the Air Act, violations are:

> punishable with imprisonment for a term which shall not be less than one year and six months but which may extend to six years and with fine, and in case the failure continues, with an additional fine which may extend to five thousand rupees for every day during which such failure continues after the conviction for the first such failure.[106]

Continued failure beyond one year after conviction attracts a penalty of imprisonment of not less than two years, and up to seven years with a fine.[107] Section 40 of the Air Act provides for the criminal prosecution of companies (including

101 The Air (Prevention and Control of Pollution) Act (Act No. 14/1981). (Air Act)

102 Air Act, Preamble.

103 National Ambient Air Quality Standards, No. B. 29016/20/90/PCI-I (Nov. 18, 2009).

104 Ghosh, Shibani. *Reforming the Liability Regime for Air Pollution in India*, 6 available at http://www.cprindia.org/articles/reforming-liability-regime-air-pollution-india.

105 Air Act, §22, & §22A.

106 Air Act, §37(1).

107 Air Act, §37(2).

incorporated bodies, firms, or other associations) for offenses under the Act, holding "every person who, at the time the offense was committed, was directly in charge of, and was responsible to, the company for the conduct of the business of the company, as well as the company" to be deemed guilty of the offense and liable for penalty unless they can prove that the offense was committed without their knowledge, and despite the exercise of all preventive due diligence on their behalf.[108] The S.P.C.B. or authorized officers can initiate criminal prosecution by filing a complaint before the respective criminal court,[109] or any other person can file a criminal complaint against an offense if they have given a minimum notice of sixty days to the concerned S.P.C.B. regarding their intention to approach the judiciary.[110]

Despite introducing criminal legislation against both water and air pollution in 1974 and 1981 respectively, the National Crime Records Bureau of India began to collect data on environmental crimes only in 2014, with the latest report being for the year 2016. Through an application to the N.C.R.B. made under the Right to Information Act, 2005, this author sought data specifically on crimes registered under The Water (Prevention and Control of Pollution) Act, 1974,[111] and the Air Act; specifically § 47 and § 40 of those Acts respectively referring to offenses by companies.[112] The N.C.R.B. responded to the request by letter[113] confirming that they did not possess records or data of environmental crimes registered in India prior to 2014, and all of the data available was limited to that available in the public domain as part of the N.C.R.B.'s annual "Crime in India" statistics reports. Therefore, this chapter will be using the very small sample of data currently released by the Indian government on environmental crimes from 2014 to 2016. The following table combines the data on crimes registered under the Water Act and the Air Act, along with crimes under the Forest Act[114] and Wildlife Protection Act[115] (W.P.A.), to offer an important frame of reference.

While the evaluation of criminal law enforcement in environmental crimes in India is difficult given the relatively short period of data covered, the data itself as listed in Table 10.2 is a confirmation of the conflict theory in criminology discussed earlier, which asserts that there are opposing interest groups in society depending on how they possess social power, and this social power manifests itself in the criminalization of behavior observed in groups lacking that social

108 § 40, Air Act.
109 § 43(1), Air Act.
110 § 43(1)(b), Air Act.
111 The Water (Prevention and Control of Pollution) Act (Act No. 6/1974).
112 The text of the application to N.C.R.B. under the Right to Information Act, 2005, and the confirmation of receipt of the application dated August 7, 2018 are attached as Appendix 1 and Appendix 2 respectively.
113 The response dated August 31, 2018 by the N.C.R.B. to my application under the Right to Information Act, 2005, is attached as Appendix 3.
114 Indian Forest Act (Act No. 16/1927).
115 Wildlife Protection Act (Act No. 53/1972).

TABLE 10.2 An illustration of the conflict theory in criminology

Year/Statute	*Forest Act*	*Wildlife Protection Act*	*Air Act*		*Water Act*	
2014	4901	770	48		15	
2015	3968	829	50		10	
2016	3702	852	25	Jharkhand (2) Maharashtra (21) Uttar Pradesh (2)	11	Gujarat (3) Himachal Pradesh (2) Madhya Pradesh (6)

Source: N.C.R.B. Data

power.[116] In a nation of twenty-nine states, it seems highly improbable and unbelievable that the N.C.R.B. data shows that crimes were registered in only three states under the Air Act and the Water Act, which could be attributed to the state boards for pollution control being unwilling to register criminal cases against polluters, as criminal prosecution is seen as time-consuming and cumbersome, and to the boards preferring to issue showcase notices instead. The unwillingness to prosecute industrial pollution can also be attributed to the governments' tacit focus on states not being perceived as litigious, because that would drive way investment in the state economy.[117]

The N.C.R.B. data also shows that crimes committed under the Forest Act and the Wildlife Protection Act (W.P.A.) are over a hundredfold and tenfold, respectively, compared to crimes registered under the Air Act and Water Act. This glaring disparity could be attributed to multiple factors. First, the majority of accused under the Forest Act and W.P.A. belong to poor communities who reside around forest areas, and were prosecuted for gathering minor forest produce or firewood, or for carrying out shifting cultivation on forest land.[118] This is an illustration of how criminal law, and environmental criminal law, tends to criminalize the poor, while industrial pollution is not prosecuted with similar enthusiasm, suggesting a pro-corporation bias of criminal environmental justice enforcement. The excesses of the Forest Service in prosecuting forest-dwelling communities has been well-documented, given that they have the discretion to make arrests without warrants.[119] This aligns with the premise in green criminology that the state and its agents are indirectly controlled by economic interests

116 Vito, *supra* note 33.

117 Sahu, Geetanjoy. *Off the mark: National Crime Records Bureau data gives a misleading picture of environmental crime*, Scroll (Jan. 02, 2018, 06:30 AM) available at https://scroll.in/article/863087/off-the-mark-national-crime-records-bureau-data-gives-a-misleading-picture-of-environmental-crime.

118 *Ibid.*

119 Kodiveri, Arpitha. *Conflicting Laws Are Criminalising Forest Communities for Exercising Their Rights*, The Wire (May 02, 2017) available at https://thewire.in/environment/conflicting-laws-criminalising-forest-communities-for-exercising-their-rights.

and decide that which is to be prosecuted and perceived as environmental crime. It also substantiates the hypothesis that the state–corporate nexus criminalizes the poor having access to nature, even if it is for subsistence outside of the liberal market economy.[120] This is also supported by the earlier finding that pollution control boards prefer opting for showcase notices against polluters, to avoid being seen as litigious and to ensure the entry of investment into their economies.

Another reason for the disparity in crimes recorded between forest-related crimes and pollution-related crimes claims that the Forest Act, a colonial law, was created in 1927 to maintain forest reserves for Britain, and it created a forest service, with the officers highly trained in policing and therefore willing and able to work in a criminal law enforcement capacity.[121] Furthermore, the pollution control boards created in the 1970s are primarily tasked with the issuance of permits and have not been trained as policing or enforcement officers and therefore tend to opt for showcase notices while implementing the pollution control statutes.[122]

However, for the purpose of comparison, since this is the available data provided by the government of India, I will be using the three-year data period for the purpose of comparison in this chapter. While in 2014 there were forty-eight crimes registered under the Air Act, a number that marginally increased to fifty in 2015, 2016 saw a sharp decline of 50 percent, with only twenty-five crimes registered under the Air Act, resulting in a negative linear trend line for the number of prosecutions in this short span of time.

10.8.3 Germany

In Germany, criminal liability is based on personal culpability so that only natural persons can be criminally liable.[123] However, § 14 of the *Strafgesetzbuch* (StGB), the German Criminal Code,[124] provides for criminal liability of persons

120 Lynch, *supra* note 8 at 148.

121 Bhuyan, Ragini. *The gap in environmental crime statistics*, LiveMint (Dec. 26 2015, 12:35 AM) available at https://www.livemint.com/Opinion/aE8vmBDeBKOJCODKhBwctN/The-gap-in-environmental-crime-statistics.html.

122 *Ibid.*

123 Sina, Stephan. "Fighting Environmental Crime in Germany: A Country Report. Study in the framework of the EFFACE", 2015, p. 9.

124 Strafgesetzbuch, § 14: Acting for another: (1) If a person acts: 1. in his capacity as an organ authorised to represent a legal entity or as a member of such an organ; 2. as a partner authorised to represent a partnership with independent legal capacity; or 3. as a statutory representative of another, any law according to which special personal attributes, relationships or circumstances (special personal characteristics) form the basis of criminal liability, shall apply to the representative, if these characteristics do not exist in his person but in the entity, partnership or person represented. (2) If a person, whether by the owner of a business or somebody delegated by him, has been 1. commissioned to manage the business, in whole or in part; or 2. expressly commissioned to perform autonomous duties which are incumbent on the owner of the business, and the person acts on the basis of this commission, any law, according to which special

acting in their authority for or on behalf of legal entities, and special personal characteristics which would form the basis of criminal liability in a person if they do not exist in him or her per se, but in the entity or person that he or she is representing. Since it is often difficult to pinpoint the exact personal liability of an individual within a company,[125] the *Bundesgerichtshof* (BGH), the German Federal Court of Justice, has established that all members in management can be held jointly responsible,[126] and a senior member on whose orders a crime is committed is not just guilty as an instigator, but as an indirect perpetrator, and sufficiently senior members of a corporation can be found guilty as indirect perpetrators by virtue of their control over the organization.[127] Germany introduced environmental crimes under the StGB in 1980. One of the findings in the German environmental criminal law system is that the prospect of standing trial, and not the probability of conviction or magnitude of fines, positively supports the attempts at deterrence, and that environmental compliance in Germany can be partly attributed to this deterrence.[128]

When the StGB was amended in 1980 to include environmental crimes, the goals of this extension were "to raise the level of general deterrence on account of harsher sanctions being available through the criminal justice system" and "to harness the additional policing and prosecutorial resources available in the criminal justice system."[129] Under Directive 2008/50/EC of the European Commission,[130] all Member States of the E.U. are required to achieve ambient quality standards specific to each pollutant and the air quality standard for Member States for annual $PM_{2.5}$ ($\mu g/m^3$) levels is 25 $\mu g/m^3$.[131]

Section 325 of the StGB contains the penal provisions for air pollution offenses detailing the circumstances in which criminal prosecution could be pursued and the individual penalties including fines and imprisonment for those offenses. The

personal characteristics give rise to criminal liability shall apply to the person commissioned, if these characteristics do not exist in his but in the person of the owner of the business. Within the meaning of the 1st sentence above an enterprise shall be the equivalent of a business. If a person acts on the basis of a similar commission for an agency performing public administrative services, the first sentence shall apply mutatis mutandis. (3) Subsections (1) and (2) above shall apply even if the act of commission intended to create the power of representation or the agency is void.

125 Sina, *supra* note 123 at 18.

126 Bundesgerichtshof [BGH] [Federal Court of Justice] Jul. 06, 1990, Neue Juristische Wochenschrift [NJW] 2560, 2008 123, 131–32 (Ger.).

127 Bundesgerichtshof [BGH] [Federal Court of Justice] May 13, 2004, Neue Juristische Wochenschrift [NJW] 2248, 2004 163–64 (Ger.).

128 *Ibid.*

129 Almer, Christian Alexander, "The Economics of Environmental Crime: Theoretical Aspects and Econometric Investigations," (Dr. rer. pol. Inaugural Dissertation, Heidelberg University, Jan. 2009), p. 11.

130 EC Directive 2008/50/EC, OJ L 152, (11.6.2008) available at http://data.europa.eu/eli/dir/2008/50/oj (last visited Nov. 12, 2018).

131 European Commission, *Air Quality Standards*, available at http://ec.europa.eu/environment/air/quality/standards.htm (last visited Nov. 12, 2018).

TABLE 10.3 Cases recorded under Section 325, StGB: 2001–2017

Year	*No.*	*Year*	*No.*
2001	303	2010	204
2002	261	2011	256
2003	308	2012	165
2004	184	2013	139
2005	176	2014	137
2009	211	2015	119
		2016	127
		2017	135

section criminalizes particular alterations made to the air including by release of harmful substances in significant amount in violation of duties under administrative law.

Using the database maintained by the B.K.A., the table above has been compiled using the data contained in annual reports and datasets of police crime statistics, from 2001 to 2005 and 2009 to 2017.[132] Table 10.3 contains the number of crimes recorded under §325. Germany has seen a decline in criminal prosecutions under §325, with 303 in 2001, and 135 in 2017, which is a decline of 55 percent in a period on seventeen years, and the general negative trend can be observed in Table 10.3.

10.8.4 Analyses

This chapter seeks to answer two questions: first, how the U.S.A., India, and Germany differ in their treatment of air pollution using criminal enforcement provisions, and, second, whether there is a relationship between the frequency of criminal enforcement and quality of air in these countries. The preceding part of the chapter has established the essential differences in the legislation, and data on crimes in each country and the following part of the chapter attempts to infer the relationship between criminal enforcement and air quality, through analysis of the available data.

The World Bank through its open data platform has published the levels of $PM_{2.5}$ air pollution, and for the purpose of this analysis I will be considering the data of annual mean exposure (μg/m^3) for each individual country.[133] All of the air quality data used below has been extracted from the dataset obtained from

132 Bundeskriminalamt, *supra* note 71.

133 The numerical calculations for the regression analysis of the data are on file with the author.

TABLE 10.4 Mean Annual Exposure to PM2.5 (ug/m3) from 1990 to 2016

Year	*U.S.*	*India*	*Germany*
1990	11.38	59.85	17.45
1995	10.88	60.59	15.2
2000	10.7	61.49	14.08
2005	10.43	65.66	13.39
2010	8.62	64.6	13.7
2011	8.65	63.87	13.68
2012	8.68	66.07	13.3
2013	8.53	68.95	13.66
2014	8.5	71.61	13.87
2015	9.17	75.58	13.48
2016	9.2	75.8	13.46

Source: Bundeskriminalamt; World Bank

the World Bank,[134] and is available for the years 1990, 1995, 2000, 2005, and 2010–2016.

Using the data for criminal prosecution of air pollution crimes in the U.S.A. listed in Table 10.4, the $PM_{2.5}$ readings for the U.S.A., we can observe that over time there has been an increase in prosecution of crimes under the C.A.A., and that there is a positive and direct relationship between crimes prosecuted over consecutive years. Along the same time period, there has been a small decline in annual exposure to $PM_{2.5}$ air pollutants in the ambient air, indicating a direct and inverse relationship. According to the N.A.A.Q.S. standards set by the U.S.E.P.A., the annual mean exposure to $PM_{2.5}$ pollutants is 12 $\mu g/m^3$ and all of the readings in the U.S.A. in the given period have fallen within the prescribed limit, in a consistently declining trend. There is an inverse and negative relationship between the two variables, indicating that an increase in the number of crimes prosecuted coincided with a decrease in exposure to $PM_{2.5}$ pollutants in the U.S.A. from 1990 to 2016.

Thus, not only has the U.S.A. achieved its goal completely, in terms of exposure to $PM_{2.5}$ pollutants being limited to under 12 µg /m3, but the analysis also indicates an inverse relationship between the prosecution of crimes and exposure to pollution, and implies a direct relationship between prosecution of crimes and decline in exposure to pollution. Using the limited N.C.R.B. data on crimes registered in India under the Air Act, and the values of mean annual exposure to $PM_{2.5}$ pollutants, a sharp decline in the number of prosecutions over time is indicated, while the exposure to $PM_{2.5}$ pollutants is a direct and positive relationship,

134 Brauer et al., *PM2.5 air pollution, mean annual exposure (micrograms per cubic meter)*, The World Bank available at https://data.worldbank.org/indicator/EN.ATM.PM25.MC.M3.

indicating that the annual mean exposure to PM$_{2.5}$ pollutants has been steadily increasing over the passage of time.

The National Ambient Air Quality Standards in India for PM$_{2.5}$ (μg/m^3) is 40 μg/m^3[135] and, as we can see from the data, the mean annual exposure to PM$_{2.5}$ has been steadily increasing from 59.85 μg/m^3 in 1990 to 75.80 μg/m^3 in 2016, which is nearly double the permissible limit. When plotted against the number of crimes prosecuted under the Air Act, the analysis reveals a negative and inverse relationship between the variables, indicating that, in India, increase in mean annual exposure to PM$_{2.5}$ pollutants has coincided with a decrease in crimes recorded. While India has completely failed to maintain its desired ambient air quality with respect to mean annual exposure to PM$_{2.5}$, its increasing mean annual exposure to PM$_{2.5}$ is inversely related with its declining use of criminal provisions of the Air Act.

Plotting the number of crimes recorded and the annual mean exposure to PM$_{2.5}$ (μg/m^3) over the years in Germany, using the B.K.A. data and the World Bank data on PM$_{2.5}$, reveals a decline in the number of crimes under §325 of the StGB recorded over time.

Additionally, given that the goal for mean annual exposure to PM$_{2.5}$ (μg/m^3) levels is 25 μg/m^3,[136] and from 1990 (17.450204 μg/m^3) to 2016 (13.4577961 μg/m^3) Germany has not only achieved this air quality goal, but seen a decline of exposure to PM$_{2.5}$ pollutants, one could also possibly explain the decline in number of crimes recorded, if the causes of action for crimes themselves have lowered over time after having achieved the air quality targets. In Germany, one cannot therefore observe a significant relationship between the variables of crimes recorded and exposure to PM$_{2.5}$ pollutants.

After accounting for the limitations of the data as previously discussed in this chapter, we find that in two of the three jurisdictions, India and the U.S.A., which are different in terms of development status, an inverse relationship between instances of criminal prosecution or registration and the exposure to PM$_{2.5}$ pollutants is clearly observed, while Germany shows no significant relationship, thus answering both the research questions in this chapter as to how the three jurisdictions differ in their treatment of air pollution crime and if there exists a relationship between criminal enforcement and the exposure to pollution in each of these jurisdictions. While it would be incorrect to imply causation, especially when comparing the U.S.A. and India, which show clearly identifiable relationships, it is possible that criminal enforcement is associated with greater technical vigilance over violations, indicative of a more robust environmental enforcement regime overall, leading to better performance on environmental indicators.

135 National Ambient Air Quality Standards, *supra* note 103.
136 European Commission, *supra* note 131.

10.9 Conclusions

This chapter has offered qualitative and quantitative arguments in favor of criminal prosecutions for environmental harms, especially against corporate perpetrators. The quantitative section presents a significantly representative, if not exhaustive, discussion on the extant theories in green criminology and corporate liability in environmental offenses, and compares cultural responses to criminal environmental law enforcement in the U.S.A., India, and Germany. Using available data, one can glean that there is a direct and positive relationship between instances of criminal prosecution and quality of air. Without implying causation, it makes a case for justifying the continued use of criminal justice mechanisms for the prevention and control of air pollution, and possibly other environmental crimes. This analysis would have undoubtedly benefitted from having access to data on not just prosecution of crimes in each country, but also the nature of the defendants (whether natural or legal persons), as well as a longer data period specifically for India. An important inference is that India is not just the only developing country in the set, and therefore the only one not having achieved its air quality goals, but its visible paucity of data on environmental crimes is possibly illustrative of a weaker enforcement regime overall. Additionally, it is a massive generalization to use air quality data averaged for entire countries and a more accurate picture could be drawn if individual cities or rural areas are compared to gauge the performance on air quality with respect to criminal enforcement.

If behavior that harms society and the environment is seen as *de rigeur* when engaged in by corporations, it is no longer seen as deviant. If everyone is a criminal, nobody is, and if all corporations are essentially engaging in criminal enterprise, making it difficult to prosecute each one for violation, an industrial structure based on environmental harm is created. It is crucial in the attribution of corporate criminal liability to not merely impose strict liability, or to attribute liability to individual persons, but also to establish how legal entities and corporations as responsible actors display criminal behavior. This is especially difficult in a political landscape where liberalism and the individual occupy the center of the legal imagination, and corporate criminal behavior conceptually can shift this perspective toward greater social welfare.[137] A change in language in how we document, report, and ultimately legally treat corporations is essential for the protection of the right of the public to access its environment.

Literature

Almer, Christian, and Timo Goeschl. "Environmental Crime and Punishment: Empirical Evidence from the German Penal Code." *Land Econ.* 86 (2010): 707, 709.

137 Quaid, Jennifer. "The Assessment of Corporate Criminal Liability on the Basis of Corporate Identity: An Analysis." *McGill L.J.*, vol. 43, 1998, p. 71.

Barett, Kimberly L. "Monetary Penalties and Noncompliance with Environmental Laws: A Mediation Analysis." *Am J Crim. Just.* 3 (2017). 1 Oct 2018. doi:10.1007/S12103-017-9428-0.

Billiet, Carole M., and Sandra Rousseau. "How Real is the Threat Of Imprisonment For Environmental Crime?" *Eur. J. L. Econ.* 37 (2014): 183, 196.

Brisman, Avi. "Crime-Environment Relationships and Environmental Justice." *Seattle J. Soc. Just.* 6 (2008): 727, 731.

Cao, Anh Ngoc, and Tanya Wyatt. "The Conceptual Compatibility between Green Criminology and Human Security: A Proposed Interdisciplinary Framework for Examinations into Green Victimisation." *Critical Criminology* 24 (2016): 413, 419.

Cassidy, Kevin M. "The Role of Motive in White Collar Environmental Crimes." *Nat. Resources & Env't.* 23 (2009): 37, 38.

Cochran, Joshua C. et al. "Court Sentencing Patterns for Environmental Crimes: Is There A "Green" Gap in Punishment?" *J. Quant. Criminology* 34 (2018): 37, 51.

Jarrell, Melissa L., and Joshua Ozymy. "Few and Far Between: Understanding the Role of the Victim in Federal Environmental Crime Prosecutions in the United States." *Crime L. & Soc. Change* 61 (2014): 563, 565.

Lazarus, Richard J. "The Greening of America and the Graying of United States Environmental Law: Reflections on Environmental Law's First Tree Decades in the United States." *VA. Envtl. L.J.* 20 (2001): 75, 77.

Lynch, Michael, et al. "The Weak Probability of Punishment for Environmental Offenses and Deterrence of Environmental Offenders: A Discussion Based on U.S.E.P.A. Criminal Cases, 1983–2013." *Deviant Behavior* 37 (2016): 1, 29. 12 Oct 2018. doi:10.1080/01639625.2016.1161455.

Martin, Ana M., et al. "The Relationship between Moral Judgments and Causal Explanations of Everyday Environmental Crimes." *Soc. Just. Res.* 26 (2013): 320, 335.

Mullikin, Nancy. "Holding the Responsible Corporate Officer Responsible: Addressing the Need for Expansion of Criminal Liability for Corporate Environmental Violators." *Golden Gate U. Envtl . L.J.* 3 (2010): 395, 423.

Ruggiero, Vincenzo, and Nigel South. "Green Criminology and Crimes of the Economy: Theory, Research and Praxis." *Critical Criminology* 21 (2013): 359, 360.

Sahramäki, Iina, et al. "Prevention of Environmental Crime through Enforcement—Finland and Sweden Compared." *J. Scandinavian Stud. Criminology Crime Prevention* 16 (2015): 41, 50.

Sina, Stephan. Fighting Environmental Crime in Germany: A Country Report. Study in the Framework of the EFFACE Research Project 9. (2015). 1 August 2018 Https://Efface.Eu/Fighting-Environmental-Crime-Germany-Country-Report.

Uhlmann, David M. "Environmental Crime Comes of Age: The Evolution of Criminal Enforcement in the Environmental Regulatory Scheme." *Utah L. Rev.* 4 (2009): 1223, 1235.

Uhlmann, David M. "Prosecutorial Discretion and Environmental Crime." *Harv. Envtl.* L. Rev. 38 (2014): 193–194.

Walters, Reece, and Diane Solomon Westerhuis. "Green Crime and the Role of Environmental Courts." *Crime, L. & Soc. Change* 59 (2013): 279, 286.

EPILOGUE

Conclusions Across Cultures

Kirk W. Junker

Although functionalism cures some of the ills of encyclopedism and formalism in comparative law, one must still justify—beyond the scope of the functionalist enterprises—why these particular topics, and not others, were chosen. We attempted a balanced presentation from north and south, as well as east and west, and hope the topics do not suggest a bias in the direction that Reimann and David noted when they commented that most comparative law literature is a U.S.A.–Europe comparison. Our ability to use law to negotiate our sustainable relationship to the material world is a concern for all of humanity, not just certain regions of the earth. We hope to have presented these topics in such a light.

Supporters and critics of comparative law projects alike admit that too often comparative law texts are academic abstractions written for an audience of academics. We hope this book will not have fallen prey to that trap. Based upon our collective environmental law study and practice experiences, we have endeavored to theorize from practice, and thereby offer comparisons for practice.

One of the most common pieces missing from discussion of law in general and environmental law in particular is the piece that discusses expectations and behaviors of those subject to the law in any given culture. That is a major difference between comparative law being treated as comparative *laws* (as in statutes) and not as comparative legal cultures. A discussion of legal cultures certainly must include the perspective and expectations of the person for whom and against whom the laws are legislated and implemented. These perspectives, attitudes, and practices cannot be reduced to the politics of voting for legislators. Citizens as legal actors are not reducible to citizens as political actors. That is one reason that we have explicitly distinguished the two in Chapter 6, "The Interface of Law and Politics." Using the perspective of citizens under the rubric "culture" is not just a clever label or intellectual embellishment. Social scientific studies demonstrate—in all cultures—direct correlations between a citizenry's compliance with sources of law and its belief in, and support of, that law. While some might package that phenomenon under the label "legitimacy," it is more. It is indispensable for lawmakers, judges, police officers, administrators, and practicing lawyers to know the citizen's expectations and attitudes. When the world's cultures first began to realize some of the harms of industrialization, questions had to be asked and answered in all industrial cultures to distinguish the services, processes, and products that we felt that we needed, despite pollution and its harmful impact on air, water, and earth. We eliminated a few things to which we had grown accustomed, but had become accustomed to too many of most things to eliminate them, so we prohibited them by statute, but then allowed them as exceptions to statutory prohibition through issuing permits to continue production but to conform that production to necessary statutory limits after a bridge period of permitting. Electrical energy is clearly not just one of, but the, leading industry that received that treatment. But there are other things that the world's salespeople would have us believe are indispensible, and yet just one generation ago, we did not have, thus demonstrating they are very likely dispensible.

Too frequently, when presented with the facts of science, those new to considering the environment around them feel overwhelmed and ask whether anything really can be done. So incredulous are the facts they learn that they feel they cannot be true and must be someone's political agenda. The slow evolution to which I referred in Chapter 1 that brought our eyes to the front of our heads, and that gave us the ability to see stereoscopically and thereby have depth perception and make choices, is a tool that has proven beneficial to the human over the course of our existence. Our attitude toward the technological fix to our problems, especially our self-imposed problems, is not a tool that will prove to be beneficial. We are capable of making very long-term changes and even irreversible changes to the world in which we live. We do so with a mild laissez-faire sense that we can technologically reverse any mistakes, as we sense them imminently threatening us. Unlike stereoscopic vision, that attitude is not a tool for survival. Employing a metaphor from technology, Marshall McLuhan reminded us that we are sitting in the driver seat. Some are looking in the rear-view mirror, but we can all look through the window before us instead and see what is coming. We hope that the tools presented in this book offer a selection of provocations and applications that enable citizens and lawyers to look forward through that window with the comparative abilities of stereoscopic depth perception and to make choices that not only enable us to survive, but thrive.

INDEX